HISTORIES OF SCIENCE

Histories of Science

*Natural Philosophy in the
Eighteenth-Century Atlantic World*

Edited by David Alff and Danielle Spratt

University of Virginia Press
Charlottesville and London

The University of Virginia Press is situated on the traditional lands of the Monacan Nation, and the Commonwealth of Virginia was and is home to many other Indigenous people. We pay our respect to all of them, past and present. We also honor the enslaved African and African American people who built the University of Virginia, and we recognize their descendants. We commit to fostering voices from these communities through our publications and to deepening our collective understanding of their histories and contributions.

University of Virginia Press
© 2025 by the Rector and Visitors of the University of Virginia
All rights reserved
Printed in the United States of America on acid-free paper

First published 2025

9 8 7 6 5 4 3 2 1

LIBRARY OF CONGRESS CATALOGING-IN-PUBLICATION DATA
Names: Alff, David, editor | Spratt, Danielle, editor
Title: Histories of science : natural philosophy in the eighteenth-century
    Atlantic world / edited by David Alff and Danielle Spratt.
Other titles: Natural philosophy in the eighteenth-century Atlantic world
Description: Charlottesville : University of Virginia Press, 2025. | Includes
    bibliographical references and index.
Identifiers: LCCN 2025000452 (print) | LCCN 2025000453 (ebook) |
    ISBN 9780813951683 hardback | ISBN 9780813951690 trade paperback |
    ISBN 9780813951706 ebook
Subjects: LCSH: Science—History—18th century | Natural history—
    18th century—Historiography | Literature and science—History—18th century |
    Philosophy and science—History—18th century | LCGFT: Essays
    Classification: LCC Q126.8 .H578 2025 (print) | LCC Q126.8 (ebook) |
    DDC 509—dc23/eng/20250214
LC record available at https://lccn.loc.gov/2025000452
LC ebook record available at https://lccn.loc.gov/2025000453

Publication of this volume has been supported by the Walker Cowen Memorial Fund and by the University at Buffalo Humanities Institute (a unit within the College of Arts and Sciences) and the University at Buffalo Office of the Vice President for Research and Economic Development.

Cover art: Flower, *A Poppy in Three Stages of Flowering,* Johanna Helena Herolt (Metropolitan Museum of Art, New York; The Elisha Whittelsey Collection, The Elisha Whittelsey Fund, anonymous gift, in memory of Frits Markus, and Karen B. Cohen Fund, 2005); vials, *Two Glass Containers for Scientific Experiments with Liquids,* 1666 (Metropolitan Museum of Art, New York; The Elisha Whittelsey Collection, The Elisha Whittelsey Fund, 1952); background, ceiling paper (Cooper Hewitt, Smithsonian Design Museum, gift of Victorian Collectibles)
Cover design: Elke Barter

CONTENTS

Introduction   1

Part I. Rhetoric

Science for the Birds: Figurative Language and
   The History of the Royal Society   13
Jess Keiser

Centlivre's Frankenstein: Science, Authorship, and Monstrous
   (Re)Productions in *A Bold Stroke for a Wife*   33
Melissa Bailes

Picturing Air: The Rhetoric of Nondescription in Robert Boyle's
   New Experiments Physico-Mechanicall and Daniel Defoe's *A Journal
   of the Plague Year*   46
Vivian Zuluaga Papp

Romancing the Placebo   65
Jayne Lewis

Part II. Reception

"The Eye of Mr. *Anson* Himself": Art and Evanescence in *A Voyage
   Round the World* (1748)   89
Anne M. Thell

Literary Technologies of the Sextant in Eighteenth-Century Britain   110
Aaron R. Hanlon

Newtonian Legacies in William Hogarth's *A Scene from "The Indian Emperour," or "The Conquest of Mexico by the Spaniards"* 125
Laura Miller

"Fully Prov'd by the Plates": The Unauthorized *System* of J. T. Desaguliers 140
Al Coppola

Part III. Embodiment

Plants, Principles, Strata: John Woodward's Improbable Corpuscles 165
Helen Thompson

"Mice in a Barn" or "Every Little Miss"? Figurative Imagination and Demographic Narratives of the Long Eighteenth Century 188
Lisa Forman Cody

Jane Barker and Virgin Anatomy 209
Frank Boyle

Margaret Cavendish, a Sensitive Witness 227
Kristin M. Girten

Maria Edgeworth's Avian Entwinements: Experimental Science and the Cultivation of the Female Mind in *Practical Education* and *Belinda* 243
Adela Ramos

Part IV. Environment

The Hoe and the Plow: Plantation Labor under the Somatic Energy Regime 261
Ramesh Mallipeddi

Infrastructural Inversion at Clarens: St. Preux in the Garden 278
Eric Gidal

Taxonomic Subversion and Vegetal Expansion in Charlotte Smith's "Beachy Head" 289
Anna K. Sagal

Spicy Forests and Amboyna Burl: Dryden and the Ecology of Disaster 305
Rajani Sudan

Geomythography: A Genealogy 325
Tobias Menely

Notes on Contributors 341
Index 345

Introduction

This volume convenes eighteen literary historians to discuss the representation, reception, and application of natural philosophy in the eighteenth-century Atlantic world. Drawn from a cohort of emerging and established scholars, these essays demonstrate the current breadth of early modern science studies. Rather than impose a topical outline that presumes to know the boundaries and subdivisions of our evolving field, we invited contributors to pursue topics of their own choosing. Just as Thomas Sprat encouraged his Royal Society colleagues to "heap up a mixt Mass of *Experiments*, without digesting them into any perfect model," our collaboration locates via miscellany areas of common concern among scholars working in eighteenth-century science studies today.[1]

Inspired by recent monographs that assign critics a formative role in interpreting scientific culture, *Histories of Science* conceives of literature not simply as a category of writing that records and answers empirical inquiry but as a capacious makerly mode that enables and structures experimental knowledge.[2] As Tita Chico argues, literariness—the aesthetic organization of language to produce possibility—constitutes both a vehicle for transmitting findings and a "way of ascertaining truth about the world" that facilitates an "imagining of the proper object of natural philosophical inquiry as well as the proper subject to carry it out."[3] Although eighteenth-century studies has long appreciated rhetoric's role in organizing natural philosophical endeavors, much recent work shares a propensity for advancing new artifacts, genres, pedagogies, and tools as worthy and rewarding of literary critical attention. Following the late twentieth-century rise of cultural poetics and Larry Stewart's specific 1992 call to "move beyond the rhetoric and rationalizations into the action of the proponents of the new science," scholars have drawn a raft of practices and artifacts under the rubric of science studies.[4]

As a demonstration and endorsement of such pluralism, this volume collates essays on placebos, engraving, lecture plates, irrigation systems, novels, sextants, plows, births, and myth. Our contributors show how even seemingly disparate source materials raise field-relevant questions over the status of witnesses and witnessing, the legitimacy of evidence, the affordances of representation, and the formation of scientific publics through print circulation, visual aids, curricula, and lecture hall theatrics. By examining the "actions of proponents" as well as the experiences of those less commonly associated with the rise of the new science—women, enslaved African people, ship hands, patients, artists, children, and craft laborers who studied nature or experienced the impact of such inquiries—this collection returns a messier portrait of eighteenth-century science and a fuller account of its present-day reverberations.[5]

To chart eighteenth-century science's ramifying networks of analytical objects and historical subjects, our writers avail themselves of various overlapping methodological frames, including formal criticism, new historicism, object-oriented ontology, earth system theory, mythology, plant studies, the medical humanities, performance studies, and infrastructure studies. This convergence of approaches reflects Al Coppola's conception of science as a "dynamic amalgam of ideas, technologies, and procedures."[6] Through its panoply of topics and methods, our volume's essays show how science studies—the study *of* the study of nature as "amalgam"—has become a methodological intersection and site of continuous interpretive adjustment.

This collection focuses on the Atlantic world in an age of revolutions, between the English Civil Wars in the mid-1600s and the abolition of the slave trade in the early 1800s, a time and place often credited with (and blamed for) the invention of dualist strategies for classifying and instrumentalizing nature—a utilitarian ethos and posture of epistemological confidence that predominates many sectors of society today. It is the eighteenth century, argue Christina Lupton, Sean Silver, and Adam Sneed, to which we attribute a contradictory "break between, on the one hand, a confidence in the integrity of the modern subject, and, on the other hand, a serene satisfaction with the concreteness of the modern object."[7] Our collective attempt to revisit this epistemological rupture finds urgency in a present afflicted by its afterlives: the caloric aftermath of a carbonated atmosphere, the explosive growth and seepage of human populations, the sixth mass extinction of plants and animals, and the relegation of the earth as a fungible resource to capital and its clients.

The "chronological coincidence" of Enlightenment empiricism and the era of coal-fired industrialization that followed James Watt's improvement of the Newcomen steam engine makes the 1700s an especially pivotal period for understanding subsequent intellectual and ecological conditions, including the present.[8] As Alan Mikhail observes, the eighteenth century "invites both humanists and scientists to the table of a long tradition of trying to explain the emergence of the modern world."[9] Our contributors take up this explanatory work by showing how different media forms registered scientific exploits during the long eighteenth century and contributed simultaneously to the production of the present and the sometimes hopeful, sometimes apocalyptic imagination of alternative pasts and futures.

The authors in this collection focus most explicitly on representations of science found in popular print. These include genres typically associated with entertainment and art—novels, satires, plays, performances, and portraits—along with trade manuals on navigation, colonial husbandry, demography, and medicine. It is perhaps no wonder, then, that Henry Stubbe, one of the Royal Society's most vociferous critics, referred to its fellows not just as virtuosi but as "novellists," a word that captures the period's obsession with the new and unreal.[10] Like the new science itself, these media forms were largely novel sources of information and entertainment for the increasingly literate populace, and the boundaries between these two genres then, like now, were more often porous than fixed.

Histories of Science therefore documents how popular media created, disseminated, abstracted, and applied new scientific knowledges to local and global effects across the long eighteenth century. Part 1 shows how poets, novelists, chemists, and physicians employed the resources of language and rhetoric to represent scientific practice. Jess Keiser opens the collection by assessing the contested use of metaphor within "plain style" scientific writing. Keiser shows how Sprat's *History of the Royal Society* and Cowley's "To the Royal Society" advocate different uses of figurative language—the former values its capacity to illustrate concepts, while the latter grants tropes the power to bring about new knowledge—instantiating a broader deliberation over the affordances of words in the study of nature. Melissa Bailes shows how Susanna Centlivre's *A Bold Stroke for a Wife* draws from accounts of biological reproduction an analogy between heredity and inheritance that allows her to critique the patrilineal models of scientific and literary originality. She argues that Centlivre's ideas of innovation and heredity not only influenced Mary Shelley's *Frankenstein* but also welcomed other authors to participate in knowledge creation through a model grounded in Whiggish notions of property that were created and maintained through labor rather than birthright.

Vivian Zuluaga Papp considers the figure's capacity to represent the limits of representation through her study of "nondescription," the narrative staging of events that defy articulation. Papp's essay compares instances of nondescription in Robert Boyle's accounts of the air pump and Daniel Defoe's *Journal of the Plague Year*, in which the narrator fails to describe both minute details and enormous losses in the face of an invisible epidemic that destroyed families and altered urban and rural communities. Jayne Lewis contemplates another kind of absence that embroiled eighteenth-century medical correspondence. The placebo effect's "experience of nothing" derives therapeutic value from its deliberate concealment of fact. Lewis traces this form of scientific storytelling through readings of William Godwin's translation of Benjamin Franklin's report on Franz Anton Mesmer and Matthew Lewis's *The Monk* alongside his failed attempt to manage birth rates on his colonial plantation.

Part 2 examines how natural philosophical acts of reception—witnessing, observing, perceiving—implied different entanglements of authors, audiences, and nature. Its essays show how humankind's empirical faculties came to license the ontological distinction between seeing subjects and seen objects. What Francis Bacon called "simple sensuous perception" turns out to have been anything but self-evident according to our contributors, who reveal the strenuous cognitive feats it took to become modest witnesses.[11] What happens to the empiricist's authority when nature proves unobservable? The speed of light and slowness of glaciation escape human perception. Mirages dazzle it. Faulty and incomplete witnessing compromised not only empiricism's power to describe but its fundamental imagination of subjects that could stand apart from the world they studied.

Anne M. Thell describes a rivalry between visual and written accounts of travel in George Anson's *A Voyage Round the World*, which incorporates biophysical theories of optics to elevate the draftsman's depictions of landscape over piecemeal narratives gathered from senior crew members. Thell argues that "pictorial facticity" complemented and ultimately displaced written representation as the most authoritative means of observing, compiling, and disseminating empirical observations from sea voyages. Complementing Thell's study, Aaron Hanlon considers the race to discover an accurate measure of longitude through various optical instruments, most specifically the sextant. Where we might think of instruments as detached engines of empirical evidence, tradesmen and satirists alike emphasized the imbrication of tools with their error-prone users. While trade manuals depicted the uses and failures of such navigational technology as a means of trafficking reliability for a specialist audience, literary satire often used sextants to evoke the quixotically unattainable.

Laura Miller's examination of William Hogarth's *A Scene from "The Indian Emperour," or "The Conquest of Mexico by the Spaniards"* shows how another visual mode, portraiture, captured the crises of scientific posterity that emerged in the wake of Isaac Newton's death. Hogarth depicts Newton's closest heirs, his niece Catherine Barton Conduitt; her husband; John; and members of their social circle viewing their child's performance of Dryden's titular play, a scene conspicuously framed by a bust of Newton. For Miller, this portrait indicates the failures of Conduitt to inherit a place in both the Royal Society and Whig politics. Through his satirical representation of familial theater and sculpture, Hogarth reveals a contradiction between the new science's professed commitment to disavowing scholasticism and cult iconography with its continuous generation of memorial artifacts.

Al Coppola considers the propertization of scientific practice through his study of Theophilus Desaguliers, a natural philosopher whose lecture notes and visual plates were purchased and reprinted without his authorization. Coppola maps the nexus of visual media and print transactions that allowed others to market Desaguliers's "system" as their own, surveying the institutions that described scientific concepts to general audiences and the legal terms under which public knowledge could be subject to exclusive possession. Helen Thompson shows how a similarly multifaceted convergence of writing and performance challenged the reliability of representation. Her essay examines the barbed public debates between physician and author John Woodward and the Scriblerians, specifically fellow doctor John Arbuthnot. Using the term "corpuscular induction," a theory that provides "a representational mode beyond verisimilitude," Thompson shows how Woodward's theories of micromaterial matter, from the human digestion of vegetables to the transmutational potential of the earth, became a ripe satirical target for Arbuthnot and friends via the Scriblerian drama *Three Hours after Marriage* (1717). Beyond its merciless lampoon of Woodward, the play demonstrates art's capacity to adjudicate competing epistemological claims.

Explanatory theories of the body inform part 3, which concentrates on gendered scientific subjectivities. A focus on embodiment as a starting point and precondition for observation endows "passive" objects of scientific inquiry—including reproductive bodies, women, animals, and children—with the capacity to produce equally valid and valued contributions to scientific knowledge. Lisa Cody reveals how early modern demography analogized the sexual appetites and "breeding bodies" of human populations to the construction of buildings and the propagation of small animals. Her essay claims that such comparative tropes proved "essential to conveying meaning and making arguments" about human totalities. Frank Boyle also addresses metaphors of gender and sex. His reading of Jane Barker's "A Farewell to Poetry with a Long Digression on Anatomy" and its later revision in *Patch-Work Screen for the Ladies* alights on feminine forms of knowledge production that appropriate the rhetoric of masculine scientific celibacy. Poetic diction establishes the epistemic authority of Barker's narrative persona while developing a system of collective knowledge grounded in women's domestic work. Together, these gestures upend traditional and often violent heteropatriarchal conceptions of scientific inquiry.

Kristin M. Girten approaches the familiar dichotomy of embodiment and observation from the radically novel vantage of Margaret Cavendish's Epicurean materialism. Girten shows how Cavendish's "sensitive witness" privileges the scientific observer's "feeling and affective openness" as a more reliable instrument than optical technologies like microscopes and telescopes. The result is a fundamental reconfiguration of the witness, no longer domineeringly "modest" and male but ideally humble and potentially female. Adela Ramos traces Maria Edgeworth's own questioning of dispassionate objectivity in *Belinda*, which depicts scientific education in the domestic sphere. Ramos concentrates specifically on *Belinda*'s "avian entwinements," scenes in which women respond affectively to the objects of experimental observation, including birds, among colonial imports and curiosities. Ramos's essay helps us reevaluate Joseph Wright's famous depiction of the *Experiment on a Bird in the Air-Pump*, decentralizing the austere scientist and elevating the role of both the family of observers and the bird itself.

The final part of the collection shows how people applied empirical knowledge projects to alter their physical surroundings. Ramesh Mallipeddi identifies the seemingly counterintuitive resistance of colonial agrarian manuals to technological advancements. Mallipeddi argues that the unwillingness of planters to employ tools like the plow stemmed not just from unfavorable conditions of soil and land but from their commitment to the hoe, with its capacity to objectify and instrumentalize enslaved Africans, whom colonists supposed were at once disposable and unruly labor in need of mechanization and constant surveillance. While abolitionist tracts from the period identified the whip as the most brutal and dehumanizing tool of enslavement, Mallipeddi establishes the hoe's vast ties to oppression and violence. Eric Gidal considers another scene of problematic cultivation. His essay presents the concept of "infrastructural inversion," or the social scientific attempt to foreground overlooked technologies and protocols, as a means of considering the utopic possibilities and limits of scientific mastery over nature in Jean-Jacques Rousseau's *Julie*. Rousseau depicts feats of earthmoving and hydroengineering that elevate estates into spectacle, ultimately teaching audiences how to distinguish between those phenomena that are perceptible to their senses and the unseen moral interventions that landscaping sublimates. The artificial garden represents both an application of natural knowledge through the practice of engineering and an ontological challenge to the category of nature that shows how land itself can be convincingly simulated.

Anna K. Sagal contends that Charlotte Smith's poetic and citational form in *Beachy Head* contests prevailing norms about binomial classification that privilege and prioritize the mammal (and the human) above other forms of nonhuman animal life, performing what she calls "taxonomic inversion." Sagal traces the poem's "ambivalence in human-plant relationships" by showing how Smith incorporates and then marginalizes examples of taxonomic nomenclature developed to impose pristine order on botanical life. In so doing, Sagal suggests, Smith recasts poetry as an alternative observational method attuned to the agential capacities and unknowable ontologies of plants. Smith thus critiques the masculinist and colonially minded methods of depicting the environment. In her essay, Rajani Sudan traces one such imperial imagination of vegetative growth, the "spicy forest" setting of John Dryden's tragedy *Amboyna*, to Restoration practices of religion, pharmacology, and land improvement. While holding out the potential for nationalistic world-making, Amboyna's abundant woods become a site of trauma, where the English mercantile project "collapses into an ecology of disaster."

In the collection's final essay, Tobias Menely unearths Baconian empiricism's overlooked indebtedness to ancient myth. Menely evaluates Robert Hooke's "geomythmaking" in order to push readers to question fixed and fast notions of disciplinary differences between science and literature. For Hooke, myth offered a better record of earth's history and transformative potential than the messy metaphors of science writing that Keiser unsettles in his opening essay. Over time, researchers have come to treat mythic earth writings as either articles of geological history or artifacts of a human past. Such ontological distinctions may collapse as the Anthropocene binds natural and social history together. As Menely shows, myth's connection to more environmentally and ethically sound Indigenous intellectual systems, such as Māori knowledge, has the potential to heal our increasingly damaged earth in more radically restorative ways than a reliance on Western technologies and traditions of science.

Sprat's conception of science writing as "unfinish'd History" aptly characterizes the broader generic, epistemic, and methodological junctures of this collection. As J. Ereck Jarvis notes, Sprat's *History* "is always compromised, not in any derogative regard but in the term's most basic etymological sense, *sent forth together*. Multiple—perhaps even contradictory—authorities are always under negotiation in the work and its composition."[12] This collection aims to stage what Robert Markley calls a "collective—and dialogic—attempt to develop new forms of critical intervention in the discourse of history, theory, and culture."[13] Both our title and design emulate the "unfinish'd History": the generic hybridity of, and ongoing inquiry about, scientific discourse across the long eighteenth century. In so doing, *Histories of Science* registers popular media's role in realizing and resisting the capacious and often contradictory possibilities and practices of Western science's emergence in the eighteenth century: its ability, then as now, to create and destroy, to liberate and to oppress, to heal and to harm. Our hope is that by reflecting on science's unfinished plural histories, this collection will help expand our record of the past, our understanding of the present, and our ability to imagine the future.

Notes

1. Thomas Sprat, *The History of the Royal Society of London, For the Improving of Natural Knowledge* (London: 1667), 115.
2. See Al Coppola, *The Theater of Experiment: Staging Natural Philosophy in Eighteenth-Century Britain* (Oxford: Oxford University Press, 2016); Helen Thompson, *Fictional Matter: Empiricism, Corpuscles, and the Novel* (Philadelphia: University of Pennsylvania Press, 2017); Courtney Weiss Smith, *Empiricist Devotions: Science, Religion, and Poetry in Early Eighteenth-Century England* (Charlottesville: University of Virginia Press, 2016); Jess Keiser, *Nervous Fictions: Literary Form and the Enlightenment Origins of Neuroscience* (Charlottesville: University of Virginia Press, 2020); Laura Miller, *Reading Popular Newtonianism: Print, the "Principia," and the Dissemination of Newtonian Science* (Charlottesville: University of Virginia Press, 2018); and Rajani Sudan, *The Alchemy of Empire: Abject Materials and the Technologies of Colonialism* (New York: Fordham University Press, 2016).
3. Tita Chico, *The Experimental Imagination: Literary Knowledge and Science in the British Enlightenment* (Stanford: Stanford University Press, 2018), 13. For additional work on the generic porousness of science, history, and literature across the eighteenth century, see Ruth Mack's *Literary Historicity: Literature and Historical Experience in Eighteenth-Century Britain* (Stanford: Stanford University Press, 2008), 12.
4. Larry Stewart, *The Rise of Public Science: Rhetoric, Technology, and Natural Philosophy in Newtonian Britain, 1660–1750* (Cambridge: Cambridge University Press, 1992), xxxiii.

5. In this way, the collection builds on studies like Deborah E. Harkness's *The Jewel House: Elizabethan London and the Scientific Revolution* (New Haven: Yale University Press, 2007).
6. Coppola, *Theater of Experiment*, 16.
7. Christina Lupton, Sean Silver, and Adam Sneed, "Latour and Eighteenth-Century Literary Studies," *Eighteenth Century: Theory and Interpretation* 57, no. 2 (Summer 2016): 165–79.
8. Alan Mikhail, "Enlightenment Anthropocene," *Eighteenth-Century Studies* 49, no. 2 (Winter 2016): 218.
9. Ibid., 212. In a similar vein, Darren Wagner and Joanna Wharton claim that eighteenth-century studies has long proven "particularly encouraging of research on literature and science." "Literature and Science in Eighteenth-Century Studies: Mountain Gloom or Mountain Glory?," *Journal of Literature and Science* 10, no. 1 (2017): 15.
10. Michael McKeon, *The Origins of the English Novel, 1600–1740* (Baltimore: Johns Hopkins University Press, 2002), 71, 441n13.
11. Francis Bacon, "Novum Organum," in *The Works of Francis Bacon*, ed. James Spedding, Robert Leslie Ellis, and Douglas Denon Heath (London: Longman, 1858), 4:40.
12. J. Ereck Jarvis, "Thomas Sprat's 'Mixt Assembly': Association and Authority in 'The History of the Royal Society,'" *Restoration: Studies in English Literary Culture, 1660–1700* 37, no. 2 (Fall 2013): 57, 69.
13. Robert Markley, *Fallen Languages: Crises of Representation in Newtonian England, 1660–1740* (Ithaca: Cornell University Press, 1993), 260. Indeed, these essays join in the ongoing and provocative conversations found in recent collections on early modern science by Paula Findlen, Kristin M. Girten and Aaron R. Hanlon, and others.

PART I

Rhetoric

Science for the Birds

Figurative Language and The History of the Royal Society

JESS KEISER

Plainness rarely catches the eye, but for much of the twentieth century, scholars interested in early science and its language noticed little else. Convinced that seventeenth-century writers had developed an austere style suitable, somehow, to the needs of the new science, these scholars trained their focus on the unadorned, the nonrhetorical, the literal.[1] Their fixation on the period's apparent "plain style" was the result, in part, of taking some early advocates of the new science at their word.[2] In doing so, they missed, or at least dismissed, what we now see so clearly: that early natural philosophers ornamented their work with sometimes baroque figures, that they did so purposefully, and that they thought seriously about the nature of figurative language and, in doing so, sought to make rhetoric part of scientific work.[3]

For all its insights, however, contemporary scholarship's ready recognition of the rhetoric of early natural philosophy has obscured an important nuance in science and language: that with the adoption of figure came ambivalence. This ambivalence manifests *not* as an attempt to somehow strip away rhetoric from writing (as an earlier generation of scholars sometimes suggested). Rather, it is evinced in efforts to institute a distinction between different *kinds* of tropes, between, on the one hand, figures that adorn some already established idea and, on the other, figures that actively shape ideas or even reveal new ones. Courtney Weiss Smith is one of the few scholars to stress this difference and insist on its importance. She draws a contrast between early scientific tropes that were "mere decorations or after-the-fact appeals to the reader" and those that "accessed real knowledge, generating new links, questions, and applications."[4] Building on Smith's work, we might call the first type of figurative language "instrumental" and the second kind "creative." Instrumental figurative language functions as a pedagogical tool. Like a thought experiment or visual illustration, it helps us see something clearly by making it more vivid or concrete (and, in this respect, it is easily discarded or ignored once we grasp the core concept). Early science readily and forthrightly embraced this type of figurative language. Creative figurative language undertakes more original thinking. It helps us perceive new things rather than simply ornamenting old ones. Although this sort of rhetoric certainly appears in early scientific thought, its presence there is more controversial and is often accordingly disavowed by natural philosophy's more stringent advocates.[5]

While early scientific thinkers recognized the distinction between "instrumental" and "creative" figurative languages, modern scholars tend to deal with this difference only implicitly or not at all. This is a mistake, since at stake in these different visions of figurative language are different accounts of how science and language work (or ought to work). Just how integral were figure and rhetoric to early science? Did tropes only help along science by making it more accessible to a general audience, or did they play a more fundamental role in early scientific thinking? We can't begin to answer these questions until we begin to appreciate the different, even conflicting, *kinds* of figurative language at play in early science.

To do just that, this essay considers what remains one of the most popular, and most misunderstood, documents in early science, Thomas Sprat's *History of the Royal Society*. While citing the *History* is requisite in discussions of science and language, I read this work differently than most by focusing on its split nature. That split becomes manifest once we appreciate that Sprat's is not the only voice resounding through the *History*. Not only does this text showcase documents from the early years of the Royal Society—its charter and some of its first scientific reports—but it also begins with a poem, "To the Royal Society," written by Sprat's friend, Abraham Cowley.[6] I'll show that, though the writing of Sprat and Cowley appears literally on the same page, these thinkers weren't on the same page when it came to figure.[7] Despite a still prevalent belief that it advocates against rhetoric, Sprat's *History* offers a canny explanation and defense of the instrumental uses of figurative language.[8] Conversely, Cowley's poem trafficked in tropes that (to borrow Smith's phrasing) accessed knowledge rather than just accessorizing it. In short, Cowley makes a case for creative figurative language. Seen side by side, Sprat's and Cowley's writings illustrate the two contrasting kinds of figurative language at play in early science.

* * *

The History of the Royal Society's best-known passage—by which I mean its most cited, most often anthologized, and most given to distortions, rumors, and misprisions—appears roughly a third of the way into that voluminous text.[9] It is the second part, section 20, entitled "Their Manner of Discourse." It's here that I want to begin teasing out Sprat's vision of language and the scientific work it supports. Sprat begins this section by noting that, though the Royal Society had gained ground in purely natural philosophical matters—having "directed, judg'd, conjectur'd upon, and improved *Experiments*"—the society soon worried that their advances would be undermined by a failure to communicate their findings clearly.[10] And so, the society came to dwell on their "*Discourse:* which, unless they had been very watchful to keep in due temper, the whole spirit and vigour of their *Design,* had been soon eaten out, by the luxury and redundance of *Speech.*"[11] This profuse language represents one of the early Royal Society's signal obstacles. According to Sprat, such language obscures the information it ought to convey ("Who can behold, without indignation, how many mists and uncertainties, these specious *Tropes* and *Figures* have brought on our Knowledg?"), and it indulges imagination and affect at the expense of intellect and truth (it is in "open defiance against *Reason;* professing, not to hold much correspondence with that; but with its Slaves, *the Passions*"; 112). Thankfully, Sprat explains, the solution to such excess is simple enough: it is simplicity itself. To combat luxurious speech, the Royal Society pares down its discourse, hewing away the flowers of rhetoric so as to uncover the most plainly primitive roots of language lying underneath: "a close, naked, natural way of speaking; positive expressions; clear senses; a native easiness: bringing all things as near the Mathematical plainness, as they can" (113). Thanks to the society's "constant Resolution, to reject all the amplifications, digressions, and swelling of style," they have managed to return language to something like its prelapsarian state, "back to the primitive purity, and shortness, when men deliver'd so many *things,* almost in an equal number of *words*" (113).

In his account of these efforts to curtail figurative profusion, Sprat's own linguistic theory stands starkly revealed. Put simply, for Sprat, the purpose of "words" is to "deliver" "things."[12] This formula turns language into a medium through which messages move. Words are like the postal service or plumbing: they serve to serve up something else (letters, water, referents). Or, less metaphorically, language is re-creative, not creative; it transmits a message but does not originate its content. This conception of language is anodyne enough. But it institutes an asymmetry that organizes much of Sprat's thinking: words are less important than things. That is, language, understood as an enveloping form or medium, is only a way of dressing up the core content or message it imparts, which means that words can be stripped away or swapped out with little of their substance lost.

Once we see this asymmetry at work in the *History*, Sprat and the Royal Society's fear of the "luxury and redundance of Speech," of "amplifications, digressions, and swelling of style," becomes more intelligible in turn. Such excess would mean the overturning or inversion of the aforementioned asymmetry. The linguistic medium, grown dense through rhetorical overgrowth, would become more substantial than its message, and the very weight of words would threaten to outweigh the content they are to carry along. Language would serve itself rather than its purpose (to convey extralinguistic information). It would swell into a tool made inoperable by its own ornateness. The Royal Society's insistence on curbing linguistic extravagance, then, represents an effort to reassert the right balance of words and things, to institute an equilibrium where language wouldn't oppress its referents. As a result, when Sprat describes the Royal Society's writing, he often makes it sound as if they strived for a style where language nearly disappears in its use.[13] Their choice, he implies, seems to have been a language marked by obscurity (think, again, of those "mists and uncertainties" dispersed by "*Tropes* and *Figures*") or a discourse bright as day (a style made "naked," "as near the Mathematical plainness" as possible).

In fact, the continual recourse to imagery of light and dark, transparency and opacity, in Sprat's account of language is no accident. Throughout his *History*, Sprat treats linguistic processes as if they were acts of perception, as if every time a word denominated a thing, we somehow saw *through* a signifier to the referent behind or beyond or beneath it. And for good reason. For Sprat, language interposes itself between mind and world. We are given to understand that when words deliver things, these things are delivered to the mind.[14] This follows from the *History*'s implicit but prevalent philosophy of mind.[15] Like so many of his contemporaries, Sprat seems to have delineated the psyche along certain well-worn lines. In this account, the mind's primary task is to represent or mirror the world.[16] This involves reproducing a model of external reality in the mind's interiority—"out there" is re-created "in here"—through the intercession of various mediums: sense organs, ideational images, and (especially important for Sprat) language. It is as if the mind stood at a remove from its surround, as if it were an eye poised behind a pane of glass. No wonder, then, that Sprat and the Royal Society worry about the relative "opacity" of the linguistic medium. In their vision, language becomes an optical tool, a means of seeing what lies beyond the confines of the psyche. And like any optical instrument, in order to work correctly, it must be *seen through* rather than *seen*, lest we find ourselves taking a smudge on a lens for an empire in the stars.[17]

Given this model of the psyche, with its focus on the veil between mind and world, terms like *transparency* and *opacity* become the key values around which the entirety of Sprat's *History* turns.[18] In Sprat's writing, these concepts not only determine what counts as language fit for the new science; they also come to define scientific work itself. For instance, a "naked" style can help communicate scientific findings by making them plain, evident. But natural philosophy, according to Sprat, operates similarly: it, too, works to disclose or unveil the world beyond the mind (and beneath language) by obviating any obstacles obscuring our way. And just as transparent language is paired with science, opaque language finds a parallel in the study of rhetoric. If science strives to see and to know the external world, then rhetoric blocks that view by viewing language itself rather than extralinguistic reality, by focusing on words rather than things.

The stakes of this struggle are exemplified in a short passage in the third part of the *History*. In the course of discussing educational reforms, Sprat explicitly pits rhetoric against science. He argues there that with the adoption of the Royal Society's methods, much of what the ancient and medieval periods claimed about nature and physics must be abandoned. Thankfully, this is no great loss because those ages have largely produced jargon—empty words denominating nothing real: "Perhaps there will be no more use of Twenty, or Thirty obscure Terms, such as *Matter*, and *Form*, *Privation*, *Entelicihia*, and the like. But to supply their want, an infinit variety of *Inventions*, *Motions*, and *Operations*, will succeed in place of words. The Beautiful Bosom of *Nature* will be Expos'd to our view: we shall enter into its *Garden*, and tast of its *Fruits*, and satisfy our selves with its *plenty*: instead of Idle talking, and wandring, under its fruitless shadows" (327). The passage draws a series of contrasts. On the one hand, we have modern science, associated with the work of experiment ("an infinit variety of *Inventions*, *Motions*, and *Operations*"). Such experiments "succeed in place of words," a success that is cast as a moment of enlightenment. Thanks to scientific insight, we move from darkness to light, from "Twenty, or Thirty obscure Terms" to the "Beautiful Bosom of *Nature* . . . Expos'd to our view."[19] As a result, we seem to slip past the veil of language screening nature from our view and thereby "enter into its *Garden*, and tast of its *Fruits*, and satisfy our selves with its *plenty*." On the other hand, we have rhetoric, which, for Sprat, is what most ancient/medieval natural philosophy amounts to. Whereas modern science produces "fruit" by interacting directly with nature—even prying into its substance with experiments—rhetoric remains resolutely "fruitless." Its work involves purposelessly playing with words ("idle talking"), an activity that leads nowhere (it is "wandring"), since it is groundless, its only bedrock the "shadows" cast by more substantial things. Elsewhere, Sprat makes this contrast even starker—especially in those moments where he discusses the prehistory of the Royal Society. In the course of surveying ancient and medieval philosophy, Sprat consistently draws attention to a similar pattern. These periods began to study nature directly—only to find themselves blocked by undue attention to the words standing in for (or even in front of) things.[20] Luxuriant speech, idle talking, rhetoric—which all comport with words rather than the things beyond them—confine us. We become bound to a world hollowed out by simulacra, where semblances loom larger than the reality they represent (hence "shadows") and where real progress is just a species of "frictionless spinning" (hence "fruitless").[21]

Transparency and opacity, then, organize much of *The History of the Royal Society*. Important aspects of language, mind, knowledge, and reality are brought into focus by these values; otherwise disparate things—style and science, the natural world out there and the mental space in here—come to connect depending on their place in this play of light and dark. So Sprat tells us of transparent language (plain, naked, natural, referential) and opaque language (ornamented, extravagant, distracting, deceptive). But he derives this distinction, in turn, from an account of the mind that hopes for transparency (e.g., the psyche should see the world through a clear glass) but worries about opacity (e.g., the world could lie veiled from view; words might obstruct things). And these dynamics in language and mind are tied, in kind, to certain intellectual pursuits, which are marked by their capacity to reveal or disclose (science) or to deceive and confine (rhetoric). Indeed, at his most capacious, Sprat even treats nature and culture as stand-ins or exemplifications of the clear and the obscure, with nature figured as a glittering country beyond the confines of our subjectivity and culture cast as the shell enclosing us in a world of words where we are kept deluded by shadows.

Once we survey the wider sweep of Sprat's thinking, some of its finer details—and even some startling features—come into view. Most surprising is Sprat's insistence that figurative language can aid scientific research. Why might this claim be surprising? It is tempting to assume that the transparency-opacity dichotomy that determines so much of Sprat's thought maps onto the distinction between literal and figurative: as if "plain" were simply another word for "literal," as if all rhetoric necessarily casts "fruitless shadows." And so it is tempting to assume as well that science—which seeks to see the natural world clearly—must have little use for the obscurity of figure. To be sure,

Sprat himself sometimes suggests as much. As we have seen, when he needs an example of extravagant, obstructive language, he often reaches for metaphor. Nevertheless, it's also important to recognize that Sprat is not being absolute in these instances. Certainly, he thinks that figurative language is *more likely* to be "opaque." And yet he never argues that *all* figures must be so. Opaque language is just language that "gets in the way": it draws attention to itself rather than to the extralinguistic world beyond; it blocks the intercourse between subject and object; and as a result, it screens us from nature by confining us to culture. But if figurative language could avoid this impasse, then there is nothing in Sprat's thought barring it, in principle, from serving scientific work. In this respect, even metaphors could be "plain."

Sprat explains how certain kinds of figurative language could aid knowledge: "[Ornaments of Speaking] were at first, no doubt, an admirable Instrument in the hands of *Wise Men:* when they were onely employ'd to describe *Goodness, Honesty, Obedience;* in larger, fairer, and more moving Images: to represent *Truth,* cloth'd with Bodies; and to bring *Knowledg* back again to our very senses, from when it was at first deriv'd to our understandings" (111–12). Sprat's description of rhetoric as an "admirable Instrument" that can "represent *Truth*" or "bring *Knowledg* back again to our very senses" ought to put us in mind of similar statements about language's power to deliver things through words. Understood as an instrument, figure can accomplish the same task as more literal discourse. In fact, Sprat implies it can do more. Presumably, an entirely plain rendering of "*Goodness, Honesty, Obedience*" would fall flat, given the stolid nature of these concepts. But by outfitting such ideas in "Bodies" and thereby transforming inert abstractions into "larger, fairer, and more moving Images," rhetorical ornaments are capable of making vague or abstruse ideas into concrete or vivid ones. Sprat's gambit is that sometimes figurative "clothing," rather than concealing things, can amplify an underlying substance.[22] In this sense, metaphor becomes a means of empiricism, a way of reproducing or reviving an originary sensible scene.

Invested in these figures, Sprat even explains how writers can create them. "*Truth,*" he writes, "is never so well express'd or amplify'd, as by those Ornaments which are *True* and *Real* in themselves" (414). This verges on tautology (i.e., truth should be clothed with true ornaments). But elsewhere Sprat explains what he has in mind. Wit—the ability to perceive a "resemblance of one thing to another" and so the engine of metaphor—is "founded on such images which are generally known, and are able to bring a strong, and a sensible impression on the *mind*" (413). In other words, when we create a figure by drawing together two things ("X is like Y"), one term must strike us with an intense lucidity, as if it were seen and understood in the blink of an eye. Modern, more technical terms can help here: we sometimes say that in a metaphor, the "vehicle" (the known or intuitive part) illuminates the "tenor" (the otherwise unknown or unclear part). In these terms, Sprat argues that especially bright, vibrant, and conspicuous vehicles make for "*True* and *Real*" ornaments. These are ornaments that use their vehicles to make evident their tenors, to bring "*Knowledg*" back again to our very senses."

So while Sprat resists consigning figurative language as such to obscurity, he nevertheless continues to draw on his generative transparency-opacity distinction when describing the mechanisms of rhetoric. After all, a figure can be analyzed into the clarifying (vehicle) and the obscure (tenor), and once we grant that division, other elements soon come into focus. For instance, Sprat insists that when making metaphors, we cultivate true or real ornaments by reaping the "*Works of Nature,* which are one of the best and most fruitful Soils for the growth of *Wit*" (415).[23] Lest that point appear too abstract, we can see how it manifests in Sprat's own writing. The *History's* metaphors follow a distinct pattern. Tenors are plucked from culture; vehicles are derived from the natural world. Sprat seems to think that nature is evidently clearer than culture and that culture, usually obscure, must be illuminated by the light of nature. So the Scholastic's language is like feathers; the writing of alchemists is like smoke; wit is like salt, and so on.[24] In Sprat's metaphors, nature—always clear, ever enlightening—helps us make sense of culture. When Sprat needs to explain how exemplary cultural products like "*Tropes* and *Figures*" make our ideas hazy, indistinct, and vague, he naturally likens them to "mists." In this metaphor, encountering figurative language is like peering into a foggy landscape—a natural scene that is surely obscure but *vividly, concretely* so.

This defense of figurative language, it is worth stressing, is hardly odd or unique. Sprat's claim that tropes can convey more abstract information in "larger, fairer, and more moving Images" follows from the idea, evident especially in classical rhetoric, that tropes trade in *enargeia:* a force for enlivening—and so making more intuitive and apprehensible—certain ideas. Nor was Sprat the only science-minded writer at the time who thought *enargeia* could be of service to natural philosophy. As Alexander Wragge-Morley notes, Sprat's account of figure is arguably the most mainstream one in scientific writing of the time.[25] As we can see, this theory instrumentalized figures and, in doing so, tamed them. In this account, metaphor could move from a force that obstructed science's ends—by making communication difficult or even occluding nature—to a means to an end, to a tool that magnifies the work of natural philosophy. And as Sprat's writing in particular demonstrates, this tamed, instrumentalized vision of metaphor was itself grounded on some deeper assumptions about the substance of the mind, the relationship between language and reality, and the nature of nature and culture, science and rhetoric.

*   *   *

Rumors to the contrary, early advocates of science never sought to strip away all figurative language from their writing in some mad quest to speak in pure, flat literalities. In fact, such ornaments were purposefully placed in scientific discourse, their luster serving to animate otherwise dull material. So long as these figures remained *only* ornaments, so long as they amplified without distorting some more primary element, so long as they could be discarded like pleasing if inessential trimming, so long as they revealed the fruits of nature, thereby freeing us from the confines of culture, they could serve as valuable tools in the work of natural philosophy—which, in Sprat's writing, was itself figured as a stripping away of various surfaces and excrescences in an effort to see nature more plainly. Appreciating the proper place of rhetoric in early scientific work clears up an apparent contradiction. If we read early natural philosophers (or one of their advocates) as absolutely opposed to any kind of ornament, then it's simply hard to know what to make of those moments when a metaphor inevitably manifests itself in their writing. And while it's always possible to attribute such slips to hypocrisy or to the fact that figure can evade the strictest censorship, the discovery that such tropes are used purposefully and are even justified by a sophisticated rhetorical theory represents a real advance in insight.

With that in mind, we might now consider what is arguably the most elaborate ornament in *The History of the Royal Society*: not one of Sprat's many metaphors but rather Cowley's "To the Royal Society," the poem standing at the head of that text. As it tracks the rise of the Royal Society and comes to praise the group for some of the same accomplishments underlined by Sprat, Cowley's work evidently complements the *History* by rehearsing that work's narrative. And given this, we might assume that the poem serves as an ideal ornament (in Sprat's sense): that it makes vivid, without disfiguring, the *History*'s plainer prose.[26] But "To the Royal Society" is much stranger and more startlingly subversive than it first seems. Rather than merely versifying Sprat's claims, it is a work that, precisely in the course of augmenting and animating its underlying material, comes to critique Sprat—and even to offer alternatives. More specifically, Cowley's poem rejects the schematic transparency-opacity distinction that organizes Sprat's thinking. In its place, Cowley offers a more complex, playful, and dynamic conception of nature, culture, science, and mind. That Cowley's poem refuses to serve complacently as a "mere ornament," that it offers a new vision of science and language rather than a pleasing but faithful revision of Sprat's, makes sense once we appreciate how the poem itself uses figurative language. Where Sprat wants to "tame" metaphor—making it a means of magnifying, without warping, some underlying reality—Cowley's poem employs a freer use of figurative language. "To the Royal Society" opts for tropes that are creative rather than simply re-creative.

Over the course of nine stanzas, "To the Royal Society" tracks the growth of knowledge from its first, halting stages to its maturity, a period marked by the rise of the Royal Society itself. The poem commences with the story of "Philosophy," Cowley's term for all of human culture writ large. In its opening stanza, the poem personifies "Philosophy," transforming it into a young "Heir" who is set to inherit a "vast Estate": the natural world and its "endless Treasurie." Even though the heir is "three or four thousand" years old, he remains in an unnatural "nonage," thanks to the work of his "Guardians and Tutors," cultural elites who have stunted his growth to maintain their power. The narrative shifts in the poem's second stanza, when Francis Bacon enters the scene to take up "the injur'd Pupils caus." Bacon exorcises the "guardians"—now revealed as a ghost and a scarecrow—who have kept the heir scared off from the natural world. With their gothic presence gone, the heir (and indeed all humanity) is now welcomed by Bacon into a nature transformed into a lush garden: "Come, enter, all that will / Behold the rip'ned Fruit, come gather now your Fill." At this point—the third stanza—the poem's allegorical energies grow profuse. The original story gives way to a series of biblical allusions. The heir's "vast Estate" becomes the garden of Eden. Bacon himself transforms into a Moses who leads humanity to the frontier of the promised land but who, barred from entering himself, must leave his inheritors—the Royal Society—to breach its borders.

Again, much of this narrative parallels the history Sprat conveys more straightforwardly. Furthermore, the poem also appears to amplify some of Sprat's more general philosophical claims. For example, Sprat's worry that language and culture will cut off humanity from nature, that we will be confined to slippery simulacra rather than rock-solid reality, is also entertained in Cowley's allegory. Philosophy's guardians prevent the heir from realizing his nature by plying him with the distractions of words and wit:

> That his own business he might quite forgit,
> They amus'd him with the sports of wanton Wit,

> With the Desserts of Poetry they fed him,
> Instead of solid meats t' encreas his force;
> Instead of vigorous exercise, they led him
> Into the pleasant Labyrinths of ever-fresh Discours:

Still later, Cowley illustrates humanity's propensity to mistake words for things by likening our mistake to the birds taken in by Zeuxis's falsely realistic painting of grapes:[27]

> From Words, which are but Pictures of the Thought,
> (Though we our Thoughts from them perversly drew)
> To Things, the Minds right Object, he [Bacon] it brought,
> Like foolish Birds to painted Grapes we flew.

Such passages can appear as near perfect poetic refigurings of Sprat's points. And sure enough, read in isolation, they give the impression that Cowley accepts much of Sprat's thinking: that language is a screen or veil interposed between mind and nature, that it should be as transparent as possible, that an ideally plain language is aligned with the work of (Baconian) science, which pierces through the shadows wrought by rhetoric to reach the natural reality beyond it.

But to see Cowley's poem, as some of his modern critics have, as simply Sprat's work tricked out in tropes is to look past its other aspects. Cowley's differences from Sprat become clearest when we consider how the former's poem plays with the figuration of nature. We will remember that for Sprat, nature is consistently portrayed as a place of utter clarity, transparency, and plainness. Sometimes Cowley's poem, as if following Sprat to the letter, presents nature similarly. For instance, in the poem's concluding stanza, Cowley compares Sprat's style to limpid water: "His candid Stile like a clean Stream does slide, / And his bright Fancy all the way / Does like the Sun-shine in it play." Though we find ourselves wading in water, we seem to be on safe ground in these lines, since this is precisely how Sprat would want the natural world—not to mention his own writing—to be reproduced. Like a clear current—a synecdoche for the whole of nature—Sprat's language is pleasing, smooth, transparent. And so Cowley's figure ornaments something cultural by likening it to a nature marked in its lucidity; a vehicle drawn from the natural world (the stream) enlightens a tenor taken from culture (Sprat's style). But that dynamic hardly holds steady throughout the poem. Indeed, just two stanzas before this passage, Cowley describes nature in different, entirely un-Sprat-like terms when he comes to characterize the work of the Royal Society.

> Natures great Works no distance can obscure,
> No smallness her near Objects can secure.
> > Y' have taught the curious Sight to press
> > Into the privatest recess
> Of her imperceptible Littleness.
> She with much stranger Art than his who put
> > All th' *Illiads* in a Nut,
> The numerous work of Life does into Atomes fit.
> > Y' have learn'd to Read her smallest Hand,
> And well begun her deepest Sense to Understand.

In other words, nature—far from appearing in the guise of a bright, blatant stream—now stands cloaked by a kind of facade, beneath which are "private" recesses and bits of "imperceptible Littleness" that must be pried into by the Royal Society's "curious Sight." All of this is thanks to nature's "art," its ability to fit its secret mechanisms into the smallest interstices (such as atoms that conceal the "numerous work of Life").[28]

That nature evinces artifice, that it can hide and deceive, is already a striking departure from Sprat's work. Perhaps even more striking, though, is that this transformation finds nature taking on a new role in Cowley's metaphorics. It moves from vehicle to tenor. Consider that to capture this vision of an obscure or deceptive nature, Cowley draws on the resources of culture, as he likens natural philosophy to the work of reading crabbed handwriting. Hence nature now stands in the tenor position (it is the obscure thing in need of borrowed light), while something cultural (namely, handwriting) takes the place of the vehicle.[29] This shift neatly turns Sprat's system of tropes—nature as enlightening vehicle, culture as obfuscating tenor—on its head.

If we track Cowley's imagery throughout his poem, we soon discover that the verse puts into play terms like *nature* and *culture*, *transparency* and *opacity*, *tenor* and *vehicle* (terms that were arranged in strict, stable hierarchies in Sprat's writing). In Cowley's writing, nature and culture can be clear, obscure, tenor, vehicle, and so on depending on their place in the verse: a transparent stream at one moment, a tricky artificer the next. In fact, at its most complex, Cowley's poem makes every effort to make the distinction between nature and culture undecidable. It does so not by collapsing or confusing the two categories altogether. Rather, nature and culture come to take on the qualities of their opposites—only to swap them again. Here, for instance, is how the poem concludes. This passage follows the description of Sprat's style I quoted earlier:

> It [Sprat's writing] does like *Thames*, the best of Rivers, glide,
> Where the God does not rudely overturn,
>   But gently pour the Crystal Urn,
> And with judicious hand does the whole Current guide.
> T' has all the Beauties Nature can impart,
> And all the comely Dress without the paint of Art.

To translate, Sprat's style (something cultural) is like a river (something natural), and because his style has taken after nature, because it has learned to "glide" like a river, and because, as a result, it appears with "all the Beauties Nature can impart," it is *also* like a "comely Dress" (something cultural) that appears *as if* its beauty were entirely natural, as if it lacked any ornament dressing it up ("without the paint of Art"). So Sprat's style is so like nature that it appears like culture again, insofar as it seems to erase its own nature (as culture). Nature knows the art of erasing art, and culture becomes capable of taking on the appearance of an artificial nature. In other words, the qualities that Sprat himself rivets to nature and culture—transparency and opacity—float free, turn about, transfer, and are transformed thanks to the play of Cowley's metaphor.[30]

I want to suggest that if Sprat's metaphors exemplify the "instrumental" vision of figurative language, then Cowley's figures represent the more "creative" alternative. Cowley's figures are not beholden to some already instituted truth about, say, the nature of language or the essence of nature. They are not simply casting new and brighter light on old but abstract truths. Indeed, these metaphors are just as likely to question—precisely by probing different aspects of nature and culture, reality and art, the transparent and the opaque—as they are to clarify. This tendency to stir up uncertainty is not a sign of Cowley's commitment to some variety of skepticism but a mark of his poetic (in the widest sense of the word) natural philosophy. Like the work of more traditional natural philosophy, Cowley's figures undertake experiments: experiments *with* the resources of rhetorical language (think of his playing with tenor and vehicle) and experiments *upon* conceptions of nature, culture, language, mind.[31]

Notes

1. The belief in a rigid "plain style" is usually attributed to the influence of a series of essays by R. F. Jones (esp. "Science and English Prose Style"). For a helpful overview of writing on the plain style, see Guillory, "Mercury's Words." For an examination of the ideology fueling calls for plain style and further linguistic reform more generally, see Markeley, *Fallen Languages*.

2. See especially Brian Vickers's careful and critical examination of the evidence Jones adduces for the "plain style" ("Royal Society and English Prose"). Vickers demonstrates, thoroughly and convincingly, that the various arguments for "plainness" Jones locates in early scientific writing are more complex, ambivalent, and ambiguous than they seem. If Jones's work established the very idea of a "plain style" among contemporary scholarship, then Vickers's critique of Jones signaled the waning of that idea's influence. Nevertheless, for an essay focused on some of Vickers's own oversights, see Stillman, "Assessing the Revolution."
3. While the work of Vickers and others casts doubt on Jones's "plain style" thesis, it seems to me that we've only recently begun to get a grip on early science's reliance on nonplain (figurative, rhetorical, imaginative) language. On uses of metaphor, analogy, and other figures in scientific writing, see Rogers, *Matter of Revolution*; Smith, *Empiricist Devotions*; Goldstein, *Sweet Science*; Preston, *Poetics of Scientific Investigation*; Thompson, *Fictional Matter*; Wragge-Morley, *Aesthetic Science*; and Chico, *Experimental Imagination*. As its title implies, Chico's study considers not only tropes but the imagination more generally—and if, following her work, we consider not just the importance of figurative language for early science but imagination more generally, we might also add Bender and Aït-Touati on fiction (*Ends of Enlightenment; Fictions of the Cosmos*) and Al Coppola on theater (*Theater of Experiment*) to this list. To be drastically sweeping, though, we could say that scholarship on science and literature went through three stages: (1) a claim that early natural philosophers tried to institute a "plain style"; (2) a counterclaim that evidence for a sweeping "plain style" was misunderstood, overstated, or distorted; and then (3) the recognition that early natural philosophers purposefully used figurative language and other imaginative/literary devices in their work. The books listed in this note belong to this third, current period. My own essay suggests, perhaps, a fourth stage: (4) that while early scientists embraced *some aspects* of figurative language, they also feared and sought to limit rhetoric's more extravagant qualities.
4. Smith, *Empiricist Devotions*, 60. For a similar distinction, see Taylor, *Language Animal*. Lisa Forman Cody's essay in this collection turns on a similar division. As I read her, Cody shows how the various figures for population growth (e.g., people breeding like various exponentially reproducing animals) didn't just illustrate known truths but drove scientific research in deeper ways.
5. Do contemporary thinkers recognize the distinction between "instrumental" and "creative" uses of figurative language? Richard Moran, in one of the best surveys of recent work on metaphor, suggests a similar distinction when he divides metaphor theorists into two camps. For Moran, one camp, exemplified

by the philosopher Donald Davidson, insists that metaphors make us notice surprising or unseen aspects of things, but for all that, such figures do not impart wholly new meanings or information. The other camp, exemplified by Friedrich Nietzsche, argues that figures perform real conceptual work. With a great deal of caution, we could lump advocates of "instrumental" metaphors into the first camp (i.e., metaphors simply dress up, and so present differently, preexisting knowledge) and advocates for "creative" metaphors in the second (i.e., metaphors impart a wholly new significance). The "great deal of caution" is in order here, though, since, as we will see in the section on Thomas Sprat in this essay, early writers on metaphor relied on a theory of language and mind that most contemporary philosophers (in both the analytic and Continental traditions) reject. Moreover, as Moran himself demonstrates, a proper understanding of figurative language necessitates moving past this divide and finding a synthesis of the two camps (*Philosophical Imagination*).

6. On the cooperative character of the Royal Society and the authorship of the *History*, see Jarvis, "Thomas Sprat's 'Mixt Assembly.'"
7. There are now a handful of essays focusing on Sprat's *History* as well as a handful on Cowley's poem. But while an essay on Sprat sometimes mentions Cowley (or vice versa), no one (to my knowledge) has compared or contrasted Sprat's and Cowley's work at great length. The assumption is that they're basically in agreement and that Cowley is merely recasting Sprat's history in meter and metaphor—an assumption I challenge in this essay. I mention scholarship on Cowley in note 26.
8. This belief is now changing thanks to essays like Skouen's "Science versus Rhetoric?" To my mind, one of the best accounts of Sprat's rhetoric—and one that focuses on its sophisticated defense of rhetoric's ornamental nature—is Ryan J. Stark's reading of the *History* in his *Rhetoric, Science, and Magic*. For the rhetoric of the Royal Society more generally, see Dear, "Totius in Verba"; Nate, "Rhetoric in the Early Royal Society"; and Lynch, *Solomon's Child*.
9. This passage serves as the centerpiece of Jones's (now widely critiqued) work.
10. Arguably the same worry fueled the period's many attempts at imagining a more perfect language and at developing an artificial alternative to our messier natural languages. See Markeley, *Fallen Languages*.
11. Sprat, *History*, 111. All further citations from this text appear parenthetically.
12. For advocates of early science, the distinction between "words" and "things" was at the heart of linguistic theory. See Howell, "Res et Verba," for more on these terms. For just how ambiguous that distinction could be, see Lynch, *Solomon's Child*, esp. 20–22. See Mann, *Outlaw Rhetoric*, 201–18, for a helpful description of the contrasting early modern and Enlightenment accounts of this pairing.

13. Cf. Sprat, *History*, 105.
14. This is certainly how Cowley understood Sprat. See his poetic recasting of this point in his poem: "From Words, which are but the Pictures of the Thought, . . . To Things, the Minds right Object."
15. It seems to me that much of what Sprat implies about the nature of mind and language is said, more explicitly and more systematically, by Locke. I'm thinking particularly of Locke's conception of the mind as a representational tool (which squirrels away sense data in an internal alcove) and his insistence that language serves as both an ordering and clarifying mechanism. See Locke's *Essay*.
16. For more on this model of mind, see Rorty, *Philosophy and the Mirror of Nature*; and Dreyfus and Taylor, *Retrieving Realism*. See Brandom, *Articulating Reasons*, for representationalism and language; for representationalism in science and naturalism more generally, see Hacking, *Representing and Intervening*, and Price, *Naturalism without Mirrors*; and in literary studies, see Silver, *Mind Is a Collection*; and Kramnick, *Paper Minds*.
17. For an excellent account of how these optics followed from the words/things distinction, see de Grazia, "Words as Things."
18. For a (much) more wide-ranging consideration of these terms and their interplay with language and truth, see Blumenberg, "Metaphorics of the 'Naked' Truth."
19. On figurations of nature as a veiled and then naked woman who must be exposed to the penetrating gaze of the male scientific community, see Merchant's classic account, *The Death of Nature*, as well as Hadot's more wide-ranging examination of such imagery from antiquity to the present, *The Veil of Isis*. Katharine Park's more fine-grained examination of this imagery in medieval and early modern works corrects some of Merchant's claims ("Nature in Person").
20. Cf. Sprat, *History*, 5–22.
21. This striking phrase is from McDowell, *Mind and World*.
22. Think, for instance, of the difference between clothing that suits you, that seems to reflect and bring out your personality, and clothing that is more like a costume, that transforms you into something else.
23. Indeed, not just nature but a nature explored and experimented on by natural science produces the best vehicles; cf. Sprat, *History*, 415.
24. Sprat, *History*, 15–16, 74–75, 419.
25. See Wragge-Morley, *Aesthetic Science*, esp. 126–33. Wragge-Morley provides one of the best accounts of the rhetorical and emotional theories lurking behind so much theorizing about science and language in the seventeenth century.
26. This seems to be the current consensus on Cowley's relationship to Sprat. Indeed, most critics tend to look past Sprat—to Bacon or perhaps Hobbes—in

order to locate the real source of Cowley's thinking. See Butler, "Stagirite and the Scarecrow"; Green, "Poet in Solomon's House"; Malpas, "In No One Thing"; Martin, "Two Baconian Poets"; Guibbory, "Imitation and Originality"; and Hinman, *Abraham Cowley's World*.

27. For an extraordinary reading of Cowley's take on these birds—and some of the paradoxes of nature and culture, reality and art in which they find themselves ensnared—see Festa, *Fiction without Humanity*, 40–47.
28. The image of nature evident at this moment in the poem—a nature that hides the "work of Life" in atoms, wholly unlike what we see on its surface—is more in tune with the Royal Society's actual conception of the natural world than Sprat's idealized visions of pleasant gardens filled with abundant fruit ripe for the picking. After all, the Royal Society was fascinated by what came to be called the primary/secondary quality division (the idea that the actual makeup of material things was nothing like the phenomenal effects these things produce in us), and in a quest to make sense of such dynamics, natural philosophers like Boyle found themselves hunting for the invisible corpuscles or atoms in the "privatest recess" of nature's "imperceptible Littleness." For more on this conception of the natural or physical world and its connection to literary and aesthetic matters, see Thompson, *Fictional Matter*; and Norton, "Aesthetics, Science, and the Theater of the World."
29. On Cowley's experiments with metaphor, see Trotter, *Poetry of Abraham Cowley*, 45–47; and Stogdill, "'Pindaric Way,'" 488.
30. That nature, in Cowley's poetry, can be both plain and obscure, blatant and veiled, is one reason his relationship to the imagistic tradition dissected by Merchant and Hadot—a tradition that finds a male-coded science prying into the confines of a half-naked female nature—is so complex. As we can see, Cowley certainly indulges in such imagery at certain points in his poem (e.g., "the curious Sight" of the Royal Society comes "to press" into nature's "privatest recess"), and Frank Boyle, in his essay in this collection, has adduced another example of this dynamic in Cowley's poem of praise to William Harvey. But at other moments, the figurative energy of Cowley's verse grows so profuse, and the clothed/naked imagery so baroque, that Cowley almost seems to be mocking or ironizing this tradition (e.g., note that the same Sprat who longs to see the "Beautiful Bosom of *Nature* . . . Expos'd to our view" in his own prose finds himself wearing a "comely Dress" "impart[ed]" by "all the Beauties [of] Nature" in Cowely's poetic refiguring).
31. We might add that the purpose of such experiments is both to probe our conceptions of nature and culture and to make sense of the way these different domains interact and overlap. In this respect, Cowley's poem is interested in one of the key questions raised by modern science: Just what is the relationship between nature and culture? For more on this question and its

importance to contemporary conceptions of science, see Sellars, "Philosophy and the Scientific Image."

Bibliography

Aït-Touati, Frédérique, and Susan. Emanuel. *Fictions of the Cosmos: Science and Literature in the Seventeenth Century.* Chicago: University of Chicago Press, 2011.

Bender, John B. *Ends of Enlightenment.* Stanford: Stanford University Press, 2012.

Blumenberg, Hans. "Metaphorics of the 'Naked' Truth." In *Paradigms for a Metaphorology,* 40–51. Ithaca: Cornell University Press, 2016.

Brandom, Robert. *Articulating Reasons: An Introduction to Inferentialism.* Cambridge, MA: Harvard University Press, 2000.

Butler, Charles. "The Stagirite and the Scarecrow: Stanza 3 of Cowley's Ode 'To the Royal Society' (1667)." *Restoration: Studies in English Literary Culture, 1660–1700* 21, no. 1 (1997): 1–14.

Chico, Tita. *The Experimental Imagination: Literary Knowledge and Science in the British Enlightenment.* Stanford: Stanford University Press, 2018.

Coppola, Al. *The Theater of Experiment: Staging Natural Philosophy in Eighteenth-Century Britain.* New York: Oxford University Press, 2016.

Dear, Peter. "Totius in Verba: Rhetoric and Authority in the Early Royal Society." *Isis* 76, no. 2 (1985): 145–61.

de Grazia, Margreta. "Words as Things." *Shakespeare Studies* 28 (2000): 231–35.

Dreyfus, Hubert L., and Charles Taylor. 2015. *Retrieving Realism.* Cambridge, MA: Harvard University Press.

Festa, Lynn. *Fiction without Humanity: Person, Animal, Thing in Early Enlightenment Literature and Culture.* Philadelphia: University of Pennsylvania Press, 2019.

Goldstein, Amanda Jo. *Sweet Science: Romantic Materialism and the New Logics of Life.* Chicago: University of Chicago Press, 2017.

Green, Mary Elizabeth. "The Poet in Solomon's House: Abraham Cowley as Baconian Apostle." *Restoration: Studies in English Literary Culture, 1660–1700* 10, no. 2 (1986): 68–75.

Guibbory, Achsah. "Imitation and Originality: Cowley and Bacon's Vision of Progress." *Studies in English Literature, 1500–1900* 29, no. 1 (1989): 99–120.

Guillory, John. "Mercury's Words: The End of Rhetoric and the Beginning of Prose." *Representations,* no. 138 (2017): 59–86.

Hacking, Ian. *Representing and Intervening: Introductory Topics in the Philosophy of Natural Science.* Cambridge: Cambridge University Press, 1983.

Hadot, Pierre. *The Veil of Isis: An Essay on the History of the Idea of Nature.* Cambridge, MA: Belknap Press of Harvard University Press, 2006.

Hinman, Robert B. *Abraham Cowley's World of Order.* Cambridge, MA: Harvard University Press, 1960.

Howell, A. C. "Res et Verba: Words and Things." *ELH* 13, no. 2 (1946): 131–42.
Jarvis, J. Ereck. "Thomas Sprat's 'Mixt Assembly': Association and Authority in 'The History of the Royal Society.'" *Restoration: Studies in English Literary Culture, 1660–1700* 37, no. 2 (2013): 55–77.
Jones, Richard F. "Science and English Prose Style in the Third Quarter of the Seventeenth Century." *PMLA: Publications of the Modern Language Association of America* 45, no. 4 (1930): 977–1009.
Kramnick, Jonathan Brody. *Paper Minds: Literature and the Ecology of Consciousness*. Chicago: University of Chicago Press, 2018.
Lynch, William. *Solomon's Child: Method in the Early Royal Society of London*. Stanford: Stanford University Press, 2001.
Markley, Robert. *Fallen Languages: Crises of Representation in Newtonian England, 1660–1740*. Ithaca: Cornell University Press, 1993.
Malpas, Simon. "In No One Thing, They Saw, Agreeing: Communicating Experimental Philosophy in Cowley and Butler." *Restoration: Studies in English Literary Culture, 1660–1700* 43, no. 2 (2020): 49–74.
Mann, Jenny C. (Jenny Caroline). *Outlaw Rhetoric: Figuring Vernacular Eloquence in Shakespeare's England*. Ithaca: Cornell University Press, 2012.
Martin, Catherine Gimelli. "Two Baconian Poets, One Baconian Epic." In *Milton and the New Scientific Age*, 17–52. Routledge, 2019.
McDowell, John. *Mind and World*. Cambridge, MA: Harvard University Press, 1994.
Merchant, Carolyn. *The Death of Nature: Women, Ecology, and the Scientific Revolution*. San Francisco: Harper & Row. 1980.
Moran, Richard. 2017. *The Philosophical Imagination: Selected Essays*. New York: Oxford University Press.
Nate, Richard. "Rhetoric in the Early Royal Society." In *Rhetoric and the Early Royal Society: A Sourcebook*, edited by Tina Skouen and Ryan J. Stark. Leiden: Brill, 2014.
Norton, Brian Michael. "Aesthetics, Science, and the Theater of the World." *New Literary History* 51, no. 3 (2020): 639–59.
Park, Katharine. "Nature in Person: Medieval and Renaissance Allegories and Emblems." In *The Moral Authority of Nature*, edited by Lorraine Daston and Fernando Vidal, 50–73. Chicago: University of Chicago Press. 2004.
Preston, Claire. *The Poetics of Scientific Investigation in Seventeenth-Century England*. Oxford: Oxford University Press. 2015.
Price, Huw. *Naturalism without Mirrors*. Oxford: Oxford University Press, 2010.
Rogers, John. *The Matter of Revolution: Science, Poetry, and Politics in the Age of Milton*. Ithaca: Cornell University Press, 2018.
Rorty, Richard. *Philosophy and the Mirror of Nature*. Princeton: Princeton University Press, 2008.
Sellars, Wilfrid. "Philosophy and the Scientific Image of Man." In *In the Space of Reasons: Selected Essays of Wilfrid Sellars*, edited by Kevin Scharp and Robert B. Brandom, 369–408. Cambridge, MA: Harvard University Press, 2007.

Silver, Sean. *The Mind Is a Collection: Case Studies in Eighteenth-Century Thought.* Philadelphia: University of Pennsylvania Press, 2015.
Skouen, Tina. "Science versus Rhetoric? Sprat's History of the Royal Society Reconsidered." *Rhetorica* 29, no. 1 (2011): 23–52.
Smith, Courtney Weiss. *Empiricist Devotions: Science, Religion, and Poetry in Early Eighteenth-Century England.* Charlottesville: University of Virginia Press, 2016.
Sprat, Thomas. *The History of the Royal Society.* London, 1667.
Stark, Ryan J. *Rhetoric, Science, and Magic in Seventeenth-Century England.* Washington, DC: Catholic University of America Press, 2009.
Stillman, Robert E. "Assessing the Revolution: Ideology, Language, and Rhetoric in the New Philosophy of Early Modern England." *The Eighteenth Century* 35, no. 2 (1994): 99–118.
Stogdill, Nathaniel. "Abraham Cowley's 'Pindaric Way': Adapting Athleticism in Interregnum England." *English Literary Renaissance* 42, no. 3 (2012): 482–514.
Taylor, Charles. *The Language Animal: The Full Shape of the Human Linguistic Capacity.* Cambridge, MA: Harvard University Press, 2016.
Thompson, Helen. *Fictional Matter: Empiricism, Corpuscles, and the Novel.* Philadelphia: University of Pennsylvania Press, 2017.
Trotter, David. *The Poetry of Abraham Cowley.* London: Macmillan, 1979.
Vickers, Brian. "The Royal Society and English Prose Style: A Reassessment." In *Rhetoric and the Pursuit of Truth: Language Change in the Seventeenth and Eighteenth Centuries,* edited by Brian Vickers and Nancy S. Struever, 3–63. Los Angeles: William Andrews Clark Memorial Library, UCLA, 1985.
Wragge-Morley, Alexander. *Aesthetic Science: Representing Nature in the Royal Society of London, 1650–1720.* Chicago: University of Chicago Press, 2020.

Centlivre's Frankenstein

Science, Authorship, and Monstrous (Re)Productions in A Bold Stroke for a Wife

MELISSA BAILES

Much modern scholarship about the early eighteenth-century poet and playwright Susanna Centlivre focuses on the extent to which she should be considered a plagiarist.[1] Regarding this phenomenon, Laura Rosenthal remarks on the gendered nature of such critiques, noting that they exemplify and perpetuate a broader trend in Centlivre's own time of women writers being denied the capacity for originality and accused instead of various forms of literary borrowing to explain their writing's success.[2] As Centlivre's authorial choices demonstrate, she was very aware of the gender dynamics involved in contemporary standards of literary borrowing. Nevertheless, I argue that she also helped catalyze an important shift in understandings of plagiarism and originality that continued to develop throughout the century and into the Romantic period by interrogating and transforming these literary concepts' analogies with new political and scientific ideologies.

Restoration-era notions of authorial borrowing often relied on comparisons between monarchical and literary lines of succession so that imitation could be justified through adherence to the work of a writer's aspirational literary "father." I argue that in her play *A Bold Stroke for a Wife* (1718), which remained popular well into the nineteenth century, Centlivre models a break from this earlier mode of imitation patterned on sociopolitical functions of absolute monarchy. As other scholars have asserted, this play champions Whig politics and the social promise of constitutional monarchy, with its implied leveling of relationships through contracts. I show that Centlivre additionally thereby reformulates approaches to literary originality, adhering to Whig ideals of liberty and universal rights in her conceptions of authorship and authorial property.

Moreover, as I will demonstrate, Centlivre's strategy to adjust ideas about authorship draws literary comparisons not only with politics but also with contemporary science, as ideas about inheritance were espoused simultaneously through these different discourses. Invoking earlier notions of originality through "familial" lines of succession and thus the analogy of biological reproduction, she suggests that such expectations for literary reproduction can only end in monstrous results. Centlivre employs variations of the term *monster* throughout her play, associating its sociobiological implications with the creation of hybrid works containing unsuccessful literary borrowings and imitations. In this way, she anticipates and influences standards of plagiarism that would become prevalent in the latter half of the eighteenth century and Romantic era, when the incorporation of another poet's style, identity, or voice into one's literary work, if not sufficiently improved, would appear as a hybrid monstrosity, vulnerable to charges of aesthetic plagiarism. In thinking about these notions of monstrosity, reproduction, and authorship, I conclude the essay by briefly exploring the play's surprising afterlife during the Romantic period, including its similarities with and possible influence on Mary Shelley's *Frankenstein* (1818). Centlivre's text thus signals an important transition, determining how changing perceptions of science and politics would inform and reshape subsequent ideas about authorship and literary originality.

Dryden's Literary Monarchy: Legitimized Borrowing through Hereditary Lineages

During the Restoration era, John Dryden championed a line of literary inheritance and reproduction that would encourage and justify the imitation of particular authorial predecessors. On the one hand, he wanted to reinforce the legitimacy of monarchy and its line of succession, and on the other hand, he felt daunted at the prospect of trying to surpass the giants of the stage in the previous age. Dryden thus presented himself as the literary inheritor of "Shakespeare's adoptive parentage," but as Richard Terry points out, "the danger was always that . . . the right of sonship might shrink to mere filial replication of the father in a manner that, in literary terms, would be indistinguishable from plagiarism."[3] Despite this authorial risk of imitation, Dryden not only wished to determine his own literary parentage but also hoped to father some poetic sons of his own. Both Alexander Pope and Charles Churchill prided themselves "on having taken up Dryden's mantle, but [William] Congreve appears to have been the younger poet whom Dryden most coveted as his natural heir."[4] In fact, in Dryden's dedicatory epistle, "To My Dear

Friend Mr. Congreve" (1694), Dryden selects Congreve as his poetic heir and one true successor qualified for ascent to the "throne of wit" after his own demise.[5] Moreover, through this system of literary lines of succession, Dryden sought to bar his rivals from this literary "throne," as is famously the case in *Mac Flecknoe* (1682), which satirizes the rival playwright, Thomas Shadwell, as unrelated to the literary father Shadwell desired, Ben Jonson, and instead as embarrassingly sired by Flecknoe. In this way, Dryden employed analogies of legal rights of inheritance to police what constituted authorial genius and originality as well as who had access to them, thereby legitimizing certain literary borrowings and imitations.

Within an era of primogeniture, this idea of literary lines of succession raises the question of whether or how women might fit into this dynamic of authorial reproduction. Several critics note that women writers of the Restoration and eighteenth century were more frequently accused of plagiarism than their male counterparts and had more difficulty asserting control over their works in this era of coverture.[6] As Rosenthal argues, the poetic genealogy "trope's inscribed fantasy of male" creation and legitimation removed women from the reproductive process because it was "based on masculine lineal descent."[7] Thus, to fit into this authorial, genealogical framework, women writers sometimes expressed a masculine identity or had one imposed on them by male writers and critics. For, in this era, "women's writing was viewed by many as unseemly, as a defiance or monstrosity of nature. This sense of abnormality grew from a belief that literary composition sprang from intellectual resources that were inherently masculine, so that women who took up the pen were necessarily engaged in an act of self-unsexing or cross-dressing."[8] For example, contemporary descriptions of Aphra Behn often remarked that she "did at once a Masculine wit express," and Dryden rationalized Anne Killigrew's literary talents by suggesting, "Thy father was transfused into thy blood."[9]

Centlivre's Originality

Centlivre herself engaged in cross-dressing both literally, as when she admitted enjoying playing masculine characters on stage, and figuratively, as when she sometimes portrayed parts of her own work as written "by a Gentleman." Her prologue to *A Bold Stroke for a Wife* is presented as being written by a man but spoken by a woman.[10] Thus, she employs this gesture toward a fictitious masculine perspective to help substantiate her claims of originality within the work. In other words, this "cross-dressing" makes it appear that she does not merely assert novelty for herself, but another, masculine, authoritative voice recognizes her originality and endorses her implicit challenge to the (patriarchal) poetic line of succession as well. She opens the prologue, writing,

> Tonight we come upon a bold design,
> To try to please without one borrowed line.
> Our plot is new, and regularly clear,
> And not one single tittle from Molière.
> O'er buried poets we with caution tread,
> And parish sextons leave to rob the dead. (ll. 1–6, Prologue)

Centlivre's announcement of her method for originality is striking. She claims to have created a "new" plot, "without one borrowed line," and adds the patriotic assurance of owing nothing to French playwrights. Significantly, she also proclaims a break with the mode of poetic genealogy invoked by Dryden and other male playwrights and poets of the Restoration and early eighteenth century to legitimate their writing through alliance with literary predecessors. Centlivre instead repudiates such affiliations with "the dead." She has not "rob[bed]" these "buried poets" and conjures up an image of the author "cautio[usly] tread[ing]" around "plot[s]" in the literary graveyard to forge her own original path and "bold[ly] design" something "new."

Countering the idea that she must adhere to some line of literary inheritance by imitating or drawing on these "buried poets," Centlivre dispenses with this notion of succession. She alternatively claims a kind of unindebted authorship, an aspirational aesthetic fantasy that would become increasingly prevalent for writers within the latter half of the century and the Romantic era, theorized in, for instance, Edward Young's *Conjectures on Original Composition* (1759). In that work, biological hybrids' reputed inability to reproduce prompts Young to draw the literary analogy, "an *Original* author is born of himself, is his own progenitor, and will probably propagate a numerous offspring of Imitators, to eternize his glory; while mule-like Imitators, die without issue."[11] More contemporarily, for Centlivre, this conception of authorship also supports her Whig political beliefs in intellectual and sociopolitical freedoms for the individual as opposed to reliance on familial (or literary) inheritance and class hierarchies.

In addition to these assertions in the prologue, Centlivre emphasizes her play's originality within its dedication, affirming, "The plot is entirely new and the incidents wholly owing to my own invention, not borrowed from our own or translated from the works of any foreign poet; so that they have at least the charm of novelty to recommend 'em."[12] Despite her strong declarations of "novelty," after Centlivre's death, one of her male contemporaries, John

Mottley, declared that he "wrote one or two entire Scenes of" *A Bold Stroke for a Wife;* however, there is no evidence to substantiate Mottley's statement. As Thalia Stathas suggests, "Centlivre's dedicatory remarks cannot be disregarded, particularly since this is the only play for which she claims complete originality. That the comedy is stylistically and thematically consistent from beginning to end tends to support her statement. Presumably, she reworked any contributions that Mottley may have made."[13] Mottley's attempt to appropriate Centlivre's work after her death exemplifies a larger trend in which male writers felt entitled to assert ownership of women's texts, for as Tilar Mazzeo notes of women writers' predicament in association with coverture, "if men could assimilate her person, then why could they not assimilate her personal expressions as well?"[14] Nevertheless, my intention is not to vindicate Centlivre regarding potential borrowings in her work. Instead, I would like to posit that Centlivre claims for herself a kind of originality that does not necessarily preclude borrowings but elides their presence in a performative act of alternative authorship, approaching the relationship between novelty and reproduction differently from the standards implied in a poetic genealogy, and that she subtly demonstrates these new standards within the text of her play.

Scientific Monstrosity and Biological Reproduction

Throughout *Bold Stroke,* Centlivre critiques expectations that authors situate themselves within a literary genealogy by displaying such lines of succession as self-destructive in the sense of ungenerative or generating only imitation rather than something new. In the play's opening scene, the reader learns that the will of Anne Lovely's deceased father includes a stipulation that she can only marry and keep her inheritance with the mutual consent of four male guardians, listed in the play's dramatis personae as "an old beau," "a kind of silly virtuoso," "a changebroker," and "a Quaker." These guardians' separate interests and temperaments ensure perpetual disagreement, which is precisely why her father chose them. In describing Lovely's deceased father, other characters state, "This was his only child: I have heard him wish her dead a thousand times" because "he hated posterity . . . and wished the world were to expire with himself. He used to swear if she had been a boy, he would have qualified him for the opera" by castrating him.[15] Another character responds, "'Tis a very unnatural resolution in a father" (1.1.89). This judgment of the father's sentiments and actions as "unnatural" foreshadows the play's frequent gestures toward *monstrosity,* one connotation of which refers to "something repulsively unnatural . . . outrageously or offensively wrong."[16]

Within the context of Centlivre's dedication and prologue, this deceased father's monstrous hatred of posterity and selfish desire to control or literally cut off the possibility of reproduction and continuance of his hereditary line sets up a paralleled expectation for Centlivre's depiction of the collapse or alteration of the concept of patrilineal literary genealogy.

Centlivre repeatedly applies variations of the term *monstrosity* to the play's oppressive sociopolitical systems, generally represented by the deceased father and four guardians. In addition to its moral and aesthetic indication of something unnatural, monstrosity also had prevalent scientific connotations. Within early eighteenth-century science, monstrosity could refer to an animal or plant or some part of that organism "that is abnormally developed or grossly malformed," and the concept is also influenced by notions of hybridity or mixture, so a monster could constitute "a mythical creature which is part animal and part human, or combines elements of two or more animal forms."[17] Contemporary members of the Royal Society exhibited a fascination with natural objects that could be considered monstrous. Indeed, monstrous births, both human and otherwise, formed one of the favorite recurring topics in *Philosophical Transactions* (1665–present), the world's oldest scientific journal and England's longest-running periodical, which often contained the correspondence, theories, and observations of Royal Society members. For example, the first issue of *Transactions* contained Robert Boyle's correspondence about "a very odd, monstrous birth" of a calf "whose hinderlegs had no joints and whose tongue was, Cerberus-like, triple—to each side of his mouth one, and one in the midst," and its sternum "was perfect stone."[18] This account's combination of scientific fact and classical mythology in portraying the calf as a figure of monstrous hybridity displays both the developing state of natural philosophy at this time and the seeming outlandishness that would leave such scientific pursuits open to ridicule by writers such as Jonathan Swift and Margaret Cavendish.

Scientific forms of monstrosity appear most obviously in Centlivre's play through the interests of the virtuoso/guardian, Periwinkle, a collector of antiquities, rarities, and monstrosities of ancient civilizations and the natural world. Periwinkle's very name places him among his collection of natural objects, referring either to a species of bluish-purple flowers or to certain kinds of mollusks.[19] Although obsessed with travel narratives about different cultures and species around the globe, never having traveled himself, he gullibly relies on others' knowledge and experiences to inform his obscure interests. Therefore, to win over this guardian and attain the hand and dowry of Anne Lovely, Colonel Fainwell presents himself as a traveler and antiquarian, initially achieving Periwinkle's readiness to sign his consent for the marriage.

In convincing Periwinkle to sign his consent, Fainwell's discourse highlights the virtuoso's unnatural enthusiasm for impractical knowledge, objects, and practices from the past. To this end, Fainwell humorously explains that according to a "learned physiognomist" in Egypt, his destiny is to find and marry Lovely so that she can "bear me a son, who shall restore / the art of embalming and the old Roman manner of burying their dead, and, for the benefit of / posterity, he is to discover the longitude, / so long sought for in vain" (3.1.260–64). Periwinkle replies, "These are very valuable things. . . . If I live till this boy is born, I'll be embalmed and sent / to the Royal Society when I die" (3.1.265, 293–94). Of course, at this time, any attempt to solve the problem of longitude could seem, as William Hogarth memorably depicts in *A Rake's Progress* (1735), a ludicrous and fruitless quest, indicative of insanity. Periwinkle's hope that a child's birth may merely revive ancient practices revolving around death displays his misplaced values, emphasizing unproductive repetition and regression rather than genuine forward progress. His perverse and ungenerative views and actions become most apparent when he rejoices in the deaths of his wife and daughter:

> Pish! Women are no rarities. I never had any great taste that way. I married, indeed, to please a father and I got a girl to please my wife; but she and the child (thank Heaven) died together. Women are the very gewgaws of the creation; playthings for boys, which, when they write man, they ought to throw aside. (3.1.238–44)

Periwinkle further scorns that women's charms fall short when compared with those of natural objects such as a cockatoo, hummingbird, or butterflies and are therefore unworthy of attention. In this context, revealing science's potentially dangerous associations with isolation, death, and lunacy, Lovely later exhorts, "Ah, study your country's good, Mr. Periwinkle, and / not her insects. Rid you of your homebred / monsters before you fetch any from abroad. I dare / swear you have maggots enough in your own brain / to stock all the virtuosos in Europe with butterflies" (5.1.123–27). Periwinkle's distaste for women, and his preference instead for preserved natural objects and arcane, esoteric, solitary studies, represents a stereotype of "men of science" that would remain in place at least for the next century.[20]

However, Periwinkle's misogynistic statements also highlight an overarching theme within the play through situations that largely remove women from the process of courtship and reproduction. As Vivian Davis has shown, Fainwell's challenge in gaining each of the four male guardians' consent to marry Lovely results in her absence from much of the play so that the actions of courtship paradoxically occur within Fainwell's wooing of the guardians themselves.[21] For example, during Fainwell's pursuit of Modelove's consent, the latter fulfills the feminine role in a traditional romantic plot, responding, "We appear to have / but one soul, for our ideas and conceptions are the / same" and "I / would prefer you to all men I ever saw" (2.1.83–85, 165–66). This removal of women creates homosocial dynamics and a male fantasy of self-reproduction in which Fainwell gains the guardians' approval by replicating an ideal version of each of them. I argue that this demonstrates Centlivre's critique of literary genealogy and its similar expectations for androcentric reproduction within an authorial line of succession. Especially through Lovely's deceased father and Periwinkle, Centlivre emphasizes these male characters' antipathy to biological reproduction or the creation of autonomous individuals, demonstrating that the mindset involved in the literary line of succession may be less a hope for biological/authorial immortality than a wish that "the world were to expire with himself" (1.1.86). In both its authorial and sociopolitical applications, this egoistic obsession with controlling self-reproduction represents a kind of self-destruction that ceases actual creativity or novelty and positive progression. The burden of reproducing the past hinders the autonomy necessary to achieve authorial originality.

Through the character of Lovely, Centlivre favors instead what we might categorize as Whig principles of authorship. Lovely praises Fainwell as a soldier who "ought to be preferred before / those lazy, indolent mortals, who, by dropping into / their father's estate, set up their coaches and think / to rattle themselves into our affections" (1.2.66–69). Rather than valuing a class hierarchy and notion of authorship maintained through inheritance, Centlivre emphasizes a meritocracy in which wealth and status are earned through skill and hard work. Additionally, in Fainwell's analogized ability to manipulate these expectations of authorial genealogy, Centlivre demonstrates an approach to literary composition that encourages more autonomy in both borrowing and originality.

Literary Monstrosity: Imitation, Borrowing, and Reproduction

Significantly, in addition to the scientific, hybrid monstrosities discussed earlier as alterations of biological forms, monstrosity also acquired specific literary application within standards of plagiarism. As Mazzeo delineates, within the latter half of the eighteenth century and through the Romantic era, there were two main categories of plagiarism: culpable and poetical.[22] In this distinction, culpable plagiarism was very difficult to prove because it signified "borrowings that were *simultaneously* unacknowledged, unimproved, unfamiliar, and conscious," carrying the stigma of a moral trespass.[23] In the absence of any one of these elements, the borrowing can only be considered poetical plagiarism, which indicates an aesthetic, not moral, indictment of the author. Of course, in the earlier standards of literary genealogy, borrowings and imitations were often intentionally familiar, acknowledged, and conscious. In this new conception, successful improvement, the most important of these aesthetic components, required unity of style and seamlessness and could justify any borrowing; conversely, "unimproved texts were frequently described as *monstrous patchwork,* or unassimilated, suggesting that the evaluation of literary works depended upon precise definitions of textual unity."[24] I argue that with Centlivre's repudiation of poetic genealogy, she provides early expressions of these developing standards of plagiarism and originality, incorporating analogies of cohesive identity directly into her play's plot.

In *Bold Stroke,* the character of Colonel Fainwell thus represents an authorial persona possessing the capacity for seamless borrowing and improvement through unity of style, maintaining an original and coherent voice or identity. In order to secure Anne Lovely and her fortune, Fainwell must live up to his name and convincingly feign the expressions, interests, and personalities of each of her guardians. In this way, his actions and dialogue enact a literary analogy for developing standards of legitimate borrowings as he must successfully imitate and even improve the personae represented by each of these four men in turn. Where his borrowings or imitations are seamless, creating a unity of style that allows his ventriloquism to go undetected, he wins the unwitting guardian's approval and consent. This is the case in his first attempts at assimilating and improving the expressions of Modelove, Tradelove, and Prim by representing the ideal versions of what they each aspire to embody within their individual ideological communities.

Indeed, within his final act of feigning, Fainwell assumes the persona of the play's most enduring character, the Quaker preacher Simon Pure, whose name subsequently developed a life of its own in the English language, meaning both a person of irreproachable virtue or integrity and also that which is real, genuine, or authentic.[25] Acting as Simon Pure, Fainwell is confronted by the actual Simon Pure, forcing both men to prove their authenticity, with Fainwell humorously winning while the real Simon Pure is cast out of the house by Prim. On the other hand, when the imitation fails, as occurs at the close of Fainwell's initial attempt to ventriloquize the expressions and interests of Periwinkle when a servant inadvertently reveals the colonel's identity, then the borrowing becomes monstrous because of the now-obvious disunity of styles incorporated into the authorial persona; thus, his "borrowing" of another's "voice" becomes apparent.

Embodying Centlivre's alternative brand of authorship, in his closing speech, Fainwell refuses to allow anyone to distort his identity, declaring, "I am the person who can give / the best account of myself" (5.1.605–6). Discarding the burden of poetic genealogy, Centlivre proposes that to "force" any obvious imitation or borrowing through adherence to a line of literary succession "makes many a sinner, not one saint"; instead of such constraint to particular past literary productions or expectations, in Centlivre's Whiggish conception of authorship, "free as air the active mind does rove" and requires "liberty of choice" if originality is to be achieved (5.1.619–20, 624). Through Fainwell, Centlivre exemplifies successful borrowings that go undetected because they are improved and cohesively incorporated into his original authorial identity. Moreover, since Lovely performatively integrates Quaker beliefs and enthusiasm at least as well as Fainwell to win Prim's consent in the play's final scene, Centlivre makes it clear that women possess an equal ability to succeed in this mode of authorship.

Anticipating Frankenstein: Simon Pure, Monstrosity, and Originality

Due to its increasing popularity on stage after the mid-eighteenth century, Centlivre's *A Bold Stroke for a Wife* "could claim the status of a cult object during the decades of the Romantic period" and became bound up with ideas about authorship.[26] For example, in the *Peter Bell* controversy of 1819, John Keats's friend, John Hamilton Reynolds, found out that William Wordsworth would soon publish a poem called *Peter Bell* and so wrote his own burlesque under that title, parodying Wordsworth's poetic style and heading it with a slight misquote from Fainwell in Centlivre's play, declaring, "I do affirm that I am the real Simon Pure."[27] Of course, here that statement satirically implies "I am the real Wordsworth." Wordsworth's *Peter Bell* was published about a week later, and several reviewers mischievously claimed difficulty in discerning "the real Simon Pure."[28] Keats, S. T. Coleridge, and Percy Shelley all remarked on the significance of this reference to the integrity of authorial identity through Centlivre's play, and Keats suggested that such a parody also should be done of Lord Byron's poetic style.[29]

In light of the popularity of Centlivre's play during this era and its acknowledged implications for authorship, similarities between *Bold Stroke* and Mary Shelley's *Frankenstein* become all the more striking. Although Shelley's novel was published exactly a century later, reacting to a different historical, literary, and scientific context, a concern with authorship and creativity permeates her work. Most importantly in this regard, Centlivre's critique of a patriarchal, literary genealogy that seeks to reproduce texts and authorial lines of succession in the absence of women finds a parallel in often-repeated readings of Shelley's novel that display its interrogation of the male fantasy of removing women from the process of procreation to symbolize androcentric science and literary authorship. Further displaying similarities with Centlivre's challenges to patriarchal lines of succession, while Shelley dedicates *Frankenstein* to her novel-writing father, William Godwin, she also undermines this filial gesture by quoting John Milton's Adam from *Paradise Lost* on her title page: "Did I request thee, Maker, from my clay / To mould me man? Did I solicit thee / From darkness to promote me?"[30] And just as Centlivre sometimes practiced authorial "cross-dressing," reviewers of *Frankenstein* initially attributed the anonymous novel to Percy Shelley.

Moreover, the monster of Shelley's book, composed of biological parts from both human and animal species, arguably symbolizes the metaphorical danger of an unassimilated, monstrous hybrid text, creating a patchwork of (unsuccessful literary) borrowings. Scholars have explored Shelley's indebtedness in *Frankenstein* to numerous scientific and literary works, including those she most explicitly references as influencing the Creature, such as Milton's *Paradise Lost*, Goethe's *Sorrows of Young Werther*, Plutarch's *Lives*, and so on. In this vein, Shelley's "Giovanni Villani" essay notes the widespread complaints about "the monstrous [literary] distortions of modern times" and asserts that with the assimilation of other materials, successful borrowing is achieved through the "unifying voice of the text."[31] Ultimately, in her play, Centlivre comically satirizes Periwinkle's misogynistic, solitary scientific studies as valuing birth or creation and Fainwell's doubling only as they emphasize practices relating to death and the past, leading to a generative dead end; similarly, Shelley's darker and more serious scientific satire portrays Frankenstein's creation of life, his Creature, as a doppelgänger or double, also androcentrically born out of death through the defunct parts of other organisms, incapable of reproducing (because denied a female companion) and thus marking an end to that genealogy, so to speak.

Considering the play's familiarity within her circle as discussed earlier, Shelley undoubtedly knew of Centlivre's *Bold Stroke*. Godwin, William Hazlitt, and Leigh Hunt also praised Centlivre's talents as a playwright.[32] Byron, displaying characteristic rancor toward women writers, refers specifically to *Bold Stroke* in a letter to John Murray, questioning public taste and lamenting "that Congreve gave up writing because Mrs. Centlivre's balderdash drove his comedies off. . . . It is not *decency* but Stupidity that does all this."[33] Thus, the aristocratic Byron ironically resists the idea of Centlivre's works surpassing those of Congreve, Dryden's chosen heir and successor to the literary "throne." Nevertheless, like Centlivre, Shelley critiques masculinist science and authorship, perceiving originality as something that does not preclude borrowing from the past and does not allow the past to constrain the author in her quest to (re)produce something new. Shelley's text, in turn, of course, has influenced countless imaginative works since its creation, and it seems both women writers would view this subsequent trajectory not as a controlled and stifled line of literary succession but as a free-flowing mode of authorial originality that recognizes the importance of textual unity while exploiting the various possibilities and consequences of producing scientific and literary monsters.

Notes

1. See, for example, John O'Brien, "Busy Bodies: The Plots of Susanna Centlivre," in *Eighteenth-Century Genre and Culture*, ed. Dennis Todd and Cynthia Wall (Newark: University of Delaware Press, 2001), 165–89, esp. 169–70; and F. P. Lock, *Susanna Centlivre* (Boston: Twayne, 1979), 110–11.
2. Laura J. Rosenthal, *Playwrights and Plagiarists in Early Modern England: Gender, Authorship, Literary Property* (Ithaca: Cornell University Press, 1996), 4.
3. Richard Terry, *The Plagiarism Allegation in English Literature from Butler to Sterne* (New York: Palgrave Macmillan, 2010), 50.
4. Terry, 49.
5. Terry, 46.
6. See, for example, Rosenthal, *Playwrights*, 4; and Terry, *Plagiarism*, 121–22.
7. Rosenthal, *Playwrights*, 49, 51.
8. Terry, *Plagiarism*, 122–23.
9. Anon., "A Pindarick to Mrs. Behn," in *Kissing the Rod: An Anthology of Seventeenth-Century Women's Verse*, ed. Germaine Greer et al. (London: Virago, 1988), 261; John Dryden, "To the Pious Memory," in *Selected Poems*, ed. Paul Hammond and David Hopkins (Harlow: Pearson Educational, 2007), 361.
10. Susanna Centlivre, *A Bold Stroke for a Wife*, intro. and ed. Thalia Stathas (Lincoln: University of Nebraska Press, 1968), 6.
11. Edward Young, *Conjectures on Original Composition* (London, 1759), 68. See also Melissa Bailes, *Questioning Nature: British Women's Scientific Writing and Literary Originality, 1750–1830* (Charlottesville: University of Virginia Press, 2017), 107.

12. Centlivre, *Bold Stroke*, 5.
13. Stathas, introduction to Centlivre, *Bold Stroke*, xvi.
14. Tilar J. Mazzeo, *Plagiarism and Literary Property in the Romantic Period* (Philadelphia: University of Pennsylvania Press, 2007), 53.
15. Susanna Centlivre, *A Bold Stroke for a Wife*, in *The Broadview Anthology of Restoration and Early Eighteenth-Century Drama*, ed. Nancy Copeland, gen. ed. J. Douglas Canfield (Ontario, CA: Broadview Press, 2001), 1.1.82–83, 85–88; henceforth, all the play's quotes are from this edition and cited within the essay's text.
16. *Oxford English Dictionary Online*, s.v. "monstrosity," http://www.oed.com.
17. *Oxford English Dictionary Online*, s.v. "monstrosity," http://www.oed.com.
18. *Philosophical Transactions* 1, no. 1 (March 6, 1665).
19. *Oxford English Dictionary Online*, s.v. "periwinkle," http://www.oed.com.
20. One might think, for instance, of Jonathan Swift's *Gulliver's Travels* (1726) or William Wordsworth's famous critique of "men of science" in his preface to *Lyrical Ballads* (1802).
21. Vivian Davis, "Dramatizing the Sexual Contract: Congreve and Centlivre," *SEL: Studies in English Literature* 51, no. 3 (Summer 2011): 526–29.
22. Mazzeo, *Plagiarism*, 2.
23. Mazzeo, 2.
24. Mazzeo, 3; emphasis mine.
25. *Oxford English Dictionary Online*, s.v. "Simon Pure," http://www.oed.com.
26. Andrea Fischerová, *Romanticism Gendered: Male Writers as Readers of Women's Writing in Romantic Correspondence* (Newcastle upon Tyne, UK: Cambridge Scholars, 2008), 106.
27. Fischerová, 99–106.
28. George L. Marsh, "The *Peter Bell* Parodies of 1819," *Modern Philology* 40, no. 3 (February 1943): 267.
29. Fischerová, *Romanticism*, 99–106; Marsh, "*Peter Bell*," 267–74; Eric Lindstrom, "'To Wordsworth' and the 'White Obi': Slavery, Determination, and Contingency in Shelley's *Peter Bell the Third*," *Studies in Romanticism* 47 (Winter 2008): 556–57.
30. Mary Shelley, *Frankenstein*, ed. J. Paul Hunter (New York: W. W. Norton, [1818], 1996), 4, 3.
31. Mazzeo, *Plagiarism*, 129–30; Mary Shelley, *The Novels and Selected Works of Mary Shelley*, gen. ed. Nora Crook, 8 vols. (London: Pickering & Chatto, 1996), 2:129.
32. John Wilson Bowyer, *The Celebrated Mrs. Centlivre* (Durham: Duke University Press, 1952), 3, 181–82, 218.
33. George Gordon Byron, *Byron's Letters and Journals*, ed. Leslie A. Marchand, 12 vols. (Cambridge, MA: Belknap Press, 1977), 7:61.

Picturing Air

The Rhetoric of Nondescription in Robert Boyle's New Experiments Physico-Mechanicall *and* Daniel Defoe's A Journal of the Plague Year

VIVIAN ZULUAGA PAPP

> Knowing is seeing, and if it be so, it is madness to persuade ourselves that we do so by another man's eyes, let him use never so many words to tell us, that what he asserts is very visible.
> —JOHN LOCKE, *An Essay concerning Human Understanding*

Recalling Locke's claim that we can only trust our own eyes, we are reminded of the difficulty inherent in convincingly describing an object (whether fictional or invisible) to a potentially skeptical reader when the ability to represent it visually does not exist, such as in describing air or plague germs. Quentin Meillassoux asks us to reconsider the ways "thought" and "the absolute" interact with one another as he revisits Locke's theory of first and secondary qualities: "For there is indeed a constant link between real things and their sensations. [. . .] Whether it be affective or perceptual, the sensible only exists as a *relation:* a relation between the world and the living creature I am."[1] Cynthia Wall claims that "description in one way or another makes something visible" and that description "makes the invisible present."[2] Robert Boyle's corpuscularian philosophy points to a world in which subvisible particles, such as air, exist in a material sense, and their absence may even threaten our lives. Conversely, Daniel Defoe presents us with an invisible plague particle whose presence causes sudden and often unexpected death. Boyle's invisible air particle is necessary for our existence, while Defoe's invisible, plague-containing particle can quickly bring about sudden death. But how do these authors characterize the unseeable and tell the story of an invisible particle? How was Defoe's *A Journal of the Plague Year* influenced by

Boyle's visual rhetoric of *New Experiments Physico-Mechanicall?* In this essay, I trace the rhetorical effect of four intersecting literary and material methods that both authors use: replicated charts, distinctive font types (italics and capitals), atypical (nonprosaic) writing, and images.

Citing his inability to sufficiently portray to the reader the sounds of the "miserable Lamentations of poor dying Creatures," in Defoe's *A Journal of the Plague Year,* H.F. aspires to "but tell this Part, in such moving Accents as should alarm the very Soul of the Reader."[3] Indeed, H.F. is faced with a quandary: How does one replicate sound through language? In order to accurately portray the devastating plague vignettes, H.F. struggles to negotiate the confines of textual language: "I wish I could repeat the very Sound of those Groans, and of those Exclamations that I heard from some poor dying Creatures, when in the Hight of their Agonies and Distress; and that I could make him that read this hear, as I imagine I now hear them, for the Sound still seems to Ring in my Ears."[4] Not only does our narrator complain about the inaccuracy of aural replication, but he also alerts the reader to a similar issue with visual matters, mentioned in the epigram by Locke. Should we trust in the validity of what other people claim to see? In this essay, I will focus on how Defoe and Boyle navigate the divide between visual verification and the written word. By examining the materiality of each text in print and the context of these visual signifiers and by looking closely at incidences where visual verification is challenged, we can see the ways scientific and fictional texts were aligned in this period. I argue that it is partly through the use of nondescriptions that Defoe and Boyle achieve authenticity; in other words, both authors "show" by paradoxically "not showing."

By bolstering his fiction with verifiable data and less conspicuously embedding within it coded typography, Defoe highlights unseeable natural and supernatural objects and occurrences. H.F. uses italics to suggest a bit of distrust on our narrator's part as he describes the influx of opportunists who appear to prey on the terrified citizens of plague-ridden London, as they "made the Town swarm with a wicked Generation of Pretenders to Magick, to the *Black Art, as they call'd it.*"[5] Defoe uses italics for many reasons, but here we have an example of Defoe's use of italics to represent vocal inflection. Defoe teaches the reader to see (and sometimes hear) what might be hidden beyond the scope of ocular vision. Visual data is only one piece of the puzzle; much of the most crucial knowledge lies somewhere in the interstices within and beyond the page.

Using typographical coding, omissions, and descriptions of things that cannot be seen, Defoe creates a remarkably realistic plague-infested world—one that teaches us to know things by looking past the visible particular. More importantly, he creates a world that closely resembles our own: a world that is not survivable through scientific methods of visual observation alone. Scientific investigation has its limits as far as knowledge-seeking is concerned; like H.F., we are constantly reassessing previously held beliefs about ourselves and the world based on the latest scientific discoveries.

Through its close alignment with the meticulous observational methods established by the scientific authors of the previous century, *A Journal of the Plague Year* calls into question the trustworthiness of fictional and nonfictional sensory perception—specifically vision. By withholding visual cues that might convince the reader of the existence of the unseeable air particle, Defoe teaches the reader to take note of what is omitted and therefore unseen. For example, H.F. discusses his record-taking techniques, telling the reader how he would return home and record daily what he had seen in the streets of London. However, he does not reveal everything: "I also wrote other Meditations upon Divine Subjects, such as occurred to me at that Time, and were profitable to my self, but not fit for any other View, and therefore I say no more of that."[6] This is an example of one of the types of "nondescriptions" in the text. Nondescription is a contradictory method of describing the unseeable; it is a moment in a text where the reader expects a visual aid yet is provided none, is provided an illustration of something that is unseeable or invisible, or is provided a detailed imagistic description of something unseeable or invisible. Another feature of nondescription is the description of movement, use, or effect in lieu of an imagistic description of an object or an abstract idea, such as air pressure. Nondescription populates the negative spaces in these two texts in ways that complicate the reader's trust in sensory perception. For Defoe, this mistrust in the veracity of vision may very well save lives.

One way to better understand how nondescriptions function is to think about some distinctions between natural science and the imagination discussed in *The Culture of Diagram*: "There is a crucial division between natural science, where primary sense perception (often reinforced by instruments) organizes knowledge, and a collateral terrain where imagination reprocesses or even supplants primary sensory perception to produce surprising secret emotions. The one realm achieves authenticity—the real—by bracketing affect; the other by embracing it. Paradoxically, the realm of imagination so profoundly distrusted by Locke has become itself the object of a different order of inquiry and the locus of a new kind of knowledge."[7] In the example from Defoe seen earlier, we see H.F. negotiating between bracketing and embracing affect through his nondescriptive practices.

Where Bender and Marinan's work offers insight into the use of diagram, Alexander Wragge-Morely explains how seventeenth-century naturalists used engraved images in their work, especially to promote affective imaginative forms of understanding. He explains that "pictures were effective because they could reproduce the suddenness, ease, and pleasure of visual experience, resulting in a vivid and lasting mental image," adding that some naturalists believed "that words could sometimes be used to greater effect than pictures."[8] Wragge-Morely's explanation of the ways in which pictures and pictorial representation were employed in seventeenth-century texts such as Diderot's *Encyclopedie*, Locke's *An Essay concerning Human Understanding,* and John Ray's letters regarding the use of illustrations in his *Historia Plantarum* helps us grasp the concept of nondescriptions a bit more firmly. Visual representation can only do so much, and the imagination must fill in the gaps left behind by sensory perception. Building on these methods, I suggest that nondescription is a rhetorical tactic that engages the imagination as an epistemological tool. These are moments in the text where the author does not provide visual data for objects that can be visualized and yet (ironically) does provide visual data for objects that are not naturally visible.

In *Terra Forma: A Book of Speculative Maps,* Frédérique Aït-Touati, Alexandra Arènes, and Axelle Grégoire illustrate (quite literally) alternate ways to picture our world as a series of maps. They argue that "it is clear that the role of humanity has changed: humankind is no longer solely in control, it creates together with many other actors, it makes way for what we call animate entities: human and nonhuman, living and nonliving agents who shape space [. . .] tracking these animate beings, their movements, traces, rhythms, and affects—qualities that were once called 'secondary,' which allowed them to be erased from the map, dismissing them from the modern project of quantifying the world and localizing things based on geometric space."[9] This groundbreaking work of spatial speculation and reconfiguration is now made possible, in large part, by our own modern technology. By applying modern methodologies, technologies, and philosophies, it is now possible (necessary?) to reimagine ways of mapping by redefining what geographers have termed the "spatial turn." By reimagining the function and rhetoric of maps and reversing previous territorial boundaries put in place by early cartographers, Aït-Touati and company are discovering new worlds all around us, citing the work done by Robert Hooke's *Micrographia*: "We use the term exploration in the spirit of Robert Hooke, English scientist and member of the Royal Society, who undertook to rediscover the world around him in 1665, a world already thought to be well-understood, by means of his optical instruments. [. . .] Like Hooke, we are not discovering new countries, but we are learning to see the territories around us differently."[10] Hooke's interest in showcasing the exceptional intricacies of quotidian organisms and substances reinforces the contemporary interest in speculation and intensive study of the commonplace object, such as our geographical locations. It must not go unnoticed that the pages of *Terra Forma* contain mostly images, with a ratio of text to image that clearly favors Hooke's use of visual rhetoric that we see at work in *Micrographia*.

By relying more heavily on stunning images than text alone, the skilled author is able to achieve an approximation of truth more efficiently than text might. In her earlier book, *Fictions of the Cosmos: Science and Literature in the Seventeenth Century* (2011), Aït-Touati shows us the connection between fiction, literature, and scientific writings that existed before writing genres became more formalized, explaining that "the scholar, like the writer, is obliged to call on his imagination in order to have a working conception of the phenomenon under study." Her argument regarding the heuristic role of fiction is especially relevant to this essay, as Aït-Touati considers the use of fiction to create believable worlds and to communicate the "unknown and unprecedented."[11] She completes her study of fictional works with Bernard le Bouvier de Fontanelle's *Conversations on the Plurality of Worlds* (1686), and I likewise draw on her work to examine how Defoe's *A Journal of the Plague Year* (1722) negotiates ideas popularized through Boyle's *New Experiments Physico-Mechanicall* (1660).

This examination of the similarities and potential connections between the narrative styles of nondescription in *Journal* and *New Experiments* reveals how both authors experiment with narrative style to confer realism. Each text employs in equal measure visual elements (whether illustrations or descriptive prose) to produce knowledge. It makes sense that Defoe would not have many illustrations in his text, as we know that the early novels of this period rarely included images. However, many of the images and imagistic descriptions he does provide come from his imagination; they are not taken from actual observations such as those Boyle could have recorded from his experiments. What Defoe does provide are images and descriptive imagery of imaginary things that are not visible in his fictional history. While it is true that *Journal*, like *New Experiments*, is filled with incredibly detailed imagery, the reader is regularly alerted to take note of what cannot be visually replicated with any significant accuracy.

Defoe is particularly well suited to be put into conversation with Boyle; while many of Defoe's texts blur the boundaries that divide imaginative stories from histories, they also often call into question the value of vision as a reliable epistemological tool. For example, in 1720 (two years before *Journal* was published), Defoe's *Serious Reflections During the Life and Surprising Adventures of Robinson Crusoe: With His Vision of the Angelick World* presents Crusoe's opinions on various topics, including the truthfulness of his reports of some of the strange things he claims to have witnessed while shipwrecked. According to Crusoe, certain people are more prone to delusional visions, and they "see" things that might not be there, at least not in the material sense. While his goal seems to be clarification and validation of his outrageous claims by analyzing the nature of visions (which may or may not be real), he explains, "As I endeavour to conceive justly of these Things, I shall likewise endeavour to reason upon them clearly, and, if possible, convey some such Ideas of the invisible World to the Thoughts of Men, as may not be confused and indigested, and so leave them darker than I find them."[12]

What follows is a sometimes contradictory but always imaginative explanation of the correlation between actual seeing and the imaginative act of seeing. Are we to believe that there really was a devil at his knee, a dying goat that sounded remarkably human, or even a single footprint in the sand? Crusoe does us (or himself) no favors, as his claims regarding vision leave us feeling even more distrustful of our sense of sight; but perhaps that *is* the point: "The Fright and Fancies which succeeded the Story of the Print of a Man's Foot, and Surprise of the old Goat, and the Thing rolling on my Bed, and my jumping out in a Fright, are all Histories and real Stories."[13] In *Angelick Visions*, Crusoe regales us with descriptions of the planets as he travels through space, reporting what he claims to "see." His descriptions are remarkably vivid, suggesting the potential for unlimited vision: "When my Fancy had mounted me thus beyond the Vestiges of the Earth, and leaving the Atmosphere behind me, I had set my firm Foot upon the Verge of Infinite, when I drew no Breath, but subsisted upon pure Aether, it is not possible to express fully the Vision of the Place; first you are to conceive of Sight as unconfin'd."[14] Crusoe imagines the potential for limitless sight; this description (reinforced by biblical evidence, such as divine visions) may convince the reader that there are visible objects perceivable in both the material and immaterial worlds. However, we must remember that Crusoe's fictional existence is just that—fictional. Are we hypocrites to demand factuality and truthfulness from this fictional character's claims to a veracity of vision?

Rather than provide images, *New Experiments*, like *Journal*, leaves empty spaces for the reader's imagination to provide fitting images. Of particular relevance are the nondescriptive moments when both authors purposely omit potentially instructive visual information. In my reading, H.F. does not simply draw on the content of Boyle's work—he is an instrument that Defoe uses to create meaning. H.F. operates as a kind of air pump, or a tool to demonstrate that which would otherwise be invisible: through him, the reader can "see" what our narrator reports (or leaves out). Defoe fashions a character who reflects the practical nature of Boyle's machine.

Boyle reports what he sees happening inside the glass receptacle part of the air pump, a space designed to both contain and display seeable experiments. The glass receptacle was designed to be clear enough to see into yet strong enough to withstand the varying pressure of the vacuum created within it: "The shape of the Glass, you will find express'd in the first Figure of the annexed Scheme. And for the size of it, it contain'd about 30 Wine Quarts, each of them containing near two Pound (of 16 Ounces to the Pound) of Water: We should have been better pleas'd with a more capacious Vessel, but the Glassmen professed themselves unable to blow a larger, of such a thickness and shape as was requisite to our purpose."[15] H.F. functions as a fictional observer whose clarity and strength, like the glass receiver of Boyle's air pump, display to the reader a believable plague scenario from which we may learn many lessons regarding disease avoidance and survival. Not only does the reader learn to recognize, avoid, and recover from a plague, but Defoe also teaches us how to "see" without looking and how to seek out crucial information that lies hidden in what is neither said nor shown.

Boyle expresses the benefit to the philosopher of spending time studying the nature of things with which we have regular, life-sustaining daily interaction, like air: "So that a true account of any Experiment that is New concerning a thing wherewith we have such a constant and necessary intercourse, may not onely prove of some advantage to humane Life, but gratifie Philosophers, by promoting their Speculations on a Subject which hath so much opportunity to solicit their Curiosity."[16] This recommendation has the potential to appeal to both the scientific mind and each curious inhabitant of the earth. Is it acceptable to take the nature of air for granted so long as you can breathe? Should humans dare to look further into the operation and structure of objects and materials that had formerly been invisible—invisible due to their microscopic dimensions or even simply being so commonplace that they hardly merited notice? For some like Boyle, this was a matter of obligation (and pleasure, to some extent). Like his protégé Robert Hooke, Boyle illustrates the double reward for the contemplation of a thing "wherewith we have such constant and necessary intercourse."[17]

Many were skeptical of the existence or relevance of a subvisible world where air might be made palpable, if not visible: "By the 17th century little progress had been made on the road to truth and reason. In that historical period the revival of the spontaneous generation theory was widely accepted by most members of the scientific community. It proposed that simple life arises spontaneously from non-living matter (abiogenesis); for example mice can arise from grain and maggots from decaying meat. The idea of spontaneous generation may be tracked back in the teaching of Aristotle in 4th century BC."[18] Machines like the microscope and the air pump made visible, to differing degrees, these previously invisible objects with which we interact daily and on which our very lives often depend. "What New Experiments did do was to exemplify a working philosophy of scientific knowledge," Steven Shapin and Simon Schaffer suggest. "Boyle sought here to create a picture to accompany the experimental language-game and the experimental form of life."[19] Besides the actual illustrations in *New Experiments*, Boyle relates images encountered in everyday life to create mental pictures, or specimens if you will, of what air might look like. He says that air consists of little bodies, or corpuscles, which are in constant movement against one another and, as a result of different forces (such as pressure and gravity), are in a constant search for equilibrium.[20] According to Boyle, air is like a "Fleece of Wooll." Immediately the image of a springy fiber comes to mind. Further on in the text, Boyle discusses how these fleece-like particles attempt to maintain equilibrium by using the recognizable example of a bow:

> As the two ends of a Bow, shot off, fly from one another, whereas the Bow it selfe may be held fast in the Archers hand; and that it is the equal pressure of the Air on all sides upon the Bodies that are in it, which causeth the easy Cession of its parts, may be argu'd from hence: That if by the help of our Engine the Air be but in great part, though not totally, drawn away from the other, he that shall think to move that Body too and fro, as easily as before, will finde himself much mistaken.[21]

The image of the bow is easily conjured. However, applying this image to the air pump takes a bit more effort. But we can see that each of the descriptions of air share something in common: springiness. The fleece of wool when compressed in the palm of a hand pushes back against the hand in an attempt to regain equilibrium or return to its natural state/form. Notably, many of Boyle's metaphors for making air visible (and tactile, to some extent) have more to do with the movement of an object than an actual material object itself.

This nondescription points the reader to the fact that invisibility does not always indicate nonexistence. While air cannot be seen, its movement and impulse to maintain homeostasis can be displayed, measured, and verified. Due to air's particular tendency toward continual shape-shifting and pressure-induced movement, Boyle often evokes material and seeable images that contain and enclose the invisible air. He first mentions a tiny box that could contain a fiber of fleece in its compressed state—which is a direct allegory of the air pump itself. Imagining the tiny box created to maintain the fleece fiber in its compressed state described in experiment 2: "a Box just fit for it," the reader approaches a basic comprehension of not only what the air pump looks like but, more importantly, how it operates. We can "see" air by witnessing its reaction to forces acting against its nature to spring back into shape. We also know how a bow works and can imagine a compressed fiber being enclosed within a structure.

Since we are working from a place of both knowledge (the action of a bow) and imagination (a tiny fiber of wool in a very small box), Boyle reinforces the validity of air's existence by reporting the nature of air particles, even though he cannot replicate them in a direct visual manner, such as with an illustration. Defoe runs into similar problems while attempting to describe ghosts, angels, and the plague particle. Jayne Lewis's work on apparition novels is helpful here: "Defoe's fictions from a critical distance all seem to reduce themselves to matters of fact." Lewis points out that *Journal* is rarely considered an "apparition novel,"[22] although it features ghosts and other visions. Lewis not only considers the spectral elements of *Journal*, but like Helen Thompson's work, Lewis brings Boyle into the discussion: "Neither demon nor delusion—nor necessarily linked to the deceased—the modern apparition was often most exactly painted in the nascent language of natural philosophy, as a simple configuration of the air. This is not the crude reduction it sounds, since air's material density and attendant mediatory capacities had come under unprecedented scrutiny in the last decades of the seventeenth century, most particularly at the hands of Robert Boyle."[23]

Thompson's work on Boyle and *Journal* also must be discussed here. Her metaphor of the plague houses as leaky receptacles directly influences my concept of nondescription: "The imperceptibility of Boyle's corpuscle correlates with its capacity to induce form. 'Parts' of air that do not trigger sensible ideas like 'Heat, Coldness, or other such Qualities' portend plague's unfelt entry into persons, because the capacity to evade sense dictates how far tiny parts can 'get in.' Corpuscles miniscule enough to reach 'the most inward parts of the body' define inwardness as a site of empirically inaccessible knowledge. As Thomson remarks, the plague 'treads so softly within us, that is cannot be heard to walk.'"[24] The problem faced by Boyle and Defoe is expressed in both Lewis's and Thompson's work: how can writers communicate the reality of subvisible or invisible objects without the ability to provide convincing visual proof, descriptive or illustrative?

Boyle often blames inconclusive results on the fact that he could never totally prevent air from leaking out of or into the receiver; this inability to fully control the movement of air complicates matters of factuality. If total isolation is what he was striving for in order to present his findings in a more definitive way, the available technology prevented this. Defoe also recounts a similar problem with containing the contagions of the plague and those infected; there was no way to guarantee that isolation rules were obeyed: "It was impossible to discover every House that was infected as soon as it was so, or to shut up all the Houses that were infected."[25] Like Boyle, Defoe illustrates the inaccurate and ineffective medical "stoppers" against the spreading of the plague. If we return to my comparison of H.F. to the glass receptacle of the air pump, which both contains and displays the action of the invisible air particle, we can think of H.F.'s narrative in a similar way. Rather than attempt to describe something that exists below our unaided field of vision, H.F. chooses instead to describe the activity caused by the invisible plague particle. We cannot see the particle, but we can witness the death and destruction it causes.

Like air particles that seem to find their way in or out of the receiver from even the smallest leaks, some plague victims also were able to slip through the cracks in the receivers (houses) that were intended to enclose and control them. Without proper isolation and containment, the chance of misinterpretation of disease progression, or the nature of air particles, increases. With this rise in the potential for misinterpretation comes a higher probability of skepticism. Shapin and Schaffer tell us that this is one of the deficiencies of Boyle's experimentation that Thomas Hobbes used to counter some of Boyle's claims. Defoe also stresses to us that we cannot really understand the manner in which the plague was spread if we cannot sufficiently quarantine those inflicted. Defoe and Boyle face a similar challenge: how to make convincing claims about natural phenomena in the context of an instrument and a plague that are both leaky, a context that undermines the truth and certainty to which they both aspire. Air particles, plague particles, and plague victims are all equally difficult to contain and observe and perhaps even harder to describe during a period in time when reliance on the senses for fact-finding was paramount.

Defoe's critique of the growing reliance on visual evidence is noticeable in the scene where H.F. cannot see an angel that everyone else around him claims to see. A woman has convinced the gathering crowd that she is viewing a hovering, vengeful angel brandishing a flaming sword: "I look'd as earnestly as the rest, but, perhaps not with so much Willingness to be impos'd upon; and I said indeed, that I could see nothing, but a white Cloud, bright on one Side, by the shining of the Sun upon the other Part. The Woman endeavor'd to shew it me, but could not make me confess that I saw it, which, indeed, if I had, I must have lied: But the Woman turning upon me, look'd in my Face, and fancied I laugh'd."[26] The way that she convinces the crowd is an important point for my argument, as it mirrors the ways that many science and fiction writers employed verifiable data to confer reality to the viewer or reader.

She first tells of its existence and then proceeds to list the specific characteristics that substantiate the existence of this particular angel. It also marks a clear example of a scene of nondescription, one where H.F. cannot see what everyone else seems to see so clearly, even though according to H.F., there is nothing to see.

Defoe employs several rhetorical techniques to describe objects possessing varying levels of seeability; in this case, the visibility of an object is challenged due to invisibility or imaginary status. Since this is a fictional work, the angel could have been visible. We cannot be sure whether the visual problem is with H.F. or if the angel is nothing more than a figment of the woman's and crowd's imaginations. By taking a look at how the woman describes the angel, we can see some influence of scientific reporting. She begins with a general idea of an angel and then moves to particularized characteristics of what we might imagine Michael or Gabriel would look like (the description seems to align with the Christian idea of an avenging angel). This report is not technically completely firsthand, as H.F. is reporting what others describe seeing, while he claims to see nothing but a cloud.

The woman first describes some very general characteristics of the clothing ("an Angel cloth'd in white"), the accessories ("a fiery Sword in his Hand"), and the movements ("waving it, or brandishing it over his Head").[27] Through the use of descriptions of the actions of the angel and its clothing—not necessarily information about the angel himself—a general idea of an angel is generated in the imaginations of both the fictional crowd members and the reader. Then along with the increasingly convinced members of the mob, she moves to the descriptive particulars ("every Part of the Figure to the Life") and particulars that are demonstrative of movement ("shew'd them the Motion, and the Form"). At this point in her enthusiastic demonstration, she wins over the crowd's confidence. They begin to claim that they, too, "see" the angel: "YES. I see it all plainly, says one. There's the Sword as plain as can be. Another saw the Angel. One saw his very Face, and cry'd out, What a glorious Creature he was! One saw one thing, and one another."[28]

This unusual scene of description is of special significance, as it defies the methodology of visual description of minute particulars that Tita Chico discusses in her influential work on science and the imagination: "Literariness produces visibility, which, within an optical realm, connotes knowability," she writes. "The process of demarcating an observed particular—identifying it, justifying it—likewise utilizes the logic of figuration."[29] Chico's work on the minute particular leads the reader to think about the problems inherent in describing something bereft of any visible particulars, minute or otherwise, and I would classify the alternate narrative methodology, nondescription, as a form of "defiguration." The empty space above the crowd does not contain an actual angel, but the developing descriptions, or the particulars, fill in the empty space by refiguring a visible object (here, the angel) into the empty space. In a sense, the angel has been created by the imagination of the crowd, as a sort of backward observational experiment.

This is not the first time that Defoe will complicate the veracity of vision by referring to angelic or demonic beings. Crusoe tells us, "One would have thought such Men as they, who had the Vision of God manifest in the Flesh should not have been so much surpriz'd, if they had seen a Spirit, that is to Say, seen an Apparition; for to see a Spirit seems to be an Allusion, not an Expression to be used literally, a Spirit being not visible by the Organ of human Sight."[30] Religious faith is used here in defense of the existence of immaterial, yet seeable, beings. Defoe situates Christian faith as an optical machine of sorts, strategically placing Crusoe's unbelievers in an awkward position. Questioning the truth of what Crusoe had reported to see (whether in his imagination or not) dangerously approaches heresy. According to Crusoe, faith is needed to see angels or devils where others do not; Defoe places his faith in the ability of the reader to see what he leaves out as well. We might think of the method of nondescription here as a test of the reader's ability to discern divine visions from hallucinations in this instance.

Nondescription functions here in several ways: as a method to describe objects that escape immediate sensory detection, to provide the author a reliable rhetorical technique to create belief in the existence of sensible objects where there are none, and like religion, it relies a good bit on faith. This scene also serves to establish H.F. as a trustworthy observer, transparent and objective. Even at risk to his own safety (at one point the woman and the crowd become threatening as he appears to be mocking them when he claims to not see the angel), H.F. refuses to report what he cannot see. And yet he does. In one of his more descriptive passages, he provides the reader with seeable information (odd, since it is all imagined) that only proves how gullible people can be in the time of a plague. It is a calculated rhetorical choice to place this scene at the beginning of the novel—it reflects not only his level-headed approach to data collection but also the lengths to which he will go in the name of the truth. Indeed, H.F. is willing to hazard his very life in order to place his own eyes on the sites of the devastation caused by the plague.

A bit further on, H.F. expresses his uncontrollable urge to view the bodies in the pit at the parish of Algate, even after his friend the sexton tries to dissuade him: "It was about the 10th of September, that my Curiosity led, or rather drove me to go and see this Pit again, when there had been near 400 People buried in it; and I was not content to see it in the Day-time, as I had done before; for then there would have been nothing to have been seen but the loose Earth."[31] He convinces the sexton to let him view the pit by claiming it might be "an Instructing Sight," to which his friend exclaims, "Name of God go in; for depend upon it, 'twill be a Sermon to you, it may be, the best that you ever heard in your Life. 'Tis a speaking Sight, says he, and has a Voice with it, and a loud one, to call us all to Repentance; and with that he opened the Door and said, Go, if you will." H.F. does not face the wrath of an optically deceived crowd this time, but the danger is still clearly to his own life as he exposes himself to the possibility of infection.[32]

The earlier scene with the angel shows a description that moves from general to particular. In the scene at Algate, H.F. recounts an event that is visible. I discuss this scene because it is strategically detailed. The minute particulars are left out, and this very distressing scene is made more horrible by the lack of description, as the bodies of the man's family are reduced to undefined, unnamed, nonspecific corpses. Even though H.F.'s vision is impeded by the darkness, he still provides as much visual information as he can gather. H.F. observes a distraught man who accompanies the dead cart filled with his deceased loved ones, including his "Wife and several of his Children." This description is as particular as it will get, and this is no surprise, since this scene plays out in the dark. Very quickly, these modestly particularized characters are made quite general as they become nothing more than bodies to be dumped into the crowded pit. Notably, the bereaved man could not maintain his sober "Masculine Grief" and "went backward two or three Steps, and fell down in a Swoon" as he witnessed his family becoming nothing more than filling for the pit: "No sooner was the Cart turned round, and the Bodies shot into the Pit promiscuously, which was a Surprize to him, for he at least expected they would have been decently laid in, [. . .] no sooner did he see the Sight, but he cried aloud unable to contain himself."[33] It was the eventual sight of the bodies being dumped that evoked an emotional response from the man. Sadly, seeing really is believing in this case. Much like the experiments with the air pump where the glass shatters and needs to be replaced, H.F. does not yet "shatter" and proves that he has the bravery needed to enter into dangerous situations to provide the reader with the most accurate reports possible. Like Boyle's air pump, H.F. contains exactly the right amount of tensile strength to withstand the increasing pressure caused by the sights of the infection.

Through the inclusion of nondescriptions, Defoe keeps us guessing about the validity of what our senses grasp; additionally, he shows how what we see (and what we believe we see) cannot be taken at face value. The function of nondescription for Defoe is multifaceted: it displays his skill at description and observation, reminds us of the limitations of our sensory capabilities, and warns us of the dangers of the misguided imagination. People were reticent at first to believe in a plague germ that they could not see yet seemed easily convinced of the reality of an imagined avenging angel. Some people claimed that the plague was portended by the visitation of two comets. Not only were these comets seen, but they were also heard. Yet they had nothing to do with the plague, even though they could be observed. Blaming unusual astronomical events and the wrath of God for the devastation of the plague was commonplace in the seventeenth century. Catherine Wilson reveals to us that during this period, "traditional medical theory [. . .] was not well equipped to comprehend an illness that was not only strikingly nonselective, affecting young and old, male and female, workers and the idle rich alike, but was sudden, violent, and uniform in its manifestations."[34]

Equipped with an eighteenth-century empirical approach, H.F. counters the truth behind these seemingly convincing accounts of plague-portending comets by reminding us that we must also take into consideration that "natural Causes are assign'd by the Astronomers for such Things."[35] *Journal* reconsiders the nature of plague particles relying solely on recorded reports, a saddler's logical translation of his observations, and the imaginations of both H.F. *and* the reader. Scholars Ionut Isaia Jeican, Florin Botiş, and Dan Gheban explain, "Regarding the etiological agent and mode of transmission, the narrator creates a symbiotic theory between the effluvious theory from Antiquity, and the modern contagious theory."[36] H.F. rejects the miasmic theory, which states that the plague germ rose up as a vapor from veins of putridity beneath the earth's surface. At this point in the novel, he is not sure how it is transmitted. He eventually determines that it is passed from infected person to infected person (especially dangerous when the person is unaware that they are carrying the plague germ). H.F. is able to consider the plague from all perspectives: religious, empirical, and supernatural. It is this ability that mirrors scientific testing of hypotheses that makes him an effective witness. H.F. considers all methods of plague transmission and visitation so long as they are considered complexly. H.F. establishes a philosophical hierarchy, one that does not assign supernatural explanations to phenomena that can be understood solely through sense and reason.

Should the reader trust the report of a saddler? While *Journal* is a fictional account of an actual event, it is nevertheless a text that may teach the reader something: how to survive in a world where subvisible deadly particles may attack us at any moment without our knowledge. In 1660, Hooke horrified the contemporary reader with the close-up image of the alarming but ultimately mundane louse that bred everywhere and on everyone, Hooke included. But Defoe presents us with an even smaller threat: an invisible plague particle that can do more harm than all the lice in London. The fictional nature of this particular plague germ does not make it any less scary, and much of the information provided by H.F. has practical applications. For example, H.F. suggests that if the infected people had been confined more quickly and completely, the epidemic might have been less destructive: "Certain it is, that if all the infected Persons were effectually shut in, no sound Person could have been infected by them, because they could not have come near them."[37] Although this "certain" fact has been established through fictional means of observation, the validity of this claim is difficult to refute. First of all, it makes sense if we buy into his theory of infection method. Secondly, H.F. provides many examples to support this claim, perhaps the most relevant being his own self-imposed periods of quarantine. After all, these periods might have been included by Defoe to account for time that passes without observations; acting as another type of nondescription, H.F.'s periods of quarantine create more anticipation about what is going on outside on the streets of London.

Defoe addresses issues not only with visual accuracy through plot and examples of the plague but also by encoding his text with atypical methods of textual communication that serve to connect particular aspects of the text. Approximately one-fifth of *Journal* is composed of these rhetorical flourishes: twenty pages with twenty replications of charts, ten pages with excessive italic and capital letter usage (more than is usual for novelistic writing), seventeen pages of nonnarrative prose, and two pages composed of four images. The use of images in *Journal* seems extraordinary. Yet perhaps it is not so extraordinary if we consider what John Bender and Michael Marrinan say about the lack of patina in Jean-Baptiste-Siméon Chardin's objects in the *Encyclopedie*: "These images do not index a past time but activate the here and now of present encounter."[38] Clearly, patina is not at issue here, but the immediacy of the images makes them seem more timeless than prose, which does not age as well. Additionally, often a chart or image can convey information more quickly than prose. By applying this key to these techniques, a pattern appears:

each of the twenty charts deals with death and/or disease statistics; the ten pages of excessive italic and capital usage are confined to instances of dialogue, emphasis, and quotations; and the images all represent matters mystical, spiritual, or ineffable. Why did Defoe choose to insert charts rather than simply report the statistics? Part of what makes these charts so effective is that they allow the reader twenty opportunities for firsthand visual observation. We see the numbers for ourselves, and we normally believe what we see.

The visual impact of these charts is significant, and their replication is the most frequently used method of visual rhetoric in *Journal*. The reader participates with H.F. in a strategically archived pattern of visual observation and verification. These charts, which represent the thousands dead of the plague, serve the reader better than an illustration or descriptive prose ever could. The numbers paint the pictures, and the realism of the charts makes these pictures come to life in the mind of the reader. By including these images of data collection, H.F. further establishes his credibility as a trustworthy observer. The numbers are carefully recorded lists that provide the reader with an accurate and true accounting of the plague victims, along with their locations. Defoe creates a character whom we now recognize as a pioneer in the field of epidemiology, a field that had yet to be established. The readers could begin to consider the possibilities for disease control by careful collection and consideration of data.

Both Boyle and Defoe were involved in imagining the new empiricism. While obvious for Boyle, given his place in history, the extent to which he needed to invent tools to see (both instruments and rhetorics) and then make his ideas visible to others classifies him as an artist relying on his own imagination and supplying the imagination of others. If Defoe is more conventionally termed an artist, his work with *Journal* was not to distract or divert his readers but to create a landscape of the imagination in which a mysterious and often unfathomable world could be reimagined as ripe for reckoning and categorization, holding out the possibility that such an approach to the world might eventually lead to a system and understanding.

Notes

1. Meillassoux, *After Finitude*. Meillassoux explains this by using a few examples: "If there were no *thing* capable of giving rise to the sensation of redness, there would be no perception of a red thing; if there were no real fire, there would be no sensation of burning. But it makes sense to say that the redness or the heat can exist as qualities just as well without me as with me: without

the *perception* of redness, there is no red thing; without the sensation of heat, there is no heat."
2. Wall, *Prose of Things*.
3. Defoe, *Journal*.
4. Ibid.
5. Ibid.
6. Ibid.
7. Bender and Marrinan, *Culture of Diagram*.
8. Wragge-Morley, *Aesthetic Science*.
9. Aït-Touati et al., *Terra Forma*.
10. Ibid.
11. Aït-Touati and Emanuel, *Fictions of the Cosmos*.
12. Defoe, *Serious Reflections*.
13. Ibid.
14. Ibid.
15. Boyle, *New Experiments*.
16. Ibid.
17. Ibid.
18. Marianna et al., "From Miasmas to Germs."
19. Shapin and Schaffer, *Leviathan and the Air-Pump*. Shapin and Schaffer explain Boyle's approach to replicating his findings: "In a concrete experimental setting it showed the new natural philosopher how he was to proceed in dealing with practical matters of induction, hypothesizing, causal theorizing, and the relating of matters of facts to their explanations" (40).
20. Boyle, *New Experiments*: "That if a man should take a fleece of Wooll, and having first by compressing it in his hand reduc'd it into a narrower compass, should nimbly convey and shut it close up into a Box just fit for it, though the force of his hand would then no longer bend those numerous springy Bodyes that compose the Fleece, yet they would continue as strongly bent as before" (22).
21. Ibid. Boyle discusses the pleasure of observing visitors attempting to lift the stopper from a receiver after the air has been removed: "And if (as sometimes hath been done for merriment) onely a Bladder be tyed to it, it is pleasant to see how men will marvail that so light a Body, filled at most with Air, should so forcibly draw down their hands as if it were fill'd with some very ponderous thing" (20–21)
22. Lewis, *Air's Appearance*.
23. Ibid.
24. Thompson, *Fictional Matter*.
25. Defoe, *Journal*.
26. Ibid.

27. Ibid.
28. Ibid.
29. Chico, *Experimental Imagination*.
30. Defoe, *Serious Reflections*.
31. Defoe, *Journal*.
32. Ibid.
33. Ibid.
34. Wilson, *Invisible World*.
35. In Defoe, *Journal*, he continues, "and that their Motions, and even their Revolutions are calculated, or pretended to be calculated; so they cannot be so perfectly call'd the Fore-runners, or Fore tellers, much less the procurer of such Events, as Pestilence, War, Fire, and the like" (27).
36. Jeican, Botiș, and Gheban, "Plague."
37. Defoe, *Journal*.
38. Bender and Marrinan, *Culture of Diagram*.

Bibliography

Aït-Touati, Frédérique, Alexandra Arènes, Axelle Grégoire, and Bruno Latour. *Terra Forma: A Book of Speculative Maps*. Cambridge, MA: MIT Press, 2022.

Aït-Touati, Frédérique. *Fictions of the Cosmos: Science and Literature in the Seventeenth Century*. Translated by Susan Emanuel. Chicago: University of Chicago Press, 2011.

Bender, John. "Enlightenment Fiction and the Scientific Hypothesis." *Representations* 61 (Winter 1998): 6–28.

Bender, John, and Michael Marrinan. *The Culture of Diagram*. Stanford: Stanford University Press, 2010.

Boyle, Robert. *New Experiments Physico-Mechanicall, Touching the Air*. London: Miles Flesher for Richard Davis, Bookseller in Oxford, 1682.

Bryson, Norman, Michael Ann Holly, and Keith Moxey. *Visual Culture: Images and Interpretation*. Boston: Wesleyan University Press, 1994.

Chico, Tita. *The Experimental Imagination: Literary Knowledge and Science in the British Enlightenment*. Stanford: Stanford University Press, 2018.

Clark, Stuart. *Vanities of the Eye: Vision in Early Modern European Culture*. Oxford: Oxford University Press, 2009.

Defoe, Daniel. *A Journal of the Plague Year*. New York: Penguin, 2003.

———. *Serious Reflections During the Life and Surprising Adventures of Robinson Crusoe: With His Vision of the Angelick World*. London: Printed for W. Taylor, at the Ship and Black-Swan in Pater-Noster-Row, 1720.

Jeican, Ionut Isaia, Florin Botiș, and Dan Gheban. "Plague: Medical and Historical Characterization: Representation in Literature (Case Study: 'A Journal of the Plague Year by Daniel Defoe')." *Romanian Journal of Infectious Diseases* 16, no. 3 (2014): 124–30.

Lewis, Jayne. *Air's Appearance: Literary Atmosphere in British Fiction, 1660–1794.* Chicago: University of Chicago Press, 2012.

———. "Spectral Currencies in the Air of Reality: A Journal of the Plague Year and the History of Apparitions." *Representations* 87, no. 1 (2004): 82–101. https://doi.org/10.1525/rep.2004.87.1.82.

Marianna, Karamanou, George Panayiotakopoulos, Gregory Tsoucalas, Antonis Kousoulis, and George Androutsos. "From Miasmas to Germs: A Historical Approach to Theories of Infectious Disease Transmission." *Le Infezioni in Medicina: Rivista Periodica Di Eziologia, Epidemiologia, Diagnostica, Clinica e Terapia Delle Patologie Infettive* 17, no. 3 (2014): 124–32.

McKeon, Michael. *The Origins of the English Novel, 1600–1740.* 15th anniversary ed. Baltimore: Johns Hopkins University Press, 2002.

Meillassoux, Quentin. *After Finitude: An Essay on the Necessity of Contingency.* London: Continuum, 2009.

Shapin, Steven. *Leviathan and the Air-Pump: Hobbes, Boyle, and the Experimental Life.* Princeton Classics. Princeton, NJ: Princeton University Press, 2017.

Thell, Anne M. *Minds in Motion: Imagining Empiricism in Eighteenth-Century British Travel Literature.* Literature, Thought & Culture, 1650–1850. Lanham, MD: Bucknell University Press and Rowman & Littlefield, 2017.

Thompson, Helen. *Fictional Matter: Empiricism, Corpuscles, and the Novel.* Philadelphia: University of Pennsylvania Press, 2017.

Vickers, Ilse. *Defoe and the New Sciences.* Cambridge Studies in Eighteenth-Century English Literature and Thought 32. Cambridge: Cambridge University Press, 1996.

Wall, Cynthia. *The Prose of Things: Transformations of Description in the Eighteenth Century.* Chicago: University of Chicago Press, 2014.

Wilson, Catherine. *The Invisible World: Early Modern Philosophy and the Invention of the Microscope.* Studies in Intellectual History and the History of Philosophy. Princeton, NJ: Princeton University Press, 1995.

Wragge-Morley, Alexander. *Aesthetic Science: Representing Nature in the Royal Society of London, 1650–1720.* Chicago: University of Chicago Press, 2020.

Romancing the Placebo

JAYNE LEWIS

Error is endlessly diversified; it has no reality, but is the pure and simple creation of the mind that invents it.
—WILLIAM GODWIN, *Historical Introduction to Report of Dr. Franklin, And other Commissioners Charged . . . with the Examination of the Animal Magnetism* (1785)

Science, meet storytelling.

In the dying days of the ancien régime, the French state "charged" a "regular and authentic committee" with investigating a "universal magnetic fluid" that the itinerant German healer Franz Anton Mesmer professed himself able to channel into an agent of healing.[1] The committee fashioned dummy versions of Mesmer's favorite tools and techniques and then applied them to select unsuspecting invalids, whose symptoms frequently receded as if they had experienced the real thing. Except there was no real thing. Or was there? The committee's "conclusion"—"the imagination [was] the true cause of the effects attributed to the magnetism"—has attracted the attention of scholars eager to discover the material presence of "the imagination" in eighteenth-century science or to find the limits of sentimental empiricism's axiom that "feeling [is] the ultimate test of truth." How better to do so than via a felt fluid that "did not exist"?[2] The so-called Franklin report's English translator, William Godwin, however, spied not just an empirical demonstration of "the influence of the imagination upon the animal frame" but the formal realization of what "has no reality."[3] As if to taunt present-day historians of the scientific imagination whose skepticism is founded on an underexamined materialism, Godwin—soon to become a writer of fiction, after all—took the drawing of nothing into form as a realization of "things as they are."[4]

The Mesmer trials themselves represent the first large-scale manipulation of what we now call the placebo effect—the effect of what "has no reality" on bodies that supposedly do. The trials also coincide with the emergence of the placebo as a respected medical therapy. Akin to the nonexistent causes that generate multiple sensational effects in the then-ascendant genre of gothic romance, an "agent" proves "useful and curative, when in reality it has no action at all."[5] As one present-day skeptic reckons, any placebo "test is an ordeal of darkness," a shady business in which the very "process of determining legitimacy often 'involves [unexamined] prior agreements about what is to count as admissible evidence.'"[6] Why should eighteenth-century healers and scientists have legitimated their knowledge practices by way of such underhanded deeds? And why should they have done so in plain sight?

The placebo put on its modern face at a well-publicized climacteric not just in the histories of fiction and evidence but in that of mediation: the late Enlightenment consolidation of a global and print-based communication system.[7] Despite its obvious materiality, a centerless network whose predicates were the distance and even absence of communicants from one another had become sufficiently familiar—indeed ubiquitous—to disappear from notice. Yet as historians of medicine have shown, modern physicians relied on their publications to compel popular belief in their modernity as well as to create the professional communities that were a mark of that modernity.[8] Reseeing the print infrastructure through which all of this was achieved raises the suspicion that, like that of early gothic literature, Enlightenment science's story has too long been told so as to flatter the belief games and imaginative forms characteristic of midcentury sentimental fiction. To be sure, placebos like Mesmer's—or like the other placebos that exposed them—seemed to "work" because they made belief. The first explicitly medicalized placebo stories, however, captivated a generation of scribbling physicians grown rather sick of sentimental empiricism's prestige as the presumed condition of such belief. Like the readers who consumed their work, they were more likely to be spellbound by sensation's inscrutable temporality, especially as it was simulated, all but infinitely, in the ever more formalized experience of thrill reading. Edgy new aesthetic theories cast the terror of what isn't there as a time-sensitive stimulant of pleasure. At the same time, scientists like Joseph Priestley, Thomas Beddoes, and Humphry Davy—empirics turned projectors as the historian Mike Jay puts it—drafted new criteria of scientific authority by recording themselves as the subjects of the slow-motion sublimations of newly isolated gases such as carbon dioxide and nitrous oxide, gases that were as obscure as they were rarefied.⁹ For its part, the Franklin report pleased amateurs and experts alike by activating—then endlessly replicating across absent bodies—the elusive experience of a hitch in time, a switch point between what's perceived and the about-to-be. In Godwin's terms, what matters isn't so much matter, or the illusion of it, as the "active energy" that pushes something that "has no reality" to "endlessly diversif[y]" as if it did. The agent in this science wasn't simply fiction. It was also story.

This is a story about that multivalent and potentially autocritical story. Its prankster protagonist is Placebo itself as it traveled, often incognito, through the nervous communication systems that bound enlightened physicians and their readers into a community transcending space and time. After offering a character sketch of the traveler and charting some of its travels, my story pursues the "modern" placebo to its hiding places in presumably archaic faith traditions that in turn revealed ways in which physicians who thought of themselves as modern were not especially so at all. Insofar as this gave rise to a way of unknowing the sentimental body—or at least of knowing it, often critically, as in part a nonbody—our tale ends with Placebo's migration to the pages of gothic romance. "Knowledge," as Deidre Lynch has put it with appealing assonance, "is a topic of gothic [literature]."¹⁰ It has also been a topic of recent reevaluations of "evidence-based medicine" that find that quantitative, purely biomedical descriptions of illness and cure "correspond to reality" only within the system of "knowledge" that those descriptions generate.¹¹ In any event, the antics of a medicalized placebo in the gothic moment teach us to take "topic" not just as a set theme but as an open-ended experience. Playfully, but far from innocently, Placebo asks, *What if this were an experience of nothing? And what if it were not?*

The Medicalized Placebo and Its Travels

"The history of medical treatment," it's been remarked, "is the history of the placebo effect."¹² Hardly a comforting observation. If the placebo effect is increasingly tested both by and in relation to artificial intelligence, today's standard definition of the placebo itself remains scientific and utilitarian: "Any therapy or component of therapy used for its nonspecific, psychological, or psychophysiological effect, or [. . .] for its presumed specific effect, but [. . .] without specific activity for the condition being treated."¹³ This definition took shape in the context of eighteenth-century experimental medicine. In some sense, *all* eighteenth-century medicine was experimental; even in 1750, Peter Shaw declared, "The Trade is not yet advanced to a Science," leaving "the Physician [. . .] like a Man married to a Phantasical Lady."[14] Much of what might "work" as medicine could at best be seen as removed one tantalizing step into a future as yet unknown. Paving the way to the rise of the placebo, experimental medicine doubled as an experience-based trial and an immersive venture into possible error.

Placebos themselves often seem to "work" because they "inspire expectant trust in a patient." Such "agents [. . .] exert their effects primarily by symbolizing the physician's healing powers," thereby isolating for scrutiny the role that rhetoric, ritual, and representation play even in modern healing.[15] Not quite an object of knowledge, never really an article of faith, the placebo flatters modernity's self-defining claim to possess both without damaging its pride in its capacity for self-mistrust and autocritique. This is as true of the word *placebo* as it is of the "medication" it seems to designate. Considered strictly as a noun, *placebo* only half reveals its own part of speech. In point of etymology, it's not a noun at all but rather a Latin verb conjugated in the optative first person: "I will please."

Who might this first person be? Whom might that person be pleasing? Evading the question, George Motherby's 1785 *Medical Dictionary* was the first text of its kind to define the placebo as "a commonplace method or medicine"—a method that by 1809 was supposed to have been "calculated to amuse for a time rather than for any other purpose."[16] Commonplace on the one hand, a toy on the other, "placebo" flirts with the unintentional in intentional healing, teasing calculation into uncertainty. And yet, notes a recent exegete, this definition has since been misquoted as specifying "a commonplace method *of* medicine."[17] Method's misleading fusion with medicine rivets our attention on matters of belief and disbelief, hiding the time-sensitive mediating techniques that not only enhance the placebo effect but carry over into the placebo's life as a verbal token—"an epithet," as Robert Hooper's 1811 edition of *Quincy's Lexicon-Medicum* has it, "given to any medicine adopted more to please than to benefit the patient.'"[18]

The medical dictionary was just one of several new genres that were busy making medicine visible as a potentially scientific practice. Linnaean nosologies and chemistry-based pharmacologies, studies of specific diseases, and published correspondence among consulting physicians and their patients all stitched private practice to public information.[19] The print system reified and integrated a dispersed community of emergent experts in Great Britain and the newly constituted American republic, making virtuous physicians seem present to common readers (and prospective clients). Now included in the display of medical knowledge, such readers were flattered by the sight of their own faces benignly reflected in doctors' own.

It was in this specious, if not spectral, atmosphere that "the" placebo became "commonplace." Literary citations of it turned into commonplaces themselves over time. If primarily "calculated to amuse for a time," bread boluses and colored ashes dissolved in water amused not as modern enchantments engineered for the suspense of disbelief but instead as high-profile "pure" and "innocent" friends of the sincerely skeptical physician. As Thomas Jefferson remarked in 1807 with reference to the protocols favored by his friend and fellow Constitution framer, the physician Benjamin Rush, "If the appearance of doing something be necessary to keep alive the hope and spirits of the patient, it should be of the most innocent character. One of the most successful physicians I have ever known, has assured me, that he used more bread pills, colored drops and powders of hickory ashes, than of all other medicine put together." Jefferson paused, bemused. "It was certainly a pious fraud."[20]

Jefferson scribbled this tongue-in-cheek "confession of faith" to please (and promote) the University of Pennsylvania anatomy professor Caspar Wistar. Like Rush, Wistar had been educated in Edinburgh's prestigious medical school; both were devoted pupils of the legendary nosologist William Cullen, best known for projecting the central nervous system as the matrix of the individual—and thus the social—body. Though far from the first physician to administer a placebo, Cullen is often credited as the first to use the word *placebo* in a medical text; its debut in his *First Lines of the Practice of Physic* (1777) sanctioned this "pious fraud" as decent (if not good) medicine and as permissible (if not necessarily best) practice. Lay readers loved *First Lines,* but it was "intended chiefly as a text-book for the use of [Cullen's] students." Ambivalent about print's authoritative stasis as well as about a wider readership he could not see or touch, Cullen would have preferred "to confine [*First Lines*] to the hands of those who attend[ed] his lectures."[21] He overcame his "diffidence" only to head off a bootleg edition of notes from his lectures.

While "placebo" was never promoted to a category in Cullen's canonical *Materia Medica* (1789), its uncertain status as "materia" gave it a natural home in *First Lines*—a text whose "lines" themselves hover between caught speech and print copy. The word "placebo" turns up just as Cullen is speculating on the delicate "management of the mind" that hypochondriacs require. Indeed, Cullen was one of the first to define the hypochondriac as an adroit somatizer of imagined maladies. "Their firm persuasion," Cullen declared, "does not allow their feelings to be treated as imaginary, nor their apprehension of danger to be considered as groundless." Hence Cullen reckoned that "if, in any case, the pious fraud of a placebo be allowable," it would be this one.[22] Cullen's equivocal syllogism places "a placebo" within an as-if "case" where apprehension, though "groundless," cannot be treated as such. His diffidently materializing "lines" make "placebo" perceptible at an elusive point of transmission and reception in a contingent sequence that tethers multiple remote, indeed absent, bodies: the physician's, the patient's, the patient's imagined one, the reader's. Meanwhile, Cullen's neurophysiology legitimated such empirically unverifiable sources of illness as "errors of sense," "false imagination," "the perception of objects other than they exist," and "perceiving in an altogether singular way." The placebo's "pious fraud" builds on these nonexistent causes of only vaguely tangible effects because it shares their virtual nonexistence.[23]

The 1770 lecture notes of an anonymous Cullen student yield two placebos already in circulation. One Mr. Gilchrist, "absolutely incurable," is "hastening fast to his fate"—is, in effect, already dead. The desperation of the case leaves the lecturing practitioner with hopes only of using him as an example, released into speculative and receptive futures that don't include a body beyond rescue: "I took him in hope of making some observations upon his case & even of learning something by his death. I prescribed therefore a pure Placebo." Whatever Cullen actually "prescribed," however, wasn't *quite* nothing: he added, "I made it a rule even in employing Placebos to give what would have a tendency to be of a use to the patient."[24] Cullen's choice of an active but irrelevant substance creates the possibility of storytelling. If the placebo in itself posed a formal problem of mimesis and a potentially ethical problem of deception and exploitation, its mobilization into narrative deflects attention from these inconveniences. In a second lecture, the good-enough doctor administers the stimulant mustard, such as "might be useful in paralytic affections," but again without faith or even any particular relevance to the problem at hand: "I own that I did not trust much to it, but I gave it because it is necessary to give a medicine and what I call a placebo."[25] "What I call a placebo" sublimates mustard—and mistrust—into a meaningful series of communications that radiates indefinitely outward and forward: the ritual giving of a medicine, an act of naming, lecturing, being heard, being transcribed, being read, being quoted in a twenty-first-century story about Cullen's placebo.

Placebo Domine

But an even longer and more capacious tale of transmission hangs in the "pious fraud" metaphor that Jefferson would copy from Cullen. Placebo's genetic code is to be found in medieval Catholic ritual, where the phrase "placebo domine in regio vivorum" (I will please the Lord in the land of the living) traditionally initiated the antiphon in vespers for the dead. Derived from the Vulgate Psalm 116, a "placebo" was originally a way of speaking formulaically at once to, for, and through the dead. The Hebrew verb is *ethalekh:* "I will walk," not "I will please."[26] An error? The iambic *placebo domine* antiphon retains the metrical footage of the Hebrew phrase. Besides, don't placating and walking share the predicate of a satisfied Lord?

Socially speaking, the Catholic "placebo" adopted the syntax of flattery and disingenuousness in the fourteenth century, when bereaved families began hiring mourners to repeat this tedious prayer ad infinitum. "Placebo" thus surfaces in Geoffrey Chaucer's *Merchant's Tale* as an insincere counselor to the doddering Januarie, newly married to healthy May. Placebo is an open toady, admitting, "What that [Januarie] seith, I holde it ferme and stable; / I seye the same or elles thyng semblable."[27] In Alexander Pope's adaptation, he vows "to let my Betters always have their Will, / Nay, if my Lord affirm'd that black is white, / My word was this, your Honour's in the Right ... Your Word is mine; and is (I will maintain) / Pleasing to God, and should be so to Man." The resulting illusion speaks to human presumption in the form of a verbal "Mirrour" that depicts "figures moving by" such that the feeble geezer's "Soul [may] arrive at Ease and Rest."[28] If (the Catholic) Pope's Placebo does not try to hide the placating manipulation of form, he nonetheless retains the frisson of address to the departed soul.

Nonetheless, the placebo's history has been tailored to fit a secularization narrative that can safely deliver it to the sociable medicine of the eighteenth century, with Cullen typically hailed as the "missing link" in its mysterious "transition from religious to medical meaning."[29] Indeed, popular literacy plays a part in this transition, especially the avid consumption of medical literature produced by Edinburgh faculty and alumni. That literature sometimes took the shape of self-help manuals such as the Edinburgh-schooled William Buchan's *Domestic Medicine* (1769)—books that evened the power balance between healer and literate patient in a scene of reading that provides a paraclinical interaction. Writing about (and with) the placebo likewise reproduced a spatialized culture invested in the similarities of bodies to one another and in the transfer of sympathetic narratives that potentially resulted. The placebo's irreducibly mimetic form meant that it was naturally at home in this, its native environment. Within that environment's logic, Stewart Justman explains, individuals were constitutionally disposed to "feel what others feel." Placebos, then as now, were believed to work best when patients believed they had done so for like-minded others, flattering a social and material conception of the imagination itself.[30] But, as in Pope's revision, of Chaucer's Placebo, a nod to what's no longer quite there lingers too. The placebo of yore was still kicking around in Catholic prayer books, in antiquarian guides to empty abbeys, and in printed rules of monastic life, even though such a life was no longer lived in England.[31] Echoes of archaic ritual fostered vestigial apprehension of the absent bodies of the dead as potentially the once-more-living. This possibility animates even a sentimentalized, medicalized placebo, making the cold, apparent mirror of flattery an active site of dissimulated—technically nonexistent—feeling.

Medical satire often personified "Placebo" to this effect. Heather Keenleyside finds that the seemingly dead form of personification lingered throughout the eighteenth century because it linked "two logically interconnected but notionally segregated practices: purification . . . and translation or mediation."[32] From this perspective, Pope's redacted Placebo looks like a personification of personification itself. So does the "kind, attentive, humane, friendly" Dr. Placebo who surfaces in a 1792 volume of the *Medical Spectator*, chillingly distinguished by "the mildness and placidity of his countenance, the suavity of his attention and the unmercenary decorum of his visit." He also enjoys exceptional "influence in obtaining a stricter and more diligent attention to perseverance in regimen," an influence he chalks up to the "vis medicatrix naturae—the influence of the medical soul."[33]

If ghostly Dr. Placebo leaves "agreeable impressions on the mind of his patients and their friends," he is himself the shadow of a shadow.[34] In Alexander Sutherland's 1763 *Attempt to Revive Ancient Medical Doctrines*, itself quoting Robert Pierce's 1713 *History and Memoirs of the Bath*, "Placebo" is a mere quack who sets up for practice, though he "never saw a professor in his chair, nor never made up a Doctor's prescription." Wholly "without Knowledge chemical or practical," this Placebo is a negation ("never," "nor never," "without") within a subtle citational web uncannily devoted to the semisecret transmission of medical knowledge. The quoted Pierce: "Nor do I now care to mention his name . . . His wig was deeper than mine by two curls, he corresponded with the most eminent, as he reported they prescribed. Was it any wonder *Placebo* grew in grace?"[35]

Like these personifications, the placebo's complex first-person voicing and strictly citational substance ferries deathly nullity into the "region of the living." In the ritual whose vestige it bears, its predicate is Purgatory, whose ghostly denizens allegedly migrated into beguiling books, popular entertainments, and the chambers of the enlightened mind.[36] As actively imagined within Catholic liturgy, however, the purgatorial soul is bound within a moving discipline of suffering. The medicalized placebo thus retains spiritual relevance to afflicted or all but obsolete bodies, raising the possibility of retrieval. If it makes medicine perceptible as modern science, the placebo still presumes the power of symbol, simulacrum, and symbolic action in reaching those who suffer. And it potentially brings those same persons into a sphere of meaning that can be shared with their hopeful healers.

The medicalized placebo seemed to save the soul simply by accommodating emotion as a physiological reality and causal "influence." The physician William Falconer was a member of the Cullen fraternity. Often seen as a founding text in modern placebo literature, Falconer's prize-winning 1788 *Dissertation on the Influence of the Passions upon Disorders of the Body* embraces affective and imagining states as internal influences on the nervous body that Cullen's medicine presumed. Yet Falconer starts with a speculation about sleep—the "brother," the "neighbor or adjunct of death."[37] Here Falconer revives earlier modern and even medieval conceptions of the imagination's implication in the body's "sensitive appetite" such as those discoverable in the Catholic recusant Thomas Wright's *Passions of the Mind* (1601/3). Falconer also activates antique ideas of the dream as both a diagnostic tool and a medium of healing, one that even in Hippocrates's writings physicians tried to access by manipulating sleep.[38] In modern times, Falconer's second, insensate body, "adjunct of death," not only underlies but actively underwrites the sensate one that becomes visible in the enlightened and sociable physician's *Dissertation*. Any symbolic address to this second body solicits the syntax of mourning, a collective ritual of progressive repetition that brings the living together in the performance of hope and in the apprehension of a single universal life. Indeed, Falconer granted the dreamlike hope that physicians build with patients a special status as "both a gentle stimulant, and composing sedative."[39]

After Cullen's death in 1790, his disciples sought, in Rush's words, to "prepar[e] the way for a divine system of morals. . . . to teach mankind that they are brethren."[40] Dedicating his work to his one-time classmate Falconer, another Cullen protégé, John Haygarth, attributed the success of his own placebo devices—including waxed nails and mahogany sticks fashioned to resemble metal "tractors" recently touted as a remedy for rheumatism, palsy, and an array of "topical" inflammatory afflictions—to "the wonderful force of the Imagination."[41] Haygarth concluded that these placebos' "whole effect undoubtedly depends upon the impression which can be made upon the patient's Imagination."[42] But while that impression was found to have supported what patients "believed," Haygarth's account of the production and management of this impression stresses "the cheering affection of hope."[43] In his book, that intangible "affection" may be animated by an action—"slightly touching the skin" such that "the strong powers of the imagination" balance "the concurring influences of the soul."[44] Half-dead hope may also be revived by *transaction*, as when a "medical friend" brings his "words and looks" to "correspond with [the] other," tipping the balance of an invalid mind "strongly agitated between hope and fear."[45] During one application of Haygarth's "fictitious Tractors" at Bath, the attending physicians were "almost afraid to look each other in [the] face lest an involuntary smile should remove the mask from [their] countenances and dispel the charm." Their placebo act rehearses modern deadpan routines in which the expressionless face pleases because it pantomimes the dead, mobilizing fringe affect ("almost afraid"). In turn, by leaking the story of physicians who can barely keep a straight face, Haygarth let his own lay readers in on the joke, linking them, seemingly, to a virtual chain of experts in the know.

Such ambiguities appear to destine the placebo for the alienated familiar of Freud's uncanny. But the placebo's mobilization in a materialized communication relay whose premise is the immateriality of readers and writers with respect to one another is perhaps more relevant to what Bernard Geoghegan calls the "infrastructural uncanny." Here spectral forms naturally "emerg[e] in periods of rapid expansion in the means of technological conveyance." They arise from "a slight rupture or assonance, either internal to a networked relay or in its relationship to the embedding environment."[46] It's the anticipatory hesitation within this system—its encounter with obsolescence, even nullity—that the placebo at once figures, reveals, and mobilizes as pleasurable, even ironic self-recognition.

Newly popular studies of specific medical challenges (typically though not exclusively chronic ones such as rheumatism, syphilis, and scurvy) often pivoted on the placebo. Authors tapped its capacity to suspend a treatment narrative between what is and what is not as well as between medicine's spiritualized but often futile past and its hyperembodied but soulless future. The male midwife in William Smellie's 1752 *Treatise on [. . .] Midwifery* turns up at bedsides where, "though every thing be in the right posture," a baby is slow to arrive. The not-quite, not-yet mother is "too anxious and impatient to wait the requisite time" for it to make its appearance. Should "arguments and gentle persuasion" fail to relieve her "uneasiness," Smellie ventured, "it will be convenient to prescribe some innocent *Placemus,* that she may take between whiles, and beguile the time and please her imagination."[47] Smellie's "*Placemus*" (we will please) retains "placebo" as a personal action, requiring it to show up as a full-blown person. The "between whiles" and time-beguiling ritual of actively inactive taking eerily reorients the living body toward one that appears not to be there or that might soon be so.[48]

Gonorrhea's ambiguous nosology and moral shadiness, alongside the lack of any "specific medicine" to treat it, made it another popular site of intercession by placebo.[49] Holding that "it appears to be nearly the truth" that "time itself will effect a cure," the celebrated London surgeon John Hunter's 1786 *Treatise on the Venereal Disease* wonders "whether medicine can be of any service in this form of the disease." Deciding that the placebo can occupy the space "where [service] is not," Hunter "gave certain pills of bread that were taken with great regularity." He reports, "The patients always got well, but some of them, I believe not as soon as they would have done, had the artificial methods of cure been employed."[50]

Possibly, Hunter's bread pills went moldy, breeding the penicillin with which gonorrhea is treated now. Like other purveyors of the bread bolus, Hunter would, if inadvertently, have administered an efficacious medicine, resorting in spite of himself to one of the more "artificial methods of cure." The placebo works by *not* working, unveiling a sphere of sensation, experience, even pleasure, uncannily adjacent—"near"—to modernizing medicine's utilitarian formal realism.[51] William Blair's 1799 *Essay on the Venereal Disease* conjured a patient whose "syphlitic eruptions" were treated with a warm bath and Dover's powders, while he "likewise took a pill (ex mica panis) as a placebo, every evening and morning" until the eruptions disappeared. Even the atmospherist of heaven Thomas Beddoes treated a syphilitic sepoy to "crumbs of bread made by [Beddoes him]self."[52]

Meanwhile, Thomas Trotter's 1786 *Observations on the Scurvy* featured a placebo trial organized around the vanishing point of what "did not" transpire in the bodies of individuals already socially dead. Speculating that Africans held captive on slave ships knew to outsmart scurvy with unripe guava, Trotter devised a single-blind experiment involving nine "scorbutic" persons "kept under the deck." Only unripe guava (oh, and limes) led to remission of symptoms: "Three Negroes restricted to [. . .] ripe Guava continued in much the same situation, where others did not."[53] A chilling anticipation of the twentieth century's Tuskegee experiments, this exercise in experimental medicine yielded dark knowledge that was printed alongside the report on a very different placebo trial, courtesy of an earlier, if unnamed physician. Unlike Trotter's, that gentleman's "Observations" spoke the language of chemical enlightenment, yielding a story so stone-cold as not to count as a story at all. In it, the aboveboard bodies of members of the British Navy are favored with elixir of vitriol, an acidic gargle, on the theory that the sulfur in its base "has so strong an attraction for the oxygene that it is incapable of being acted upon" and thus "passes unchanged, through the body, without exerting any effects on the blood." Trotter interpreted this to mean that "it is thus in this way a mere *placebo*."[54] Without exerting any effect, vitriol split sensation from substance; moreover, the European sailors to whom it was administered gave full consent to be used as living confirmations of common knowledge. Yet this literally bloodless story has a real and important effect on the truly moving story that Trotter paired with it. In this second story, the reader is urged to hear the "hideous moan" of Black bodies deprived of guava's saving grace; hopelessly, those in what amounts to a control group cried out "when waking from sleep, after a dream had presented to their imagination their home and friends." For Trotter, their "hideous moan" delivered "proof that the depressing passions of the mind have a powerful effect in the production of the scurvy."[55] It did not point to a moral about forced deprivation of home and friends, or about the cruel delusion of "a dream," or about which bodies, minds, and spirits are privileged in moments when much is suspected but little is known. There is no anger—no *vitriol,* it is hard not to say—evident in Trotter's tale. But between its lines, Placebo seems to be telling a different and terrible story, one that has only become audible over time.

One need not be a medical treatise to be haunted by Placebo. Just for fun, it was possible for eighteenth-century readers to read of a "man of imposing countenance," the impersonator of a dead physician who has left several painstakingly labeled drawers of remedies behind. The impostor sends any "Stranger" he cannot diagnose to a "Placebo drawer among the others."[56] It is not hard to guess which drawer got the most traffic. In any case, such satires suggest that modern physicians were building a community of experts and a myth of expertise superficially validated by the public story of its own disenchantment. Adam Phillips identifies the post-Enlightenment "craving for experts" with modernity's unsatiated "need for belief." But a subtler truth is that "our relationship to experts is a picture of the way we need."[57] Placebo experts pictured in their readers, and patients, a need for experience itself—the experience if not of healing then at least of remission of symptoms. This is by definition a need for nonexperience, for the body to be phenomenologically absent. This absent body was one that a placebo story could touch as well as a placebo.[58]

[T]errors and Experts: The Gothic Placebo

Placebo's migration into gothic romance brings us to the final chapter of this story. The first gothic novelists were placebo experts. They honed techniques of terror premised on the uncanny ability of nothing—unless it is the mysterious relay between print and its readers—to awaken specters of sensation.[59] Small wonder that the pleasures of the supernatural are medicalized in Ann Radcliffe's oft-cited definition of "terror" as what "expands the soul, and awakens the faculties to a high degree of life," while horror "contracts, freezes, and nearly annihilates them."[60] Radcliffe suspends the perverse pleasures of gothic effects between "life" support and mercy killing, while her popular fiction solicited those same pleasures via the empty but palpable forms of printed words.

As Matthew Wickman suggests, insofar as gothic techniques of terror begin with nonexperience—or rather with the experience of mediation as it seems to recur across historical moments—they rehearse "the logic of mourning."[61] Echoing the medicalized placebo's activation of a second body that is historically gothic and notionally dead, gothic fiction posited a second, nonsensing body whose blood it could chill via the protocols of expectation it mobilized in the empty but material forms of print. Notional manipulations of this ideational body—raised hair, chilled spines, curdled blood—were as formulaic as Haygarth's placebo art. Indeed, Haygarth himself boasted of having vanquished the "very obstinate and permanent disorder" of rheumatism by "play[ing] the part of necromancer, to describe circles, squares, triangles, and other figures of geometry upon the part affected."[62]

Jane Austen tested these tricks in her deadpan gothic *Northanger Abbey* (1803/1817), many of whose scenes are set in Haygarth's Bath and thus in the very lap of fashionable medicine, where lucrative catering to wealthy patients' ambiguous fantasies had produced the very conflict that placebo resolves: physicians' increased reliance on story to validate their own authority versus the profitability of believing that patients can tell their own stories. Austen's fantasia of error and false expectation genders this dilemma. The purity encoded in the name of her robustly healthy protagonist, Catherine, evokes the placebo's innocence of intent. Having contracted a fever for gothic romance, Catherine finds herself the guest of a family, the Tilneys, whose surname winks at the suspended, not-un*til* structure of gothic pleasure as it exploits expectation and suspense. Like Haygarth's placebo artists, Catherine's tutor Henry Tilney struggles to keep a straight face when confronted with the self-stimulating errors of attribution and false expectations known to make Catherine's "blood r[u]n cold."[63]

Here, Austen turns out to be of Ann Radcliffe's party without necessarily knowing it. In Catherine's Radcliffean bible, *The Mysteries of Udolpho* (1794), Emily St. Aubert's mother, dying of smallpox, refuses to be "flattered or deceived" by physicians all too eager to do both. In Radcliffe's *The Italian* (1797), a physician summoned to the bedside of the poisoned Schedoni sees at once how little hope there is: "He was, however, willing to administer the medicine usual in such cases."[64] Expertise delivers gothic romance's signature pleasures, grounding these not in an efficacious intervention that holds communities of bodies together but rather in the sustained tension, suspense, and ambiguity that separate bodies from themselves.

For one prominent gothic novelist of the 1790s, placebo effects were literally the work of the devil. As Chaucer's *Parson's Tale* had already had it, "flatterers be the Devil's champions / That ever sing placebo."[65] As late as 1646, Thomas Browne is still saying it's Satan who "deludeth us also by [. . .] many superstitious waies in the cure of common diseases; seconding herein the expectation of men with events of his own contriving."[66] Matthew Lewis's 1796 romance *The Monk* thus naturally places the monk of his title, Ambrosio, in an echo chamber of pleasing forms and flattering voices, especially those of the cross-dressed novice Matilda, who turns out to have been the fabrication of the devil. Or at least that is who she is if we are to believe Lucifer's own boasts. Does belief matter? Sentenced to death for the crimes that have followed from Matilda's seductive cure, Ambrosio hangs on the devil's every word. But "rather dead than alive" he's lured not by faith misplaced but rather by his own groundless hopes.[67]

Ambrosio's very name ("elixir of life") identifies him with palliative medicines that sustain life by stimulating pleasure beyond the body's material boundaries. Contemporary reviews classified his story as an effort to "selec[t] what is most stimulant from the works of our predecessors."[68] André Breton would find that Lewis, "in the purest way imaginable, exercises an exalting effect only upon that part of the mind which aspires to leave the earth." Matilda flatters just that part of Ambrosio's mind into the conviction of its own existence; as Lucifer notes, his "pride was gratified by [her] flattery."[69] Matilda perhaps most flatters Ambrosio by sucking snake venom from his body. She's assisted by the nonintervention of the house physician, Father Pablos, whose name, if derived from Alain-René Lesage's *Devil upon Two Sticks* (trans. 1707), whispers the placebo. With snakebitten Ambrosio seemingly at death's door, Pablos prescribes a few inefficacious tranquilizers and then, to pass the time, conjures a fantastic cause for a fantastic effect: the nonexistent Cuban "cientropiedro."

A true placebo romance, *The Monk* couples science with storytelling. Lewis, interestingly, would engineer another such copulation on his family's Jamaica plantation. Here, confronted with plummeting birth rates among his African slaves and obviously driven by self-interest, he introduced aesthetically pleasing reward systems (scarlet girdles that would "entitle" pregnant women to "marks of peculiar respect and attention") and enlisted the authority and experience of Afro-Caribbean female "doctresses" who often drew on techniques of ritual and narrative enchantment to create community and health.[70] It is difficult not to see the irony and aptness of Lewis's own demise from yellow fever on his way home from his family's slave holdings in Jamaica. Having poisoned himself with West Indian emetics, he was heard "on his last day exclaiming every instant, 'the suspence—the suspence,' [sic] which the physician who attended him was doubtful whether he meant to allude to religious doubt or to the success of a medicine which he had taken."[71]

Notes

1. William Godwin, *Report of Dr. Benjamin Franklin, And other Commissioners Charged . . . with the Examination of the Animal Magnetism* (London, 1785), 2.
2. Jessica Riskin, *Science in the Age of Sensibility: The Sentimental Empiricists of the French Enlightenment* (Chicago: University of Chicago Press, 2002), 14. See also James Delbourgo, *A Most Amazing Scene of Wonders: Electricity and Enlightenment in Early America* (Cambridge, MA: Harvard University Press, 2006); Robert Darnton, *Mesmerism and the End of the Enlightenment in France* (Cambridge, MA: Harvard University Press, 1968); Patricia Fara, *Sympathetic Attractions: Magnetic Practices, Beliefs, and Symbolism in Eighteenth-Century England* (Princeton: Princeton University Press, 1996); Stewart Justman, *To Feel What Others Feel: Social Sources of the Placebo Effect* (San Francisco: University of California Medical Humanities Press, 2012); and Richard Sha, *Imagination and Science in Romanticism* (Baltimore: Johns Hopkins University Press, 2018), 152–53. For broad-based recent inquiries into the place of the imagination in Enlightenment science, see Sha and Tita Chico, *The Experimental Imagination: Literary Knowledge and Science in the British Enlightenment* (Baltimore: Johns Hopkins University Press, 2018).
3. William Godwin, "Historical Introduction," in *Report of Dr. Benjamin Franklin*, xvii.
4. This is not all such historians, despite the long list; see in particular Helen Thompson, *Fictional Matter: Empiricism, Corpuscles, and the Novel* (Philadelphia: University of Pennsylvania Press, 2017).
5. Godwin, *Report of Dr. Benjamin Franklin*, 37.

6. Ted. J. Kaptchuk, "Intentional Ignorance: A History of Blind Assessment and Placebo Controls in Medicine," *Bulletin of the History of Medicine* 72 (1998): 391–92.
7. Classic sources documenting and/or theorizing this phenomenon include Roger Chartier, *The Order of Books: Readers, Authors, and Libraries in Europe between the Fourteenth and the Eighteenth Centuries* (Stanford: Stanford University Press, 1992); Adrian Johns, *The Nature of the Book: Print and Knowledge in the Making* (Chicago: University of Chicago Press, 1998); Clifford Siskin and William Warner, eds., *This Is Enlightenment* (Chicago: University of Chicago Press, 2010); and most recently, Paula McDowell, *The Invention of the Oral: Print Commerce and Fugitive Voices in Eighteenth-Century England* (Chicago: University of Chicago Press, 2017).
8. Roy Porter, "Laymen, Doctors and Medical Knowledge in the Eighteenth Century: The Evidence of the *Gentleman's Magazine*," in *Patients and Practitioners: Lay Perceptions of Medicine in Pre-industrial Society*, ed. Roy Porter (Cambridge: Cambridge University Press, 1986), 283–314; Ginnie Smith, "Prescribing the Rules of Self-Help and Advice in the Later Eighteenth Century," in Porter, *Patients and Practitioners*, 249–82.
9. Beddoes, for example, reported that while inhaling a certain "mixture of airs," his own body fat "fell away rapidly," and he watched, enraptured, as a "carnation tint" crept up his fingers by degrees themselves imperceptible. Mike Jay, *The Atmosphere of Heaven: The Unnatural Experiments of Dr. Beddoes and His Sons of Genius* (New Haven: Yale University Press, 2009), 73. The novelistic qualities of such self-reportage eerily anticipate Jay's own brand of science writing; a *Lancet* review of *The Atmosphere of Heaven* quoted on its back cover hails the book as "serious historical writing that can be read like a novel." See Elizabeth Leyland, "Living, Breathing History," *Lancet* 375 (2010): 366.
10. Deidre Shauna Lynch, "Gothic Fiction," in *Cambridge Companion to Nineteenth-Century Fiction*, ed. Richard Maxwell and Katie Trumpener (Cambridge: Cambridge University Press, 2008), 49.
11. Ronald Schleifer and Jerry Vannetta, *The Chief Concern of Medicine: The Integration of the Medical Humanities and Narrative Medicine into Medical Practices* (Ann Arbor: University of Michigan Press, 2013), 37. In a book that has become a standard text in progressive medical schools, Schleifer and Vannatta seek to supplement fundamentally realist models of medicine with ones that take "a practical schema of narrative understanding" (p. 100) into account as constituents of meaningful realities. On storytelling as a key component of medical care, see also Rita Charon's classic *Narrative Medicine: Honoring the Stories of Illness* (Oxford: Oxford University Press, 2008).
12. Arthur K. Shapiro, "The Placebo Effect in the History of Medical Treatment: Implications for Psychiatry," *American Journal of Psychiatry* 116 (1959): 303.

13. Arthur K. Shapiro and L. A. Morris, "The Placebo Effect in Medical and Psychological Therapies," in *Handbook of Psychology and Behavior Change*, ed. Allen Bergin and Sol Garfield (New York: Wiley, 1978), 372. This definition remains common currency in the vast literature on the placebo that has subsequently come into existence. For a survey of its applications and debates surrounding it, see Klause Lin, Margrit Fässler, and Karin Meissner, "Placebo Interactions, Placebo Effects, and Clinical Practices," *Philosophical Transactions of the Royal Society of London B (Biological Sciences)* 366 (June 26, 2011): 1905–12. Shapiro and Morris themselves drew on Harry Beecher's seminal article "The Powerful Placebo," *Journal of the American Medical Association* 159 (December 24, 1955), 1602–6. Recent approaches to the placebo effect that posit the human mind-body's merging with technology are epitomized in Thomas Kosch et al., "The Placebo Effect of Artificial Intelligence in Human-Computer Interaction," *ACM Transactions on Computer-Human Interaction* 1c (2022): 1–32. See also Ze Hong, "The Population Dynamics of the Placebo Effect: Its Role in the Evolution of Medical Technology," *Human Ecology* 50 (2022): 11–22. Throughout this essay, brackets around ellipses indicate omissions from the original text.
14. Peter Shaw, *The Reflector* (London, 1750), 230, 226.
15. Jerome D. Frank and Julia B. Frank, *Persuasion and Healing: A Comparative Study of Psychotherapy* (1961; Baltimore: Johns Hopkins University Press, 1991), 132, 134.
16. George Motherby, *A New Medical Dictionary, or A Repository of Physic* (London, 1785), 594; Bartholomew Parr, *London Medical Dictionary* (London, 1809).
17. Arthur K. Shapiro, "Semantics of the Placebo," *Psychiatric Quarterly* 42 (1968): 653.
18. Thomas Hooper, *Quincy's Lexicon-Medicum* (London, 1811), 634.
19. Wayne Wild, *Medicine-by-Post: The Changing Voice of Illness in Eighteenth-Century British Consultation Letters and Literature* (New York: Rodopi, 2006). See also Roy Porter, *Bodies Politic: Disease, Death and Doctors in Britain, 1650–1900* (London: Reaktion, 2001).
20. Thomas Jefferson to Casper Wistar, 21 June 1807, in *The Life and Selected Writings of Thomas Jefferson*, ed. William Peden (New York: Modern Library, 1998), 534.
21. William Cullen, *First Lines of the Practice of Physic* (London, 1777), 1.
22. Cullen, *First Lines*, 136.
23. Cullen, Clinical Lectures, 1770, Royal College of Physicians of Edinburgh (RCPE) Manuscript, Cullen 1/18 130, James Lind Library. Cited in Rosalie Stott, "Health and Virtue, or How to Keep Out Harm's Way: Lectures on Pathology and Therapeutics by William Cullen, c. 1770," *Medical History* 31 (1987): 138.

24. Cullen, Clinical Lectures, 1772 February/April, RCPE Manuscript, Cullen 4/4 218–19, James Lind Library.
25. Cullen Clinical Lectures, 1772–73, RCPE Manuscript, Cullen 4/2 299–300, James Lind Library.
26. Shapiro, "Semantics of the Placebo," 655.
27. Geoffrey Chaucer, *The Canterbury Tales*, in *The Riverside Chaucer*, ed. Larry D. Benson, 3rd ed. (Boston: Houghton Mifflin, 1987), 4:1494–1500. On the gender politics of Placebo's flattery, see Amanda Walling, "Placebo Effects: Flattery and Anti-feminism in Chaucer's *Merchant's Tale* and The Tale of Melibee," *Studies in Philology* 115, no. 1 (2018): 1–24.
28. Alexander Pope, "January and May," in *The Works of Alexander Pope* (London, 1717), 196, 200.
29. C. E. Kerr, I. Miller, and T. J. Kaptchuk, "William Cullen and a Missing Mind Body Link in the Early History of Placebos," *Journal of the Royal Society of Medicine* 101, no. 2 (2008): 89.
30. Justman, *To Feel What Others Feel*, 64–65.
31. For example, Henry David, *An Historical Description of Westminster-Abbey* (1865); John Burney, *An Historical Description of the Metropolitical Church of Christ, Canterbury* (London, 1799); John Burney, *The Winchester Guide* (London, 1796); and John Burney, *Divers Liturgical Chants* (London, 1799).
32. Heather Keenleyside, *Animals and Other People: Literary Forms and Living Beings in the Long Eighteenth Century* (Philadelphia: University of Pennsylvania Press, 2016), 24.
33. "Serious Reflections on Atmospherical Diseases," in *The Medical Spectator* (London, 1792), 53.
34. "Serious Reflections," 60.
35. Alexander Sutherland, *An Attempt to Revive Ancient Medical Doctrines* (London, 1763), xxiii, xxiv.
36. Terry Castle, *The Female Thermometer: Eighteenth-Century Culture and the Invention of the Uncanny* (New York: Oxford University Press, 1995); Simon During, *Modern Enchantments: The Cultural Power of Secular Magic* (Cambridge, MA: Harvard University Press, 2002); Stephen Greenblatt, *Hamlet in Purgatory* (Princeton: Princeton University Press, 2001).
37. William Falconer, *A Dissertation on the Influence of the Passions upon Disorders of the Body* (London, 1788), 31–32.
38. On the role of sleep and dreams in ancient medicine, see Ian Hacking, "Dreams in Place," *Journal of Aesthetics and Art Criticism* 59 (2001): 245–60; and Kenton Kroker, *The Sleep of Others and the Transformations of Sleep Research* (Toronto: University of Toronto Press, 2007), 25–29. Sleep's transformations in modernizing medicine are traced in Sasha Handley, *Sleep in Early Modern England* (New Haven: Yale University Press, 2016), 18–38. A classic study

of the practice of dream incubation as a healing technique in ancient Greece and the medieval Catholic Church, where it coincides with pictorial imagery, can be found in Mary Hamilton, *Incubation: The Cure of Disease in Pagan Temples and Christian Churches* (London: Henderson, 1906). See also Sarah O'Dell, "Gothic Therapies: Literature, Sacrament and the Physic of the Imagination, 1760–1820" (PhD diss., University of California, Irvine, 2024).

39. Falconer, *Dissertation*, 22.
40. Benjamin Rush, "An Eulogium upon Dr. William Cullen," in Rush, *Essays Moral and Political* (Philadelphia, 1798), 321–22.
41. John Haygarth, *Of the Imagination as a Cause and a Cure of Disorders of the Body, as Exemplified by Fictitious Tractors and Epidemical Convulsions* (Bath, 1800), 3. In an episode too baroque to fit the limits of this essay, Haygarth was attacking the "tractors" devised by the American physician and entrepreneur Elisha Perkins as a remedy for rheumatism, palsy, and other "topical" disorders. Perkins's son Benjamin had taken the patented tractors to Great Britain and published a "tract" touting their benefits—and cutting into the lucrative health care market centered in Bath and Bristol. See Benjamin Perkins, *The Influence of Metallic Tractors on the Human Body* (London, 1798). Haygarth's elaborate placebo experiments were intended to debunk the tractors and an intervention that smacked of Mesmerism and the alternative social medicine it represented. Perkins responded to them with the indignant *The Efficacy of Perkins's Metallic Tractors, in Topical Diseases* (London, 1800), which included dozens of testimonials. Periodical culture weighed in on both sides. For an analysis of the Perkins controversy as a replay of the Mesmer debate, see Justman, *To Feel What Others Feel*, 51–58; as an event in electrical medical culture, see Delbourgo, *Most Amazing Scene*; and for a contemporary perspective favoring Perkins père et fils, see Charles Langworthy, *A View of Perkinean Electricity* (Bath, 1798).
42. Haygarth, *Of the Imagination*, 4.
43. Haygarth, 5, 25.
44. Haygarth, 28.
45. Haygarth, 31.
46. Bernard Dionysius Geoghegan, "Mind the Gap: Spiritualism and the Infrastructural Uncanny," *Critical Inquiry* 42, no. 4 (2016): 900.
47. William Smellie, *Treatise on the Theory and Practice of Midwifery* (London, 1752), 224.
48. It also complicates canonical portraits of eighteenth-century medicine that insist that male physicians turning midwifery into obstetrics sought to automate childbirth, fast-forwarding the laboring female body through a mechanical process, or spectacularly displaying male heroic prowess. For a nuanced example of the latter view, see Lisa Forman Cody, "The Politics of Reproduction:

From Midwivery's Alternative Public Sphere to the Public Spectacle of Man-Midwifery," *Eighteenth-Century Studies* 32, no. 4 (Summer 1999): 477–95; for the former, see Adrian Wilson, *The Making of Man-Midwifery: Childbirth in England, 1660–1770* (Cambridge, MA: Harvard University Press, 1995). For early readings on the protogothic elements of male midwifery, however, see Robert Erickson, *Mother Midnight: Birth, Sex and Fate in Eighteenth-Century Fiction* (New York: AMS, 1986).

49. On venereal diseases' moral ambiguity and uncertain nosology, see Noelle Gallagher, *Itch, Clap, Pox: Venereal Disease in the Eighteenth-Century Imagination* (New Haven: Yale University Press, 2018)
50. John Hunter, *A Treatise on the Venereal Disease* (London, 1786), 79–80. See also Wendy Moore, "John Hunter: Learning from Natural Experiments, Placebos, and the State of Mind of the Patient in the Eighteenth Century," *Journal of the Royal Society of Medicine* 102 (2009): 394–95.
51. Michel Foucault is the definitive archaeologist of this regime; see especially *The Birth of the Clinic: An Archaeology of Medical Perception* (New York: Vintage, 1973).
52. Thomas Beddoes, *Communications Respecting the External and Internal Uses of Nitrous Acid* (London, 1800), 19; William Blair, *Essay on the Venereal Disease* (London, 1799), 186.
53. Thomas Trotter, *Observations on the Scurvy* (London, 1786), 137–38, 184.
54. Trotter, *Observations*, 184.
55. Trotter, 37.
56. Robert Thornton, *Medical Extracts* (London, 1798), 321–22.
57. Adam Phillips, *Terrors and Experts* (Cambridge, MA: Harvard University Press, 1995), xii–xiii.
58. Drew Leder, *The Absent Body* (Chicago: University of Chicago Press, 1990).
59. Lynch, "Gothic Fiction," 47–63.
60. Ann Radcliffe, "On the Supernatural in Poetry," *New Monthly Magazine* 16, no. 1 (1826): 150.
61. Matthew Wickman, "Terror's Abduction of Experience: A Gothic History," *Yale Journal of Criticism* 18, no. 1 (2005): 180.
62. Haygarth, *Of the Imagination*, 17.
63. Jane Austen, *Northanger Abbey, Lady Susan, the Watsons, Sanditon*, ed. Claudia L. Johnson and John Davie (Oxford: Oxford University Press, 2008), 137. On the *katharsis* embedded in Catherine's name as an alternative to evidence-based medicine, see Schleifer and Vannatta, *Chief Concern of Medicine*, 5–6.
64. Ann Radcliffe, *The Italian, or The Confessional of the Black Penitents*, ed. Nick Groom (1797; Oxford: Oxford University Press, 2017), 382.
65. Chaucer, *Canterbury Tales*, 10.618–19.

66. Thomas Browne, *Pseudodoxia Epidemica*, in *Thomas Browne: Selected Writings*, ed. Kevin Killeen (1646; Oxford: Oxford University Press, 2014), 151.
67. Matthew Lewis, *The Monk*, ed. Howard Anderson and Nick Groom (1796; Oxford: Oxford University Press, 2016), 331.
68. *Monthly Review* 23 (1797): 451.
69. André Breton, "Manifesto of Surrealism," in *Manifestoes of Surrealism*, trans. and ed. Richard Seaver and Helen R. Lane (1924; Ann Arbor: University of Michigan Press, 1969), 15.
70. Katherine Paugh, *The Politics of Reproduction: Race, Medicine, and Fertility in the Age of Abolition* (Oxford: Oxford University Press, 2017), 101–7. Thanks to Danielle Spratt for this reference.
71. Thomas Moore, *The Journal of Thomas Moore*, cited in David Lorne Macdonald, *Matthew Lewis: A Critical Biography* (Toronto: University of Toronto Press, 2000, 210).

PART II

Reception

"The Eye of Mr. *Anson* Himself"
Art and Evanescence in A Voyage Round the World (1748)

ANNE M. THELL

Few voyage accounts were as popular yet as vigorously contested as George Anson's *A Voyage Round the World in the Years MDCCXL, I, II, III, IV* (1748). While Anson was celebrated as a national hero upon his return to London in 1744, his voyage was marred by catastrophic losses: scoured by storms, shipwreck, mutiny, and scurvy, the squadron returned with only 118 of nearly 2,000 men.[1] Efforts to determine a "fair and impartial Account" of these harrowing years played out in dramatic form in legal disputes over prize money; drunken brawls among returned sailors; a series of competing eyewitness accounts (at least five of which preceded Anson's, with dozens more to follow); the complicated ghostwriting of *A Voyage* (compiled by Richard Walter but likely completed by Benjamin Robins, a military engineer who did not sail on the voyage); and the various travel compilations that haphazardly combined source materials.[2] Writers like Horace Walpole and Abbé Coyer accused Anson of fantastic embellishment, while experienced naval men like James Naish scribbled across the margins of *A Voyage* to amend its failures and distortions (see fig. 1).[3] If *A Voyage* was designed to offer the official and comprehensive record of Anson's historic voyage, its publication history also demonstrates the limits of his authority and, broadly, the insufficiency of any single perspective in capturing a voyage so spectacularly multifarious in nature.[4]

In his introduction to *A Voyage,* Walter understands that establishing the stability and superiority of Anson's point of view is crucial to claiming the text's primacy. Intriguingly, he does this not by trumpeting Anson's rank or judgment, nor his commercial exploits, but by turning to the practices and theory of art. Walter differentiates *A Voyage* from "any narration of this kind

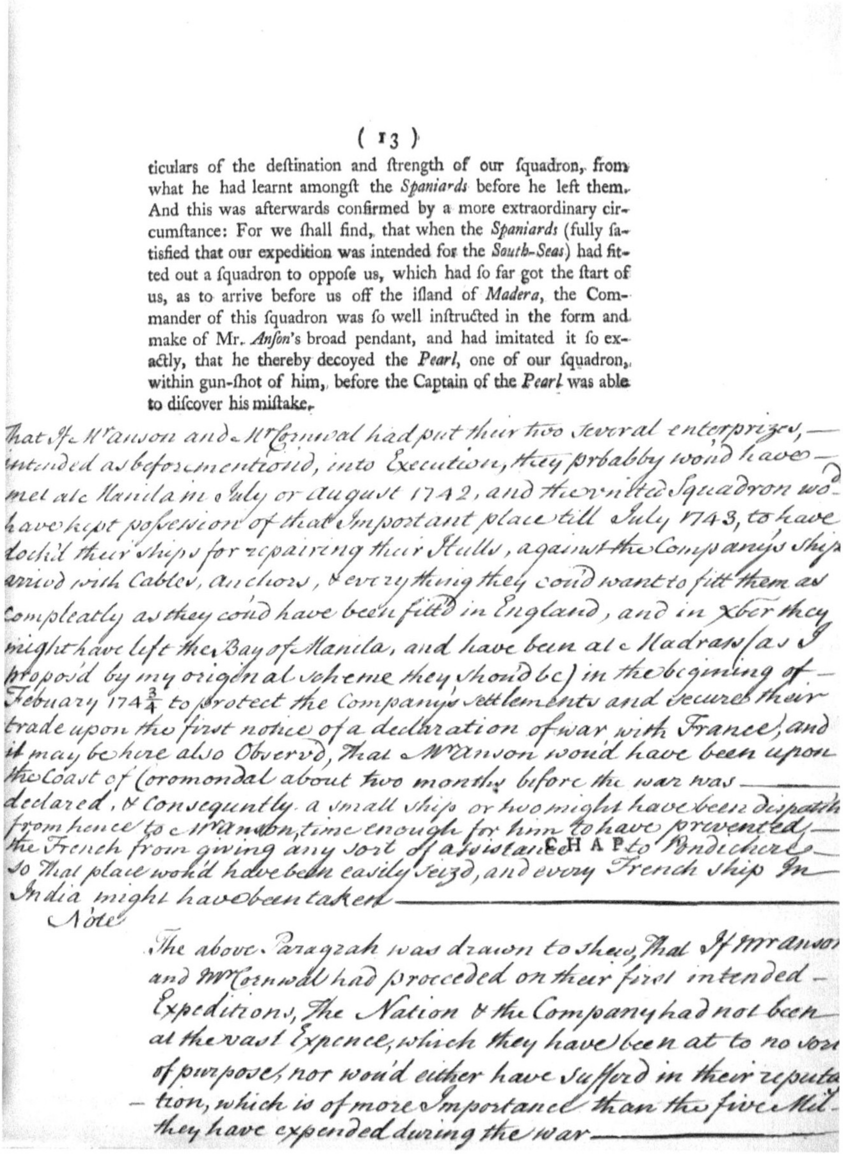

(13)

ticulars of the deftination and ftrength of our fquadron, from what he had learnt amongft the *Spaniards* before he left them. And this was afterwards confirmed by a more extraordinary circumftance: For we fhall find, that when the *Spaniards* (fully fatisfied that our expedition was intended for the *South-Seas*) had fitted out a fquadron to oppofe us, which had fo far got the ftart of us, as to arrive before us off the ifland of *Madera*, the Commander of this fquadron was fo well inftructed in the form and make of Mr. *Anfon's* broad pendant, and had imitated it fo exactly, that he thereby decoyed the *Pearl*, one of our fquadron, within gun-fhot of him, before the Captain of the *Pearl* was able to difcover his miftake.

That if Mr Anson and Mr Cornwal had put their two several enterprizes, — intended as before mentioned, into Execution, they probably wou'd have met at Manila in July or August 1742, and their united Squadron wo'd have kept poffeffion of that Important place till July 1743, to have dock'd their Ships for repairing their Hulls, against the Company's Ships arriv'd with Cables, anchors, & everything they cou'd want to fitt them as compleatly as they cou'd have been fitt'd in England, and in Xber they might have left the Bay of Manila, and have been at Madrass (as I propos'd by my original scheme they should be) in the beginning of February 1744¾ to protect the Company's settlements and Secure their trade upon the first notice of a declaration of war with France; and it may be here also Observ'd, That Mr Anson wou'd have been upon the Coast of Coromandal about two months before the war was declared, & Consequently a small Ship or two might have been dispatch'd from hence to Trincomon, time enough for him to have prevented the French from giving any sort of assistance from Pto Pondichery So that place wou'd have been easily seiz'd, and every French Ship in India might have been taken —

Note
The above Paragrah was drawn to shew, That if Mr Anson and Mr Cornwal had proceeded on their first intended Expeditions, The Nation & the Company had not been at the vast Expence, which they have been at to no sort of purpose, nor wou'd either have Suffer'd in their reputation, which is of more Importance than the five Mill. they have expended during the war —

FIGURE 1. James Naish's copy of Anson's *Voyage Round the World* (1748). (© British Library Board, shelfmark 10025.f.8, p. 13)

hitherto made public" by focusing on its "useful and instructive" engravings, most of which were drafted by junior naval officer Peircy Brett.[5] These visual depictions of distant locations, as Walter argues, bring out "a compleater and more finished delineation" of Anson's voyage from "rude well-known outlines": "No voyage I have yet seen, furnishes such a number of views" (there were forty-two in all), he explains, while their quality is "not exceeded, and perhaps not equalled" because "they were not copied from the works of others, or composed at home from imperfect accounts, given by incurious and unskilful observers," but instead "drawn on the spot with utmost exactness, by the direction, and under the eye of Mr. *Anson* himself."[6] Transforming Anson into an omniscient witness who can adopt various proxies, including his draftsman Brett, Walter goes on to theorize the special perception of the skilled draftsman, who has the capacity to perceive more faithfully and vividly than laypeople and thus to achieve an enhanced form of observation that far outstrips ordinary vision. By exulting the superior quality of Brett's drawings, Walter annexes for *A Voyage* a type of aesthetic perception that derives from practical craft knowledge and becomes its own form of inviolable testimony.

This specialized form of perception—the eye of the practical artist—connects to emerging discourses of aesthetics, a field not often associated with the century's navigational accounts. Yet Walter's turn toward the draftsman makes sense in many ways, while it also shows how concretely the practices of art and science remain entangled at this point in midcentury. His focus on skilled perception is an attempt to shore up Anson's eyewitness authority, already an imagined construct that incorporates not only his own idiosyncratic experiences but those of his entire company (across multiple ships and thousands of men). This panoptic vantage point bears a structural resemblance to that occupied by the artist, who often condenses multiple points of view. More generally, Walter is savvy in locating the synthetic "eye of Mr. *Anson*" in drawing, as Brett's professional vision obviates language and, for Walter, guarantees his neutrality and reliability. The eye of the naval draftsman corroborates *A Voyage*, then, while also creating for readers sensory experiences that are in themselves evidential. Indeed, one aspect of the text that has remained beyond question, both then and now, is the accuracy of its engravings, which include not just maps, charts, and coastlines but landscapes, encounters, and natural historical images. In reality, these share the same perspectival complexities as narrative: they are human reproductions, laden with cultural tropes and assumptions; they employ imaginative frameworks and require knowledge of specific semiotic systems; they are often composite images, drawn over hours or days.[7] What we call their accuracy is also a "tendency to beauty," in the words of William Hogarth, which is "not owing to any greater degree of exactness," necessarily, but rather "more pleasing" formal qualities.[8]

Even more specifically, Walter seems to recognize that Brett's fluency in producing externalized or allothetic points of view serves as an ideal analog to Anson's all-encompassing narrative voice. In fact, Walter proposes something more sophisticated than we might expect: he anchors Anson's authority in the allothetic—that is, in vantage points located beyond and outside of the viewing subject. In this way, points of view centered on external locations—aerial vantage points, for instance, or those rooted in locations off ship—serve as the visual equivalent of the narrative disinterestedness that official voyage accounts worked so hard to consolidate. Brett's special importance, then, has to do with his ability to disappear, as he quite literally "rises above" a given scene to adopt various positions not usually available to human observers. Walter's elision of Brett's visual acumen (his ability to see all) and Anson's narrative omniscience (the ability to know all) bolsters the account's authority and creates another unlikely parallel between Walter and Hogarth, who is similarly interested in how the trained artist can see more proficiently than ordinary observers (and thus two men in vastly different situations conceive similar solutions to a similar problem). Importantly, this conflation of Brett and Anson also constitutes a response to the tricky problem of "objective" narration—specifically, the "plain maritime Stile . . . void of Partiality and Prejudice"—that haunted scientific, legal, and navigational writing across the century.[9] Appeasing a culture of ocularity by emphasizing the evidential status of *seeing* and playing up the visual authority of the practical artist, Walter contributes to emerging discourses of aesthetics by isolating in artistic practice a rarefied form of pure or untainted (that is, then, superhuman) perceptive knowledge.

Anson's A *Voyage* marks an important moment in not only the history of navigation or maritime writing, then, but also the history of British aesthetics. What better way to view art and science at midcentury than via this massively popular, lavishly illustrated account, where elaborate foldout engravings illuminate the narrative (and the narrative, in turn, describes these images); where the eye of a naval artisan is the ultimate authority; and where the representational aims of artists and scientists dovetail in a fantasy of comprehensive vision that is usually just beyond our reach? In this brief compass, I examine this aesthetic framework in several steps. First, I discuss Walter's optimistic theorizing of the perception of the practical artist. Next, I examine how Walter approaches the interplay of text and image, as images corroborate the narrative and vice versa, thus creating a self-sustaining multimodal network that satisfies the visual tendencies of both empiricism and formal realism. I then turn to the dynamics of witnessing in several specific scenes, where Brett works to counter subjectivism by collectivizing experience and erasing his own presence. Finally, I conclude with some brief thoughts on the ephemeral nature of optical phenomena like glories and rainbows, which might offer insight into a voyage so starkly individuating that it galvanized one of the century's most explosive textual maelstroms. Throughout, I am interested in the imagined points of view produced by both written and pictorial narrative and, specifically, in the eye of the practical artist that anchors *A Voyage*.

Anson, Walter, Brett

A Voyage was designed to trumpet Anson's prizes, defend his naval and political decisions, and eclipse all competing accounts—all necessary tasks, given his political foibles and his massive losses. By the time of its publication in 1748, *A Voyage* had already accumulated a great deal of competition and was hotly anticipated: "No Expedition that ever I read or heard of," explains one officer, "has rais'd the Expectations of the Publick to [such a] very great Pitch."[10] The 1741 wreck of one of Anson's ships, the *Wager*, led to a series of publications (starting in 1743), with fuller accounts soon following. Midshipman "John Philips" published the first, *An Authentic Journal*, in 1744; that same year, "an officer" published his experiences in *The Universal Spectator*; soon after, mathematics teacher Pascoe Thomas released *A True and Impartial Journal of a Voyage to the South-Seas* (1745).[11] Here, Thomas offers an unflattering account of his superiors while also claiming that Walter and Robins suppressed dissenting views so as to claim "Authority" over experiences not theirs to "tell the world."[12] However frustrated, Anson's men did manage to be heard: travel compendiums published after 1745 nearly always include Anson's voyage but often base this section on the unauthorized narratives. For instance, the popular *Navigantium atque Itinerantium Bibliotheca* (1744–48) redacts Philips's *Authentic Journal*, while *A Complete System of Geography* (1747) bases Anson's voyage on Thomas's *Journal*. This proliferation of competitive publications, all based on firsthand experience, radically challenges the preeminence of any single account or perspective.

Indeed, Anson's endlessly contested and supplemented narrative illuminates the complexity of isolating what really happened on any such collaborative undertaking, which is not ultimately recoverable by any single individual: "It was impossible any one Man could see every Thing that happened."[13] This problem—the tricky business of establishing a point of view that is both authentic and yet comprehensive or totalizing—explains why Walter works so hard to cement Anson's authority and, specifically, why he draws so extensively on the logic of skilled artisanship. Across his introduction to *A Voyage*, he emphasizes the seemingly incontestable nature—or the "pictorial facticity," to borrow John Bender's term—of its engravings.[14] Sidestepping the complexities of composite narration to focus on the accuracy of *A Voyage*'s "useful and instructive" charts, maps, and views, Walter trumpets the medium specificity and pragmatic value of these visual materials, as well as their abundance and precision.[15] He measures Brett's drawings against both a real-world standard and those views "defective in industry and ability," which trade dangerously in "bold conjectures, and fictitious descriptions."[16] As Walter would have us believe, Brett's skilled images evade conjecture because they are exact visual representations of visual experiences and are therefore beyond language, beyond dispute, beyond judgment.

But Walter emphasizes the drawings featured in *A Voyage* not only because they are skilled and precise, capturing and allowing readers to participate in the original sensory experience in ways language cannot, but also because they symbolize for him a unique form of perception that is both sharper and more totalizing than quotidian observation. As he "lament[s]" at length, "Many of our accounts of distant countries are rendered [very imperfect] by the relators being unskilled in drawing, and in the general principles of surveying," something that might be remedied if "more of our travellers" were trained in these fields.[17] At stake here is not only "the geography of the globe" but also the faculty of human vision:[18]

> I must add, that besides the uses of drawing, which are already mentioned, there is one, which, though not so obvious, is yet perhaps of more consequence than all that has been hitherto urged; and that is, that those who are accustomed to draw objects, observe them with more distinctness, than others who are not habituated to this practice. For we may easily find, by a little experience, that in viewing any object however simple, our attention or memory is scarcely at any time so strong, as to enable us, when we have turned our eyes away from it, to recollect exactly every part it consisted of, and to recal all the circumstances of its appearance; since, on examination, it will be discovered, that in some we were mistaken, and others we had totally overlooked: But he that is employed in drawing what he sees, is at the same time employed in rectifying this inattention; for by confronting his ideas copied on the paper, with the object he intends to represent, he finds in what manner he has been deceived in its appearance, and hence he in time acquires the habit of observing much more at one view, and retains what he sees with more correctness than he could ever have done, without his practice and proficiency in drawing.[19]

Here Walter counters the innate flaws of human observation—idiosyncrasies of attention, habituation, and memory—with a fantasy of all-encompassing perception that obtains "much more at one view, and retains what [it] sees with more correctness," than any ordinary person could achieve. Importantly, this perceptive acumen is *learned* via the systematic adjudication between the scene and its representation as well as an ongoing attentiveness to repeated and therefore composite points of view: the draftsman, "by confronting his ideas copied on the paper" and comparing them with the actual scene, begins to recognize how "he has been deceived" and therefore to "rectify inattention" and sloppiness, eventually achieving a "habit of observing" that is seemingly mechanical in its completeness and precision. Echoing the comparative framework of Joseph Addison but with a focus on practical craft knowledge rather than taste, judgment, or beauty, Walter accomplishes far more than commending the quality of Brett's drawings.[20] He outlines an ideal of superior perception that allures artists and scientists alike, one that evolves from skilled draftsmanship and eradicates human distortion. Concomitantly, he argues that this particular account surpasses all others because it *sees* better.

Walter's focus on artistic practice and craft knowledge connects, surprisingly, to Hogarth, who makes a strikingly similar claim in *The Analysis of Beauty* (1753). As Hogarth strives to locate what makes certain artworks appealing—what constitutes "the sublime part," the "*Je ne sçai quoi*, or an unaccountable something to most people," or "an harmonious propriety"—he recommends study, emulation, and practical application: "A bright genius, in my opinion, who aspires to excel in the ideal, should propose this to himself, as what has been the principal study of the most famous artists. 'Tis in this part that the great masters cannot be imitated or copied but by themselves, or by those that are advanced in the knowledge of the ideal, and who are as knowing as those masters in the rules or laws of the pittoresque and poetical nature."[21] While the aspiring artist's copies may lack "invention," they bring the imitator closer to perceiving and capturing "the ideal" that sets great work apart. Like Walter, then, Hogarth promotes the perception of the artist via the *practice* of drawing, which has special resonance when applied to a naval officer making and doing in harsh and mobile conditions.[22] Like Hogarth, Walter discusses not taste or judgment but rather concrete skills and habits; his aesthetics focus on the world of objects and result in a scientific formalism that remains, of course, *formal* in its intentional configurations that aim to produce the effect of facticity. Moreover, Walter aligns with Hogarth because he seems to understand that the artist can capture more in a single frame than the human eye can ascertain in any given instant (a point to which I will soon return).

Importantly in relation to Brett, whose engravings have for centuries been praised for their painstaking fidelity, Hogarth also emphasizes that accuracy alone cannot account for the appeal of certain images; instead, we must turn to form and to aesthetics more generally. He demonstrates this by referring to three drawings of human legs—all different in "manner" yet largely natural or anatomical—and commenting,

> If in comparing these three figures one with another, the reader, notwithstanding the prejudice his imagination may have conceiv'd against them, as anatomical figures, has been enabled only to perceive that one of them is not so disagreeable as the others; he will easily be led to see further, that this tendency to beauty in one, is not owing to any greater degree of exactness in the *proportions* of its parts, but merely to the more *pleasing turns, and intertwistings of the lines,* which compose its external form; for in all the three figures the same proportions have been observ'd, and, on that account, they have all an equal claim to beauty.[23]

If we extend Hogarth's logic that "beauty proceeds from those lines"[24]—that is, from formal traits rather than accuracy alone—to Brett's engravings, we might better understand the enduring appeal of, say, Brett's "Burning of the Town of Payta" or his famously luxuriant landscape "View of the Watering Place at Tinian." Hogarth would resist explaining the allure via accuracy alone—as this might be accomplished by any number of draftsmen—and would turn, instead, to their formal qualities. Their aesthetic appeal or "grace," to borrow his term, does not arise from the exactitude of mechanical demonstration as Walter might have us believe. It relates more convincingly to culturally specific concepts of symmetry, beauty, allegory, and form, which enable collective consensus about their quality and operate over and above their politicized content. Brett might capture these scenes accurately—especially in the case of, say, navigational charts that can be tested by future navigators—but his drawings of landscapes and the company's ships exhibit formal qualities that captivate and produce beauty, which can be discerned by "the quality of attention it produces in us as spectators."[25]

As Walter describes Brett's vision as exact, neutral, and mechanical and as we pore over his precise, beautiful images, which often include his own ship, we tend to forget the human being who produced them. Indeed, Brett tends to invisibly inhabit his work, sublimated into points of view impossible to casual onlookers. But he exists—and appears occasionally in *A Voyage*. For instance, on Juan Fernandez Island, Walter describes "loose and shallow" soil, which allows "very large trees" to be "easily overturned" and which once "occasioned the unfortunate death of one of [their] sailors." Here, "Mr. Brett too met with an accident only by resting his back against a tree, near as large about as himself, which stood on a slope, for the tree giving way, he fell to a considerable distance, though without receiving any harm."²⁶ While impossible to confirm, we imagine this fall occurred while Brett was working (scrambling toward a promising vantage point, carrying utensils, drawing), and therefore we suddenly confront the physical effort of "observing much more at one view" that Walter commends (or, in Hogarth's terms, the effort of enabling "the whole figure [. . .] to be seen at one view, as at the playhouse from the gallery").²⁷ Brett's specter emerges elsewhere, but usually indirectly (in an officer gazing back at the artist, for instance, or in his imposition of an English pastoral aesthetic). Omnipresent and yet abstracted, disappearing into perspective itself, Brett provides for Walter the visual equivalent of Anson's panoptic narrative gaze and the anchoring sensory apparatus of *A Voyage*.²⁸

Text and Image

Across *A Voyage*, Walter attends carefully to the integration of image and text, often referring the reader to an image (or specific, labeled parts of an image) for a more apt depiction, thus creating a self-referential textual network that involves both reading and witnessing scenes as they unfold and accentuates embodied readerly experience. He often calls on visual media to supplement and improve verbal description: "To distinguish this bay the better at sea, I have added a very exact view of it, which will enable all future Navigators readily to find it."²⁹ Later, he announces, "The appearance of the entrance of this harbour is very accurately represented in the annexed plate."³⁰ At Juan Fernandez Island, a climax of both the written and visual narratives, he elides what the company saw and what *we* see via Brett's engravings: "On approaching it on its east side, it appears, as represented in the annexed plate, where (A) is a small Island, called *Goat Island*, to the S.W. of it; (B) a rock, called *Monkey Key*, almost contiguous to it; (C) is the East Bay, (D) *Cumberland Bay*, where we moored, and which, as will be observed, is the best road for shipping."³¹ Walter pictorializes language but also inscribes landscapes, with both tactics aiming to transform readers into witnesses. This appeal to the senses sometimes extends beyond seeing. In Brett's exquisite foldout chart of "South America," "Anson's Track" appears as a dotted line, allowing readers to follow that path with a finger as they read. Such tactility enhances our sense of progress through time and space (and unlike real life, our experience can expand to include both projected and actual routes). Again at Juan Fernandez, we find, serially, "The east prospect of the Island," a foldout "Plan" of the entire island, and then a "Survey of the North East Side" (including Cumberland Bay); we next zoom in to a closer view of Cumberland Bay with the company's ships and then "A View of the Commodore's tent." As readers, we thus reenact the gradual accrual of sensory data as we read, unfold and study materials, and refer back to the text; like the navigators, we slowly apprehend scenes as "they . . . open to our view."[32]

However, Walter's commentary in regard to the engravings is most interesting in moments when words fail—when he stumbles upon the je ne sais quoi ("an incommunicable feeling [that] covered the most painful as well as the most pleasurable interludes of navigation"[33]) and appeals to visual media to illuminate. Often, Walter implies that Brett's images are more accurate than his writing because they are uninterpretable (e.g., "a near view of the anchoring place . . . represents it so exactly, that none hereafter can possible [*sic*] mistake it").[34] But he explicitly states their advantage in scenes that elude language, or in what Jonathan Lamb calls "unregulated moments of pure sensation."[35] For instance, approaching Terra del Fuego, Walter announces, "Though the dreariness of this scene can be but imperfectly represented by any Drawing, yet the annexed plate contains so exact a delineation of the form of the country, that it may greatly assist the reader in framing some idea of this uncouth and rugged coast."[36] If overwhelming "dreariness" evades verbal description, so does "beauty," "grandeur," and "elegance," as the company discovers at Juan Fernandez:

> I shall finish this article with a short account of that spot where the Commodore pitched his tent, . . . though I despair of conveying an adequate idea of its beauty. This piece of ground which he chose was a small lawn, that lay on a little ascent, at the distance of about half a mile from the sea. In the front of his tent there was a large avenue cut through the woods to the sea-side, which sloping to the water with a gentle descent, opened a prospect of the bay and the ships at anchor. This lawn was screened behind by a tall wood of myrtle sweeping round it, in the form of a theatre, the ground on which the wood stood, rising with a much sharper ascent than the lawn itself, though not so much, but that the hills and precipices within land towered up considerably above the tops of the trees, and added to the grandeur of the view. There were, besides, two streams of chrystal water, which ran on the right and left of the tent, . . . and were shaded by the trees which skirted the lawn on either side, and compleated the symmetry of the whole. Some faint conceptions of the elegance of this situation may perhaps be better deduced from the draught of it, inserted in the adjoining plate [see fig. 2].[37]

FIGURE 2. "A View of the Commodore's Tent at the Island of Juan Fernandes." (Bodleian Libraries, University of Oxford, shelfmark AA 151 Art, plate 18)

In this famous description of the island, which has fascinated readers since Jean-Jacques Rousseau, Walter employs some of his most poetic writing in an effort to capture the overwhelming beauty of the scene, which is "better deduced from the draught."[38] Faced with profound passion—here awe and relief after tremendous suffering—Walter recognizes the inadequacy of language (a standard trope in voyage literature) but also locates this failure in a *view* and a *moment of seeing* that Brett's visual acumen comes closer to capturing (in "faint conceptions" at least). In his effort to re-create and transmit "an adequate idea of its beauty" and his simultaneous awareness of the inevitable loss of translation, Walter acknowledges that both text and image aim to convey information far beyond scientific exactitude: they strive to capture *how* the men saw and felt and, specifically, *a kind of seeing and feeling* that leads to *knowing*—an aesthetics, then, that Brett might formally articulate via his specialized observational skills (his ability, again, to apprehend "much more at one view").

Yet however much the men strive to convey such experiences to their British audience, Walter recognizes that the company's more extreme encounters are both unrepeatable and inexpressible; the pinnacle of sensation is reserved for those who dared venture on the voyage. The moments when Walter admits that both text and image fail to convey threaten to undermine his scientific project even as he employs the motif of the inexpressible to articulate that experience. For instance, after passing through the infamous "Streights *le Maire,*" he notes the importance of accurate "prospect[s]" for future navigation but then adds,

> And on occasion of this prospect of Staten-land here inserted, extremely barren and desolate, yet this Island of Staten-land far surpasses it, in the wildness and horror of its appearance: It seeming to be entirely composed of inaccessible rocks.... These rocks terminate in a vast number of ragged points, which spire up to a prodigious height, and are all of them covered with everlasting snow; the points themselves are on every side surrounded with frightful precipices, and often overhang in a most astonishing manner.... So that nothing can be imagined more savage and gloomy, than the whole aspect of this coast.[39]

Engaging tropes of both je ne sais quoi and the sublime in this description of "wildness and horror," Walter acknowledges that both narrative and visual depiction are rendered powerless in the face of nature's grandeur; in this instance, even imagination fails to conceive. Thus, although the visual and narrative elements work in tandem to authenticate and replicate Anson's voyage, both stall in the face of intense passion and sensations that might elude even a skilled draftsman like Brett. Again, this inability to signify can only be considered a failure if Walter aims at something more than navigational exactitude or pragmatic value—more than the bare strokes of Anson's logbook[40]—that is, if he strives to convey and to consolidate the company's experiences of beauty, horror, wonder, and passion. Even if they occasionally fail to capture the voluptuousness of firsthand experience, though, Walter positions Brett's engravings as the closest thing to—the most "adequate idea of"—the real.

The Virtual Eyewitness

The forty-two engravings featured in *A Voyage* are diverse: there are charts and maps; views of coastlines, landscapes, and political skirmishes; natural historical images; diagrams of Indigenous technologies. Many of the navigational materials—the maps and charts—require some specialist knowledge, especially in regard to their range of symbols (for shallow water, currents, shipwrecks, etc.). A layperson, for instance, might not know that a few abstract trees indicate dense woods or that a tiny ship marks a wreck that breaks the waterline. The engravings thus require viewers to tap into "individual reservoirs of experience to produce knowledge"[41] while they also demand feats of imagination as readers toggle between symbolic and mimetic forms of representation. We must imagine, say, to populate the land with hills (marked only in the abstract), to translate verbal prompts into visual data ("50 fa. fine dark grey Sand," "50 fa. oozy Ground," etc.), to assume an aerial perspective, or to transform a chart into a multidimensional passage as an approaching navigator would. And, importantly for my present purposes, we must learn to naturalize impossibly detailed renderings of scenes that could have appeared only fleetingly to the eye.

Most evocative in relation to Anson's quasi-omniscient narration is the point of view that Brett assumes when he removes himself from a given location in order to incorporate his own company and ship into the scene; in these moments, he creates a vantage point that exists in nature but is inaccessible to human observers. Brett's allothetic abstraction—his imagined distance from a scene that he in real life inhabits—occurs noticeably in his depiction of the company's ships, which are often a central focus and rendered in intricate detail but at times are strangely out of scale or static. For instance, in "Cape Virgin Mary," the ships stick upright in the water (no ripples, no movement), oddly perpendicular to a calm ocean; in "View of Patagonia," the beautifully engraved ships distract the eye from a distant landscape, a fact Brett must have recognized. In such scenes, Brett employs the ships to perform functions beyond the mimetic: they assist the viewer's sense of scale and familiarity, they narrativize, they symbolize British presence. But the ships also mark the limits of the artist's point of view—the location and conditions of his drawing—as well as his departure from that physical localization into the realm of artistic perception.

Despite or even because of these perspectival complexities, as shipbound Brett views himself from a distance, the most captivating images of *A Voyage* depict the ships in full sail, outrunning a storm or cruising a rugged coast. Removing himself from the scene—floating above or behind—Brett transforms an individuated experience (his own limited point of view) into a collective one (all the ships and men sailing in concert). At these moments we might recall Anson's assumed position as the collective "eye," a fictionalized construct designed to consolidate and streamline diverse experiences. Both enterprises, that of narrator and artist, rely on imaginary frameworks that conjoin spectacularly diverse perceptions into an artifact of combined experience: *A Voyage*. From a navigational perspective, the depiction of the company's actions is not always pragmatic or necessary, yet Brett populates distant scenes as a means of claiming that experience for Anson and for a British audience eager to share in his exploits. His view of Terra del Fuego, for instance, is really a portrait of the fleet in full sail, comparatively large before a distant coastline.[42] In the next engraving, "A View of Streights Le Maire between Terra del Fuego and Staten Land" (see fig. 3), the fleet again takes center stage, leaning slightly to depict motion and effort as a dramatic storm approaches; such kineticism along with the symmetry and allegory of dark storm against bright sky produces a narrative of impending doom that aligns with the tempo of Walter's writing: "We here found what was constantly verified by all our observations in these high latitudes, that fair weather was always of an exceeding short duration, and . . . a certain presage of a succeeding storm."[43] Agile, evolving, evocative, Brett's images rotate vantage points to convey motion and drama in ways that connect to fine art more than navigation while they also consolidate vantage points to produce a collective, multifocal perspective—a communal sensory experience—impossible for any individual sailor to attain.

The triumphs of the voyage in particular are portrayed in ways that surpass navigational pragmatism and both concretize and collectivize experience. For instance, "The Burning of the Town of Payta on the Coast of Santa Fee" depicts one of the company's greatest assaults on Spanish properties as billows of black smoke rise from a distant town and Spanish ships disappear into the water in the foreground[44] (see fig. 4). Fascinatingly, and unusually, Walter details Brett's location within this scene: he notes that Brett was given orders "to burn the whole town, except the two churches." "These orders [are] punctually complied with," Walter continues, as Brett and his men disperse throughout the town (they "distribute . . . combustibles . . . into houses situated in different streets of the town, so that, the place being fired in many quarters at the same time, the destruction might be more violent").[45] Brett's location *within* the city (throughout its streets and inside its buildings), in the small boats, and then back aboard the *Centurion* ensures that he has a thorough, monopolizing view

FIGURE 3. "A View of Streights Le Maire between Terra del Fuego and Staten Land." (Bodleian Libraries, University of Oxford, shelfmark AA 151 Art, plate 11)

FIGURE 4. "The Burning of the Town of Payta on the Coast of Santa Fee in the South Sea." (Bodleian Libraries, University of Oxford, shelfmark AA 151 Art, plate 24)

of the encounter. Encompassing the perspectives of town, beach, and ship, Brett embodies, in Walter's description, actor, artist, and witness. Walter emphasizes, then, how Brett accesses (and then reassembles) those perspectives in the accompanying landscape.

Perhaps these scenes that overtly reflect British interests are those that make Brett's perspective itself most visible despite its professional qualities. Like his view of Juan Fernandez, Brett's renowned "View of the Watering Place at Tinian" depicts a "luxurious" pastoral scene that is "exquisitely furnished with the conveniencies of life, and [for] the enjoyment of mankind." As in the earlier view, Brett combines familiar and exotic and populates the scene with British officers who enjoy an idyll "entirely destitute of inhabitants."[46] Here, though, one of the officers, quite possibly Anson himself, gazes back inquiringly at the artist, reminding us of Brett's presence.[47] The returned gaze conjures once again the person who produces a "tendency to beauty" that is distinctly British. But it also connects subject and artist, commander and sailor, company and individual, archiving a condensed point of view that defrays the radical individuation of these experiences, refracts subjectivism, and conceals the artist in a mirror visage, mise en abyme.

The Glory or Halo

Glories—sometimes also called coronas or halos—are optical phenomena that appear, often to sailors, when sun or moonlight passes through damp air; like a rainbow, they are real in that they are natural events perceived by the senses, but they are also immaterial, unrepeatable, and radically contingent.[48] They are perceivable only from a specific viewpoint, in a specific moment; they are vivid yet dissipate with any change in location or conditions. There is in fact "no underlying common object" that we refer to when we describe these phenomena, which exist "between Matter and Light, between Thing and Process,"[49] and produce a "dramatic differentiation of observers."[50] Of an event so unique and precarious, one must always ask, Was it real or a trick of the light? Perception or deception? Certainly, Anson's voyage was to each of his men just this kind of volatile experience; for many, it was a nightmare that eluded depiction, except perhaps in the ghastly imagery of their dreams:

> It is impossible for the Tongue or Pen of Man to give a perfect Description of the prodigious Suffering and Hardships that every person in the Fleet, as well Officers as Men, went thro' during the Voyage. Methinks I still hear the horrible Roaring of the Winds, and see the Sea rising into Mountains, the Ships clambering as it were those Hills, . . . the Rigging torn from the Masts, and the Sails split into a thousand Pieces shivering in the Wind; Wildness and Despair in every Man's Countenance, as thinking each Moment would be his last. But who can conceive, the dreadful Miseries and Hardships that those poor Sailors underwent?[51]

Wonder and terror suspend attention, as eighteenth-century thinkers believed, and synthesize seeing and feeling into knowing. Yet the knowledge gained on this voyage seems to resist narrative—is antinarrative, antidocumentary in its fracturing of experience—and threatens, at times, even the practical craftsmanship of Brett. Nonetheless, as the authors seem intuitively to recognize, synthetic, multimodal, combinational narration—and, importantly, the formal mechanisms of art—provide the only avenue of approximating a voyage that proved so challenging to their sensory, cognitive, and linguistic powers. Anson, Walter, Robins, and Brett labor to counter the individuating effects of the voyage and to consolidate ephemeral—traumatic, wondrous, passionate—experiences via a special kind of authority based on a special kind of practiced perception and to produce impartiality via a totalizing and hence inhuman viewpoint that stitches together the aims of scientific narrative and fine art.

Walter's theorizing of the perception of the draftsman precedes that of practical artists like Hogarth and should be seen as an important contribution to not just maritime history but also aesthetic discourse. Moreover, in striving to accommodate both the scientific fantasy of accuracy beyond language and the documentary ambition of capturing the depth and range of human experience, he shows us how the aims of art and science collide—sometimes smoothly, sometimes jaggedly—in voyage accounts of the period. Indeed, in its careful imbrication of text and image, individual and communal, *A Voyage* conjoins the aims of art and science via form ("*an ordering, patterning, or shaping*"); this "composite imagetext" is visual and discursive, intellectual and tactile, and constitutes an attempt to locate "Anson's voyage" via intricate formal structures and professional modes of seeing that improve but also fundamentally alter human vision.[52] Such work is intrinsically open-ended, inviting reinterpretation, reauthoring, and reimagining, as the *Voyage* saga vividly demonstrates.

In plate 2 of *The Analysis of Beauty*, "The Country Dance," Hogarth depicts "a number of people together" suspended in a single whirling moment, creating a tableau of postures, expressions, and attitudes that no human eye could ever capture within a given moment. Here Hogarth conveys not so much the actual dance—which without serial images we cannot ascertain—but the "delightful play upon the eye, especially when the whole figure is to be seen at one view, as at the playhouse from the gallery," as well as the complexity and range of all that *might* be seen in any given instant, even within a single room, from a single vantage point.[53] He therefore arranges and makes visible—he formalizes—an evanescence that could not be witnessed firsthand (no one can see so many faces, so many gestures, at once) and in so doing approximates the *experience* of an event dense with life and motion as well as the sensation of striving to see more than our faculties allow. Beautiful, carefully configured, preoccupied with vision and perspective, *A Voyage* ventures after the same kind of revelation and typifies a navigational realism that is—has always been—an aesthetic in disguise.

Notes

An MOE Tier 1 Grant (WBS R103000164115) funded archival research for this essay.

1. On this "saga of repeated disaster and appalling casualties," see Williams, "Anson at Canton," 271; and Williams, *Great South Sea*, 214–50. Only the seizure of the Acapulco galleon saved Anson's reputation.
2. Anon., *Voyage to the South-Seas*, ix. For simplicity, I refer to Walter as the narrator throughout this essay.
3. Walpole compared Anson to "his predecessor Gulliver" (*Yale Edition*, 9:55). Chinese accounts of Anson's visit to Canton further undermined his credibility; see McDowall, "The Shugborough Dinner," 9–13. On Naish, see McDowall, 12–13.
4. In this sense, the Anson saga reveals the limits of the collective writing practices that Margaret Cohen isolates as typical of voyages of the period; that is, if voyage accounts often "emphasized collectivity over originality," Anson's men refuse any such elision. Cohen, *Novel*, 36.

5. Walter, *Voyage*, sig. c[3v]. Brett served in further naval exploits, notably with Anson at Finisterre, and was eventually promoted to admiral. However, beyond a 1755 plan for a harbor at Ramsgate (see Bodleian shelfmark Gough Maps Kent 31), I have found no evidence of further draftsmanship.
6. Ibid., sig. c[3r]–c[3v].
7. See Smith, *European Vision* and *Imagining the Pacific*. On how Brett's images seep into public discourse, see Mancini, "Siege Mentalities."
8. Hogarth, *Analysis*, 58.
9. Bulkeley and Cummins, *Voyage*, v. On the imaginative dimensions of the "plain style," see Keiser, "Science for the Birds"; on the ambiguities of supposedly self-evident diagrams, see Coppola, "'Fully Prov'd by the Plates,'" both in this volume.
10. "Officer of the Squadron," iii. On "the Expectation of the Publick," see also Anon., *Authentic Account*, [A1r].
11. See Williams, *Great South Sea*, 231.
12. Thomas, *True and Impartial Journal*, 10.
13. "Officer of the Squadron," vi. Despite variations in how the author identifies himself (e.g., "Officer of the Squadron" versus "Officer of the Fleet"), the several versions of *A Voyage to the South-Seas* are largely the same, with minor exceptions. Here, the author makes special note of how he verifies "whatever [his] own Eyes were not Witnesses of" (vi).
14. Bender, *Ends of Enlightenment*, 59. That these engravings are *prints* is also important, as midcentury readers learn to associate print and facticity (63).
15. Walter, *Voyage*, sig. c[3v].
16. Ibid., sig. c[4r].
17. In earlier voyages, many surveyors were trained as draftsmen; from Cook onward, however, it became standard to travel with a retinue of professional artists.
18. Walter, *Voyage*, sig. d2[v].
19. Ibid., sig. d[3r].
20. See Addison, "On the Pleasures of the Imagination," *Spectator*, nos. 411–21. On craft knowledge and navigation, see Cohen, *Novel*, 15–55.
21. Hogarth, *Analysis*, xv–xvi. Another useful comparison is Jonathan Richardson's *Essay on the Theory of Painting* (1715), which makes similar claims about exposure to great art and practical application, although he tends to focus on the moral value of "the best Masters" (15).
22. On Hogarth's pragmatic aesthetics and his focus on form rather than theory, see Zitin, *Practical Form*.
23. Hogarth, *Analysis*, 58.
24. Ibid., 57.
25. Zitin, *Practical Form*, 24.

26. Walter, *Voyage*, 116.
27. Ibid., sig. d[3r]; Hogarth, *Analysis*, 150.
28. In this way, it seems almost too fitting that Brett's name appears on the back of Thomas Pingo's 1747 commemorative coin that celebrates Anson's exploits in the Pacific and at the Battle of Finisterre. On the front, Anson's profile gleams as he is crowned with laurels by Winged Victory; on the back, the names of six officers who served with him encircle the coin. See Object 111/1600, National Maritime Museum, Greenwich. See Mancini, "Siege Mentalities," on this and other fine objects commissioned to celebrate Anson.
29. Walter, *Voyage*, 115.
30. Ibid., 260.
31. Ibid., 114–15. These charts thus enact "the visual mechanics of diagram"; see Bender and Marrinan, *Culture of Diagram*, 63.
32. Walter, *Voyage*, 74.
33. Lamb, *Preserving*, 12.
34. Walter, *Voyage*, 315.
35. Lamb, *Preserving*, 125.
36. Walter, *Voyage*, 73.
37. Ibid., 119–20.
38. On the passions that generate this vision of Juan Fernandez as well as the broader trope of paradise, see Lamb, *Preserving*, 227–40.
39. Walter, *Voyage*, 74–75.
40. Lamb, *Preserving*, 10.
41. Bender and Marrinan, *Culture of Diagram*, 60.
42. Brett is particularly good at combining action scenes with detailed coastal renderings. For example, in "Terra del Fuego" and in figure 3, he depicts the fleet in full sail but also offers a meticulous coastal view with an accompanying key.
43. Walter, *Voyage*, 72–73.
44. Ibid., 198–99.
45. Ibid., 198.
46. Ibid., 311. It was uninhabited only because the Spanish had forcibly removed the Indigenous population.
47. Including within a scene a figure "who is both viewer and participant" was common in 1740s periodicals. See Bender and Marrinan, *Culture of Diagram*, 64.
48. On glories, subjectivism, and the "explorer," see Craciun, "What Is an Explorer?"
49. Wieting, "Cartesian Rainbow," 10. See also Fisher, *Wonder*.
50. Craciun, "What Is an Explorer?," 36.
51. "Officer of the Squadron," vi–vii.
52. Levine, *Forms*, 3; Craciun, "What Is an Explorer?," 34.

53. Hogarth, *Analysis,* 150. My thanks to Jonathan Lamb for bringing my attention to the relevance of this image. On Hogarth's interest in realism, see Miller, "Newtonian Legacies," in this volume.

Bibliography

Addison, Joseph. "On the Pleasures of the Imagination." *Spectator* 411–21 (June 21 to July 3, 1712). In *Spectator,* 1:114–28. London: Richard Eyres, 1778.

Anon. ["taken from a private Journal"]. *An Authentic Account of Commodore Anson's Expedition.* London: M. Cooper, 1744.

Anson, George. *Voyage Round the World in the Years MDCCXL, I, II, III, IV.* Edited and compiled by Richard Walter [and Benjamin Robins]. London: John and Paul Knapton, 1748.

Bender, John. *Ends of Enlightenment.* Stanford: Stanford University Press, 2012.

Bender, John, and Michael Marrinan, eds. *The Culture of Diagram.* Stanford: Stanford University Press, 2010.

Bulkeley, John, and John Cummins. *A Voyage to the South-Seas, in the Years 1740–41.* London: Jacob Robinson, 1743.

Cohen, Margaret. *The Novel and the Sea.* Princeton, NJ: Princeton University Press, 2012.

Craciun, Adriana. "What Is an Explorer?" *Eighteenth-Century Studies* 45, no. 1 (2011): 29–51.

Fisher, Philip. *Wonder, the Rainbow, and the Aesthetics of Rare Experiences.* Cambridge, MA: Harvard University Press, 2003.

Hogarth, William. *The Analysis of Beauty.* London, 1753.

Lamb, Jonathan. *Preserving the Self in the South Seas, 1680–1840.* Chicago: University of Chicago Press, 2001.

Levine, Caroline. *Forms: Whole, Rhythm, Hierarchy, Network.* Princeton, NJ: Princeton University Press, 2017.

Mancini, J. M. "Siege Mentalities: Objects in Motion, British Imperial Expansion, and the Pacific Turn." *Winterthur Portfolio* 45, nos. 2/3 (2011): 125–40.

McDowall, Stephen. "The Shugborough Dinner Service and Its Significance for Sino-British History." *Journal for Eighteenth-Century Studies* 37, no. 1 (2014): 1–17.

Naish, James. "Interlinear Notes." In *Anson's Voyage Round the World,* edited and compiled by Richard Walter [and Benjamin Robins]. London, 1748. British Library shelfmark BL 10025.f.8.

"An Officer of the Squadron." *A Voyage to the South-Seas, And to many other Parts of the World, performed From the Month of September in the Year 1740, to June 1744.* London: R. Walker, 1744.

Philips, John. *An Authentic Journal of the Late Expedition under the Command of Commodore Anson.* London: J. Robinson, 1744.

Richardson, John. *Essay on the Theory of Painting.* 2nd ed. London: W. Bowyer, 1715.

Smith, Bernard. *European Vision and the South Pacific, 1768–1850.* Oxford: Clarendon Press, 1960.

———. *Imagining the Pacific in the Wake of the Cook Voyages.* Carlton, Victoria: Melbourne University Press, 1992.

Thomas, Pascoe. *A True and Impartial Journal of a Voyage to the South-Seas, and Round the Globe.* S. London: S. Birt, J. Newbery, and J. Collyer, 1745.

Walpole, Horace. *The Yale Edition of Horace Walpole's Correspondence.* 45 vols. Edited by W. S. Lewis. New Haven, CT: Yale University Press, 1937–83.

Wieting, Thomas W. "The Cartesian Rainbow." Online essay. Portland, OR: Reed College, 2006.

Williams, Glyndwr. "Anson at Canton: 'A Little Secret History.'" In *The European Outthrust and Encounter. The First Phase c. 1400–c. 1700: Essays in Tribute to David Beers Quinn on His 85th Birthday,* edited by Cecil H. Clough and P. E. H. Hair, 271–90. Liverpool: University of Liverpool Press, 1994.

———. *The Great South Sea: English Voyages and Encounters, 1570–1750.* New Haven, CT: Yale University Press, 1997.

Zitin, Abigail. *Practical Form: Abstraction, Technique, and Beauty in Eighteenth-Century Aesthetics.* New Haven, CT: Yale University Press, 2021.

Literary Technologies of the Sextant in Eighteenth-Century Britain

AARON R. HANLON

The Scilly naval disaster of 1707 resulted in the loss of four ships and upwards of 1,400 sailors from Sir Cloudesley Shovell's fleet. The Isles of Scilly lie just off the southwestern coast of Cornwall. As stormy weather pushed Shovell's fleet off course, sailing masters strained to reorient their ships amid the squalls. Before they could determine their location, four ships had lurched perilously close to the isles and were dashed on the rocks. The inability to ascertain longitude proved fatal.

The Scilly naval disaster was both an impetus for the establishment of the Board of Longitude and a catalyst of much discussion and dispute surrounding navigational instruments in the mold of the sextant, such as the quadrant and octant, in eighteenth-century Britain. The establishment of the Board of Longitude also inspired skepticism, in part because it took decades for the board to hold its first meeting and then for applicants for its monetary reward to come up with an accurate way to find longitude at sea to arrive at promising solutions. Along with chronometers, which were expensive and not easily manufactured, sextants were choice instruments for determining longitude at sea.[1] Sextants were easier and cheaper to produce, but keeping accurate time at sea—by means of a chronometer that could remain accurate despite changes in humidity and temperature along the voyage—would become the most promising solution to the problem of longitude by the time John Harrison was designing his "sea clocks" in the 1730s. Not only was the Board of Longitude something of a quixotic endeavor, but the sextant would also become something of a quixotic instrument, simultaneously dated and optimistically relied upon.

Interest in developing reliable instruments and methods for determining longitude at sea intensified in the subsequent decade after the 1707 disaster, even if solutions were a long way coming. A pamphlet listing "Reasons for a Bill, Proposing a Reward for the Discovery of the Longitude," published on June 10, 1714, in support of the Longitude Act, includes among its reasons "Because it will prevent the Loss of abundance of Ships and Lives of Men; as it would certainly have sav'd all Sir *Cloudsly Shovel*'s Fleet, had it been then put in Practice."[2] The Longitude Act of 1714 passed a month later, creating a financial reward scheme for anyone who could develop an accurate way of determining longitude at sea, starting with £10,000 for anyone who could achieve accuracy within one degree of longitude, £15,000 for accuracy within forty minutes of longitude, and £20,000 for accuracy within half a degree.[3]

Given the importance of finding longitude for both scientific (primarily astronomical) and navigational purposes, this essay examines how quadrants, sextants, and like instruments were portrayed in eighteenth-century writing. In so doing it demonstrates three things in particular. First, there were common approaches to representing sextants and like navigational instruments across narrative genres; second, the very methods of establishing the credibility of sextants were also employed satirically to undermine their credibility; and third, literary technologies for establishing the credibility of scientific instruments relied on accounts not only of the instruments themselves but also of their use, or of their entwinement with human actors capable of error, wonder, skepticism, and deception.

Two epistemic models in particular guide my approach here. One is Steven Shapin's concept, later developed with Simon Schaffer, of "literary technology," "the expository means by which matters of fact were established and assent mobilized" in experimental natural philosophy in Restoration Britain.[4] Literary technology is for Shapin and Schaffer the basis of "virtual witnessing," "the production in a reader's mind of such an image of an experimental scene as obviates the necessity for either its direct witness or its replication."[5] Like Robert Boyle's air pump, the accuracy and reliability of the sextant as an instrument for generating positional data on which ship's masters made navigational calculations—calculations that sometimes came down to life or death—had to be widely demonstrated through literary technology. This frequently meant detailed diagrammed illustrations of the instrument and its features as well as tables of its positional data and narratives of its manufacture, storage, and use. In accordance with Shapin and Schaffer's finding that the literary technology of virtual witnessing was instrumental in the establishment of matters of fact, the establishment of the sextant as a reliable and practical navigational technology depended on credible widespread representation. At the same time, literary technologies of the sextant frequently displayed what John Rogers calls—in a study of the rhetoric of seventeenth-century natural and political philosophy—the "literary discourse of rhetorical contradiction and thematic opposition" that reflects "the cultural uses to which the systematic literary juxtaposition of incompatible models of agency and organization could be put."[6] In the case of the sextant, such incompatibility was a result of a combination of the immense practical and symbolic value conferred upon the instrument—due to the importance of navigation to so many economic, scientific, and imperial endeavors—and the instrument's obvious shortcomings. The sextant's status was similar to that of another infamous scopic technology, the microscope. In the 1660s, Robert Hooke believed the microscope would revolutionize all of science, but by the 1690s, he would lament that it had become a mere pastime for wealthy hobbyists. The sextant was a technology simultaneously valued in the present and always, from the beginning, in search of its future.[7]

The second model that underwrites my approach in this essay is Tita Chico's notion of the "experimental imagination," especially how "scientific writing requires literary knowledge but suppresses it simultaneously." For Chico, "representation of early science persistently discloses its literary status—not merely in the tropological nature of scientific writing and practice, but also through the metaphorics of science that allow writers to posit alternative models of authority and evidence."[8] In what follows, I am less interested in holding onto the category of "the literary" as Chico defines it—that is, in the figurative or aesthetic dimensions of writing that we might retroactively call "literary"—than in the more pedestrian observation that narrative accounts and descriptions of the sextant have been crucial for establishing its credibility and reliability as a means of determining longitude at sea. In other words, what matters here is not whether scientific writing is actually "literary" or narrative accounts are actually "scientific" so much as what accounts and representations of the sextant were doing to establish and bolster credibility.

Representations of the sextant across different genres of writing share common characteristics and strategies and provide a record not only of a representational history of the sextant in eighteenth-century Britain but also of how representation as such contributed to the establishment of the sextant as a reliable technology. We find, for example, autobiographical narratives of the sextant's use in the technical treatises of its developers as well as in the storytelling mode of Olaudah Equiano's *Interesting Narrative* (1789) and Jonathan Swift's satirical treatment of instruments in *Gulliver's Travels* (1726). Across genres we find modes of description and valuation that reflect not simply the desire of instrument makers to sell the Board of Longitude on the reliability of their wares but also a collective awareness of what the sextant and like instruments represent and are meant to represent: both a modern scientific culture and the ways one might overstate the prowess and practicality of such a culture.

In what follows I offer some exemplary cases from this historical record—both nonfictional and fictional—with particular attention to the relationship between the history of writing and the history of technology within the broader history of science. As Jennifer Lieberman argues, "We should formalize technology's place in the union between literature and science" on the grounds that "the study of literature and technology has been relatively neglected within Literature and Science."[9] The sextant is a particularly suitable instrument for such inquiry because of its practical value in navigation—and thus in British trade and imperialism—and the symbolic freight it carried even in minor textual mentions. This essay offers a survey examination of a written history of the sextant as a basis for demonstrating the importance of literary technology not simply within the history of science—as has been widely demonstrated—but specifically within the eighteenth-century history of the sextant as a technology of particular importance, perhaps the first mass-produced scientific instrument in Britain.

I give the sextant this designation in light of a combination of physical and literary technologies. Jesse Ramsden's development of an industry-standard dividing engine in the 1770s—which automated the marking of degrees along an arc—made mass production of sextants possible. Since Ramsden was funded by the Board of Longitude on the condition that he forfeit any patent rights to his superior dividing engine, it was not only navigational instruments that could then be manufactured at scale but also the means of manufacture—the dividing engine—that could be widely replicated.[10] The scale of the sextant's production and the centrality of navigation to so many aspects of eighteenth-century life made the sextant a recognizable technology and thus a symbol of so many scientific and navigational ambitions. As with such cultural symbols or touchstones, the mass production of the technology and its representation become mutually reinforcing.

Along with performing experiments in social spaces and encouraging replication through detailed accounts of experiment protocol, multiplying witnesses of experimental knowledge through virtual witnessing entailed, as scholars have long since observed, the inclusion of diagrams and illustrations in scientific texts.[11] The same was often the case in accounts of and advertisements for navigational instruments such as the sextant. Even where images are absent, writing about sextants in the eighteenth century frequently meant enhancing the witnessing effect through "the production in a reader's mind of... an image of an experimental scene," or an image of how the sextant was operated, under what conditions, and to what effect.[12] Whereas seventeenth- and eighteenth-century scientific atlases typically relied on illustrations of instruments used, substances or organisms studied, and other details of the experimental scene, accounts of sextants heavily relied on tables of figures—visual renderings of data recorded with the instrument—as a literary technology. Written accounts of sextants and of navigation more broadly in eighteenth-century Royal Society papers are filled with numerical tables that compare the data of one instrument with another.[13] Virtual witnessing through accounts of the experimental scene was crucial for demonstrations of both an instrument's reliability and its unreliability, but these also required the sharing of numerical data as a centerpiece of the virtual demonstration.

A 1781 dispute between the astronomer William Bayley and the mathematician and instrument maker Jesse Ramsden reflects this technique. In a brief "Account of a 15 Inch Sextant Made by Mr. Ramsden," Bayley describes how he took measurements with various sextants on a journey to the Cape of Good Hope, finding Ramsden's sextant insufficiently accurate. To demonstrate its inaccuracy such that readers could lend credence to the account, Bayley not only describes and quantifies the precise errors he found but also notes how "the sextant was carefully put into its case & kept between Decks during the interval of observations" and how he "often found the errors of adjustment to alter even during the time [he] was observing, & consequently found it necessary to examine it both before and after."[14] That is, Bayley was careful to describe an experimental scene in which the details of the storage of the sextant when not in use—when it was not generating data—become relevant to understanding and accounting for the errors he goes on to describe. In this way, Bayley uses the account of the sextant's storage to show that the variation in errors he observes—"the error of adjustment ... would differ a Minute, or a Minute and a half & that without any apparent cause"—is the fault of the instrument itself and not of its handling.[15] In this manner, credibility is a function of isolating the instrument itself from human error in the process of scientific vetting or error correction. Bayley needs not only Ramsden but also the members of the Board of Longitude to understand—by way of a transparent account—that he has been exacting in his trial of the instrument, thus all such instruments that pass scrutiny can be relied upon. In this process of virtual witnessing, the sextant must be isolated as a narrative object with a direct causal relationship to Bayley's data output.

Responding to Bayley's observations about the insufficiency of Ramsden's fifteen-inch sextant, an account from the Board of Longitude demonstrates a similar approach to setting the experimental scene. It begins by summarizing Bayley's account, familiarizing the board with it, and then proceeds to describe Ramsden's testimony about the sextants he supplied to Bayley for the Cape of Good Hope voyage. The account reports that Ramsden assured the board that "the two sextants [Ramsden supplied] were made and fitted exactly alike" and packed "each in its respective Case in the same manner," paralleling Bayley's statement about the storage of the sextants while not in use during the voyage.[16] When comparing the accounts of Bayley and Ramsden, then, a striking commonality is an emphasis on the use narrative of the instrument, not simply a presentation of "raw" numerical data. Likewise, the dispute between Bayley and Ramsden is less over the data itself than the narrative accounts of how the data from the sextant was generated, the dispute turning on the degree to which one party might convince the other of the likelihood or unlikelihood of misuse or human error.

In addition to detailed accounts of the experimental scene, including narration and numerical tables of data recorded, those presenting new and improved sextants schematized their descriptions to emphasize visuality and clarity. Benjamin Martin's *An Explanation of a New Construction and Improvement of the Sea Octant and Sextant* (1775), for example, adopts such key strategies for virtual witnessing. In addition to detailed images of the instrument he describes, Martin emphasizes the extent to which much of his intervention is about the simplification and clarification of the theory of how sextants work. As Martin boasts, criticizing the "obscure, prolix and intricate Theory" offered in John Hadley's original description of the quadrant in the *Philosophical Transactions* in 1731, "I have elsewhere given its whole Theory in *six Lines* only, from the Properties of a plain Triangle alone; these six may be reduced to *four*, as will be hereafter shewn; and is there any Instrument of a *mathematical Nature*, that can pretend to a *Rationale* so simple, so concise, so perspicuous?"[17] Here we can see not only the conventional rhetoric of visual demonstration common in Royal Society writing in the seventeenth and eighteenth centuries—"as will be hereafter shewn"—but also an awareness that distilling a mathematical theory of the instrument's efficacy into four lines makes it easier for readers to grasp and assent to Martin's claims to improvement. Martin's son, Joshua Lover (J. L.) Martin, whose signature appears on the patent illustration of the new instrument, was actually the inventor. Among the principal improvements was J. L. Martin's reduction of the number of movable lenses from three to one (whose movement is restricted), thereby reducing measurement errors occasioned by trying to coordinate too many moving pieces while on an unsteady ship at sea.[18] Simplification, demonstration, and appeals to the integrity of the visual are Martin's rhetorical strategies, but Martin's main task in the 1775 tract is to employ literary technology to explain the improvements of his son's invention, multiplying virtual witnesses by emphasizing the simplicity of the instrument, the theory behind its efficacy, and the demonstration of that theory.

Even as it offers a simplification of the theory behind the functioning of Hadley's quadrant (a simplification crucial for popular consensus), Martin's description largely follows the structure of Hadley's original presentation of "a new Instrument for taking Angles" in 1731.[19] Both descriptions open with a diagrammed figure, which is then referred to in the textual description. Accordingly, both texts lay out the theory of how the instrument should work using a series of letter references to the figure. Hadley writes, for example, "Since the Point A is in the Plane of the Scheme, the Point M will be so also by the known laws of Catoptricks. The Line FM is equal to FA, and the Angle MFA double the Angle HFA or MFH."[20] For comparison, Martin follows the same framework for annotating the figure: "Let the Plane or Surface of the Index-Glass D be denoted by ABCD; that of the first Horizon Glass E, by EFGH; and that of the second F, by IKLM."[21] Mundane as these formal observations may seem, they reflect a style common not only to scientific atlases and like forms of Royal Society writing, from Robert Hooke's *Micrographia* (1665) onward through the end of the eighteenth century, but also to textual demonstration of instruments. Visuality and clarity become the means of literary technology here, providing readers with a virtual tour of the object itself as part of a description of how theory underwrites function.

In eighteenth-century narrative accounts of related subjects—not sextants or navigational instruments as such but travelogues, sea voyages, and astronomical tracts in which one might expect to find references to astronomical or navigational instruments—sextants and quadrants tend to be marginally represented while also bearing significant symbolic weight across fictional and nonfictional texts. In Jonathan Swift's *Gulliver's Travels*, for example, the sextant and quadrant are touchstones for satirical commentary about the reliability and practicality of the Royal Society's experimental natural philosophy as well as Isaac Newton's calculations in the *Principia* (1687).[22] In part 1, the Lilliputians determine that Gulliver should be nourished with "a quantity of meat and drink sufficient for the support of 1728 Lilliputians" by employing "his majesty's mathematicians" to take Gulliver's height with a quadrant and calculate the proportion of Gulliver's size to that of the Lilliputians. In part 3, on the island of Laputa, the king orders a tailor to have Gulliver fitted for a suit of clothes, but unlike "those of his trade in Europe," the tailor measures

Gulliver's height with a quadrant. Then, "with a rule and compasses," the tailor "described the dimensions and outline of [Gulliver's] whole body," performed calculations on paper, and "in six days" produced clothing "very ill made, and quite out of shape, by happening to mistake a figure in the calculation." Gulliver notices that such mathematical errors are "very frequent, and little regarded," on Laputa, where the astronomer's cave stores "a great variety of sextants, quadrants, telescopes, astrolabes, and other astronomical instruments."[23]

As Gregory Lynall observes, such moments in the voyage of Laputa satirize not only "the cultural obsession with the new sciences, and the ubiquity of the Newtonian philosophy," but also, more specifically, some of the astronomical calculation errors that appeared in the first edition of Newton's *Principia*.[24] Further, like Martin's claim to have improved Hadley's demonstration of the sextant's efficacy by simplifying the calculations to merely four "simple" lines, Swift's invocation of the sextant draws attention to the complexity of Newton's work and the difficulty of translating it for practical use. In this context, the mention of the astronomers' store of sextants and other instruments is brief but profound, as Swift aims his satire at both the reliability and comprehensibility of instrument-based calculations.

Such brief moments in *Gulliver's Travels* use sextants and like instruments as effective satirical signifiers precisely because of their mass production and visibility. As in the dispute between Bayley and Ramsden over the accuracy of one of Ramsden's sextants and the causes of error, Swift's account of the tailor's quadrant specifies both a method (taking Gulliver's height, dimensions, and outline, then making calculations) and a cause of error (mathematical miscalculation). The frequency with which Gulliver observes such human errors on Laputa reflects Swift's understanding of the very problem Bayley and Ramsden negotiate in their exchange. All three authors identify the difficulty of separating the integrity and precision of the instrument itself—a function of the instrument's conception and construction—from the use of the instrument or the introduction of human error beyond anything possible in the instrument's construction. In a satirical moment that turns on the overcomplexity of the Laputan tailor's methods, we also see, in a kind of argument by negation, the appeal of Martin's strategy of simplifying his explanation of how the sextant works, presenting the new technology as a counter to Hadley's "obscure, prolix and intricate Theory" based on the premise that the relative complexity of Hadley's instrument was to blame for user error.

This kind of problem was particularly consequential after the Longitude Act of 1714, as an excellent instrument could fetch a substantial financial reward, but the process of verifying the instrument's excellence could come down to human error in the instrument's use or misapplication. Because virtual witnessing entailed a minute description of both components—instrument and use—even brief fictional accounts of sextants employed such a structure. Virtual witnessing in the traditional sense in which Shapin and Schaffer use the concept focuses on the credibility of the experimental process, which applies as well in this case to the development and trial of the technology; but whereas the validated product of the experimental process is conceptual—the matter of fact—the product of technological development is the technology itself. Thus, for virtual witnessing to work in the case of scientific instruments, both process and product must be validated through virtual witnessing. Because Swift's satire is in this case virtual witnessing's foil, a moment meant to undermine rather than bolster public impressions of the reliability of scientific instruments and of the Royal Society's experimental methods more broadly, it makes sense that Swift would employ virtual witnessing's techniques. Additionally, because of the importance and visibility of navigational instruments in eighteenth-century Britain amid the rise of its Atlantic empire and maritime prowess, the sextant, the quadrant, and the like were for Swift efficient reference points for satire, instruments freighted with importance but also likely to engender skepticism.

Sextants and quadrants make sporadic appearances in eighteenth-century fiction as recognizable signifiers of learnedness and experimentalism in general, whether in earnest or in mockery. In *Tristram Shandy* (1759–67), Laurence Sterne invokes the quadrant as a symbol of mechanical learning in Tristram's prefatory disquisition on wit and judgment, which comes in an effort to settle into his author's preface after his father and Uncle Toby have fallen asleep, exhausted by a discourse over how Toby's mind is like a smokejack, continually turning over thoughts: "Here the brethren of another profession, who should have run in opposition to each other, flying on the contrary like a flock of wild geese, all in a row the same way.—What confusion!—what mistakes!—fiddlers and painters judging by their eyes and ears,—admirable!—trusting to the passions excited in an air sung, by a story painted to the heart,——instead of measuring them by a quadrant."[25] In Tobias Smollett's *Roderick Random* (1748), the quadrant signifies Narcissa's learnedness and mystique—in satirical contrast with the fact that she acquires these things out of faddishness—at the point at which Random would "become enamored of" her:

> She sat in her study, with one foot on the ground, and the other upon a high stool at some distance from her seat; her sandy locks hung down, in a disorder I cannot call beautiful, from her head, which was deprived

of its coif, for the benefit of scratching with one hand, while she held the stump of a pen in the other. Her forehead was high and wrinkled; her eyes were large, gray, and prominent; her nose was long, and aquiline: her mouth of vast capacity, her visage meagre and freckled, and her chin peaked like a shoemaker's paring knife; her upper lip contained a large quantity of plain Spanish, which, by continual falling, had embroidered her neck, that was not naturally very white, and the breast of her gown, that flowed loose about her with a negligence that was truly poetic, discovering linen that was very fine, and, to all appearance, never washed but in Castalian streams. Around her lay heaps of books, globes, quadrants, telescopes, and other learned apparatus.[26]

Far from a capacious catalog of appearances of the sextant or quadrant in eighteenth-century fiction, such examples nevertheless demonstrate some of the ways these navigational instruments entered into the novelistic and popular imagination as signals of scientific authority, even in cases where such authority is being mocked or questioned. Alas, Narcissa's scientific books and instruments lay in "heaps," suggesting that for all their symbolic value, they go largely unused.

In eighteenth-century nonfictional narratives, sextants were often objects of wonder and importance. The quadrant makes one appearance in Olaudah Equiano's *Interesting Narrative* toward the beginning of the narrative, in chapter 2, when Equiano describes the horrors of being kidnapped and taken aboard the slave ship. As Equiano describes the horrific experience of being kept below deck in a state of near suffocation, he turns unexpectedly to an observation on deck, his first sighting of flying fish and the first time he sees someone using a quadrant:

> I also now first saw the use of the quadrant; I had often with astonishment seen the mariners make observations with it, and I could not think what it meant. They at last took notice of my surprise; and one of them, willing to increase it, as well as to gratify my curiosity, made me one day look through it. The clouds appeared to me to be land, which disappeared as they passed along. This heightened my wonder; and I was now more persuaded than ever that I was in another world, and that every thing about me was magic.[27]

Striking in this passage is the shift in register from an enumeration of horrors in the chapter that introduces Equiano's experience of being captured—particularly if we encounter this passage in contrast with Bayley's account of the sextant's storage, the precious instrument getting infinitely more care than the human cargo aboard the slave ship—to a reflection on curiosity, wonder, and magic at the sight of the instrument and the images of the natural world it enables. Whether or not this constitutes a key moment in Equiano's narrative—a glimpse of his unfathomable capacity for wonder in so bleak a scenario—it reflects something of the instrument's capacity to inspire wonder. Here the quadrant is not simply an observational tool but a thing of intrigue and plausible as such. Likewise, it establishes Equiano's credibility as one who recognized the object's importance and function, his reaction shifting quickly from wonder to recognition of and interest in its capacity as a technology.

If the quadrant is an object of curiosity and wonder for Equiano, by the 1790 publication of William Bligh's *A Narrative of the Mutiny*, the quadrant was something to be dispensed with in favor of the more wieldy and reliable sextant. As the mutineers begin loading the captive crew into the ship's boat to be set adrift as Captain Bligh looks on, Bligh's description of which items the mutineers are willing to part with and which are indispensable to them reflects an important hierarchy of value. Those being set off to sea in a small boat to fend for themselves are permitted to take "twine, canvas, lines, sails, cordage, an eight and twenty gallon cask of water, and the carpenter to take his tool chest. Mr. Samuel got 150lbs of bread, with a small quantity of rum and wine. He also got a quadrant and compass into the boat." We learn, however, that Mr. Samuel "was forbidden, on pain of death, to touch either map, ephemeris, book of astronomical observations, sextant, time-keeper, or any of [Captain Bligh's] surveys or drawings."[28] We see in this passage not only the importance of the technical material and instruments that would have been valuable in navigation as well as scientifically—map, astronomical observations, time-keeper, ephemeris (a data table reflecting the calculated positions of celestial objects over a regular time interval)—but also the recognition of the sextant as a technological innovation on the quadrant that was given to those set adrift. The outdated quadrant was nevertheless portrayed as a valuable instrument even by the end of the eighteenth century. In *The Journal of Captain Cook's Last Voyage to the Pacific Ocean* (1781), Cook recounts a story of an "Indian" who steals a quadrant from the astronomer's observatory. We are told the quadrant "was almost instantly missed" and later returned "very much damaged," reflecting something of the stakes of instrument care and the perils of misuse we find in Bayley's account.[29] We also see in this moment a microcosm of the anxiety of the British imperium over the prospect of their technologies falling into the hands of the colonized.

As I have suggested, this sample of narrative descriptions and references of the sextant and quadrant in nonfictional and fictional writing throughout the eighteenth century is by no means a comprehensive account but rather a modest illustration of two eighteenth-century forms of virtual witnessing. The first is virtual witnessing in the traditional sense in which Shapin and Schaffer use the concept: technical documents meant to demonstrate the function and value of the sextant used illustrations, diagrams, and verbal descriptions as well as narrative accounts of the use and storage of these instruments. Such accounts emphasized the simplicity of their design and the care with which they ought to be handled to produce accurate results. The second form of virtual witnessing, however, was dissipated over a wide range of texts, including travelogues and fiction. This was the invocation of the sextant and quadrant as lifesaving instruments of great importance, sometimes of wonder. Even in his satirical treatment of the quadrant and sextant in *Gulliver's Travels*, Swift burlesques the new science through farcical portrayals of Laputa that take up images of technical seriousness and render them ineffectual, an acknowledgment of the instrument's purchase as a symbol of technical seriousness.

My primary objective in this essay has been to call attention to the function—however brief—of the sextant and quadrant in eighteenth-century writing as a preliminary effort to show how virtual witnessing could be distributed across different genres, reverberating far beyond the epicenter of astronomical texts, navigational primers, cases entertained by the Board of Longitude, or issues of the *Philosophical Transactions of the Royal Society*. Narrative accounts of the sextant could establish credibility not only through traditional forms of virtual witnessing—images, diagrams, technical descriptions, and numerical data—but also through the generation of wonder, intrigue, anxiety, and other narrative means of creating impressions of value. They could also evince awareness of such narrative methods of generating credibility even in satirical efforts to diminish credibility. And sometimes impressions of credibility or value were simply incidental, not necessarily matters of authorial strategy but consequences of the mass production of the technology itself and the importance of navigation to the ends of transoceanic trade and British imperialism. This last point is essential to bear in mind because imperial objectives certainly drove much of the interest in representing navigational technologies as reliable objects of wonder—or, among critics such as Swift, as overhyped preoccupations. Such a preoccupation meant that audiences who were sympathetic to Equiano's *Interesting Narrative* or the cause of the "Indian" who steals the astronomer's quadrant in Cook's account might have viewed such technologies very differently from their inventors and champions, perhaps not simply as objects of wonder but as objects of horror, control, or rapaciousness. What is missing from the account I offer here, and might be promising for further inquiry, is a more capacious, quantitative study of when, how, and with what kinds of inflections navigational instruments appear across genres of eighteenth-century writing. My conjecture here is that reading across genres might give us a more thorough picture of how the credibility of navigational instruments developed and what role their popular image played in such development.

Notes

1. I use the term *sextant* in my essay title, though I refer to quadrants as well throughout this essay. The different names of these instruments reflect the size of the instrument's arc (as a proportion of a circle: a quadrant is a quarter of a circle, a sextant a sixth of a circle, an octant an eighth of a circle), though the underlying technology and purpose of each of these forms of the instrument are similar. As Margaret Schotte writes in *Sailing School: Navigating Science and Skill, 1550–1800* (Baltimore: Johns Hopkins University Press, 2019), "The sextant and the octant, invented in the 1730s and 1750s, overtook the multitudes of quadrants and sectors to become the preferred tools for observing altitudes. They used mirrors to reflect the image of the celestial object in the same field of view as the horizon, ensuring a more stable observation" (152).
2. William Whiston and Humphry Ditton, "Reasons for a Bill, Proposing a Reward for the Discovery of the Longitude," June 10, 1714, the Royal Society, London.
3. For an account of the Board of Longitude's finances and awards, see Derek Howse, "Britain's Board of Longitude: The Finances, 1714–1828," *Mariner's Mirror* 84, no. 4 (1998): 400–417.
4. Steven Shapin, "Pump and Circumstance: Robert Boyle's Literary Technology," *Social Studies of Science* 14, no. 4 (November 1984): 484. Shapin and Simon Schaffer take up this concept and develop it further in *Leviathan and the Air-Pump: Hobbes, Boyle, and the Experimental Life* (Princeton: Princeton University Press, 1985).
5. Ibid., 491.
6. John Rogers, *The Matter of Revolution: Science, Poetry, and Politics in the Age of Milton* (Ithaca: Cornell University Press, 1996), 36.
7. Tita Chico, "Minute Particulars: Microscopy and Eighteenth-Century Narrative," *Mosaic* 39, no. 2 (2006): 143.
8. Tita Chico, *The Experimental Imagination: Literary Knowledge and Science in the British Enlightenment* (Stanford: Stanford University Press, 2018), 28, 1–2.
9. Jennifer L. Lieberman, "Finding a Place for Technology," *Journal of Literature and Science* 10, no. 1 (2017): 26.

10. Maurice Daumas, *Scientific Instruments of the Seventeenth and Eighteenth Centuries and Their Makers*, trans. Mary Holbrook (London: Batsford, 1972), 200–202.
11. Shapin, "Pump and Circumstance," 481–520, 488–91.
12. Ibid., 491.
13. See, for example, the collection of correspondences and reports of and concerning the Board of Longitude, particularly regarding the dispute between William Bayley and Jesse Ramsden over the performance of Ramsden's sextants on Bayley's voyage to the Cape of Good Hope, MM/7, Royal Society Library Archive, London, United Kingdom.
14. William Bayley, "Some Account of a 15 Inch Sextant Made by Mr. Ramsden" (manuscript), ca. 1781, MM/7/22, Royal Society Library Archive, London, United Kingdom.
15. Ibid.
16. Board of Longitude, notes of a controversy between Bayley and Ramsden (manuscript), ca. 1781, MM/7/23, Royal Society Library Archive, London, United Kingdom.
17. Benjamin Martin, *An Explanation of a New Construction and Improvement of the Sea Octant and Sextant* (London, 1775), 2.
18. Ibid., 11. As B. Martin explains it, "As in the usual Construction *all the Glasses are moveable*, here they may be *all considered fixed*; for the two Horizon Glasses are absolutely riveted down to the Frame; and the Index Glass, though it be freely moveable by the Screws, yet is it immoveable from any other Cause" (11).
19. John Hadley, "The Description of a New Instrument for Taking Angles," *Philosophical Transactions* 37, no. 420 (1731): 147.
20. Ibid., 148.
21. Martin, *Explanation of a New Construction*, 4.
22. See, in this volume, Anne M. Thell, "The 'Eye of Mr. Anson Himself': Art and Evanescence in A Voyage Round the World (1748)," and Al Coppola, "'Fully Prov'd by the Plates': Desaguliers's Unauthorized System."
23. Jonathan Swift, *Gulliver's Travels*, ed. Claude Rawson and Ian Higgins (1726; Oxford: Oxford University Press, 2005), 39, 149–55.
24. Gregory Lynall, *Swift and Science: The Satire, Politics, and Theology of Natural Knowledge, 1690–1730* (London: Palgrave Macmillan, 2012), 103–4.
25. Laurence Sterne, *The Life of Tristram Shandy, Gentleman*, ed. Ian Campbell Ross (1759–67; Oxford: Oxford University Press, 2009), 157.
26. Tobias Smollett, *The Adventures of Roderick Random*, ed. Paul Gabriel-Boucé (1748; Oxford: Oxford University Press, 1999), 217–18.
27. Olaudah Equiano, *The Interesting Narrative*, ed. Brycchan Carey (1789; Oxford: Oxford University Press, 2018), 42.

28. William Bligh, *A Narrative of the Mutiny, on board His Majesty's ship Bounty, and the subsequent voyage of part of the crew, in the ship's boat, from Tofoa, one of the Friendly Islands, to Timor, a Dutch Settlement in the East Indies* (London, 1790), 3.
29. James Cook, *Journal of Captain Cook's Last Voyage to the Pacific Ocean; on Discovery; performed in the years 1776, 1777, 1778, 1779* (London, 1781), 167–69.

Newtonian Legacies in William Hogarth's *A Scene from "The Indian Emperour,"* or *"The Conquest of Mexico by the Spaniards"*

LAURA MILLER

The 1730s marked a pivotal juncture in William Hogarth's career. During this period, Hogarth moved from producing isolated satirical engravings and conversation paintings to the fusion of satire, painting, and engraving that became characteristic of his sequential art. This shift is showcased prominently in *The Harlot's Progress* (1731) and *The Rake's Progress* (1733–35). In both of these series, multilayered plots and characters come to life, narrating the tragic fates of Moll Hackabout and Tom Rakewell. Concurrently, Hogarth undertook a commissioned work titled *A Scene from "The Indian Emperour," or "The Conquest of Mexico by the Spaniards"* (1732–35). This painting was executed at the request of Catherine Barton Conduitt and John Conduitt, Sir Isaac Newton's niece and nephew-in-law. It depicts the Conduitts' opulent home, where children engage in a performance of John Dryden's *The Indian Emperour* while distinguished guests socialize in the foreground. At first glance, the painting presents endless vivacious subjects to contemplate: rosy-cheeked children; sociable, exquisitely dressed elites; intersecting bright gazes; and pervasive merriment. *A Scene from "The Indian Emperour"* has traditionally been regarded as a painting of children or as a conversation painting. Art historian David Bindman considers it "perhaps the greatest of all [Hogarth's] conversation pieces."[1] His interpretation in the *Oxford Dictionary of National Biography* underscores its significance as "evidence of the artist's connections," but the painting has not been evaluated as part of Hogarth's transition from static to sequential modes of representation.[2] Bindman is right that the painting is exceptionally accomplished as a conversation piece: it catches many sociable interactions in medias res to inspire the sustained attention characteristic of viewing Hogarth. This idea of inviting sustained attention during a time of artistic transition prompts my reevaluation of the painting's critical reception and categorization by foregrounding the painting's depiction of Newton. Hogarth's central placement of a bust of Newton offers a pointed commentary on legacies and inheritance.

In particular, Hogarth lays bare the significant differences between performing a legacy and embodying it in one's work, as found in the links between portraiture and public life. The bust of Newton in the painting sits atop a fireplace largely devoid of ornamentation—with the exception of a frieze below it, similar to that by Michael Rysbrack on Newton's tomb at Westminster Abbey.[3] These memorial elements trouble common assumptions about the painting's focus. Despite the seeming centrality of the children's amateur theatricals, the memorial prods viewers to consider how memorialization functions differently in sculpture versus repertoire. Here, the children's sprightly performance contrasts with a quieter space for critical reflection, encapsulated by Newton. Hogarth directs his viewer to focus on both.

Patricia Fara's recent book on Newton's London life uses this painting from Hogarth as an inspiration for studying Newton's time in London, when he transitioned from his academic lifestyle to a more cosmopolitan environment.[4] As in the painting, Newton's wealth and social position as Warden and then Master of the Royal Mint meant he had a far different life when he moved to London than he had in Cambridge as a professor and member of Parliament. Fara's focus on the Hogarth painting as important to reevaluate in Newton studies informs my work here, although this essay largely discusses Newton after his death in 1727. Before Fara's book, academics who analyzed nonsatirical representations of Newton generally responded in the context of hagiography, because many depictions of Newton were meant to inspire veneration. The imposing stained-glass window designed by Giovanni Battista Cipriani (1727–85; completed 1775), installed at the end of the Wren Library in Trinity College, Cambridge, shows Newton being led by the muse of Trinity to meet George III, Britannia, and Sir Francis Bacon while an angel heralds this event with her trumpet.[5] Images of Newton like this proliferated during and after his life, even compared to other often-represented nonroyal eighteenth-century figures such as Alexander Pope and David Garrick. Two hundred thirty-one images were made of Newton by 1800, as Milo Keynes's *Iconography of Sir Isaac Newton* shows. Newton outranks Pope at 81 and Garrick at 70 images.[6] One explanation for this popularity is that Newton's work connected physical objects directly to abstract ideas. Thus, his persona may have lent itself to iconography more than those of other nonroyal public figures. Another reason Newton was also widely adopted as an icon of Enlightenment

FIGURE 1. A Scene from "The Indian Emperour," or "The Conquest of Mexico by the Spaniards," William Hogarth, 1732–35. (Smith Archive / Alamy Stock Photo)

was his Englishness—he was so English he never left England and so "enlightened" he dismantled the properties of light itself.

Newton had, as with a traditional saint, a series of icons with which he was associated—including the reflecting telescope he invented, the prisms through which he revealed the spectrum, the material texts of the *Principia* and the *Opticks*, and even the apple. Newton's corpuscular theories of light and optics described light's materiality, so the content of his work was both theoretical and material. In the second edition of the *Opticks'* Query 29, Newton asked, "Are not the Rays of Light very small Bodies[?]"[7] The ability to understand the physical properties of light and the ability to use paint to highlight dimensionality on a flat canvas may separate Newton from artists, but the two still complement one another.

The three laws of motion from the *Principia*, too, present a mininarrative of objects: an object begins at rest, receives an impact from another object, and then responds to the force of that impact; then the work itself reveals the universal motion and attraction of objects in space.[8] The *Principia*'s difficulty as a text also emphasizes its status as a material object because it resists the kind of immersive reading present in other kinds of text, such as correspondence, fiction, or even sensationalized, cheap ballads—instead relying on mediation. When artists represent Newton visually, an object is often present to mediate or direct the experience. His statue in the Trinity College antechapel holds a prism, about which William Wordsworth reflected in *The Prelude*:

> I could behold
> The antechapel where the statue stood
> Of Newton with his prism and silent face,
> The marble index of a mind for ever
> Voyaging through strange seas of Thought, alone.[9]

The intellectual remoteness Wordsworth highlights, in conjunction with the relative absence of Royal Society sociability from the early years of Newton's career, contrasts with the reception of contemporaries like Robert Hooke and Robert Boyle, who inspired fewer visual representations than Newton even though Boyle was of a higher social rank.[10] Considered in this context, the use of apples, prisms, and Newton's published books in Newtonian iconography extends our comprehension of Newton and materiality in visual culture. The position of Newton in this painting by Hogarth would have been unsurprising but still would have drawn the viewer's attention. Hogarth, however, omits the familiar objects with which Newton might have been associated, including the prism, the apple, celestial/terrestrial globes, and books. Placing Newton in this domestic scene, an anchored gray bust slashing through a sea of color, generates its own contrast and interest. Although it does not seem hagiographic, it nonetheless shoulders the weight of Newton as a subject.

Hogarth's depiction of the frivolity and ephemerality of the scene links to his satirical sequential art as much as it does to his other conversation paintings. Indeed, the painting appears to criticize Conduitt, who performs the role of Newton's heir but fails to embody Newton's legacy through his own work. Newton's late life and relationships with family members like the Conduitts are described in detail in biographical works like Richard S. Westfall's *Never at Rest* and Fara's newer interpretation of his later years.[11] Robert Iliffe's recent work on Newton describes the disagreements between Newton and Conduitt regarding how each valued imagination, in particular Newton's criticism of imaginative works that indulge in "fancy" as opposed to rational thought. Iliffe extends this argument to write persuasively and comprehensively about Newton's rational approach to religious study.[12] Newton did own volumes of printed plays, which might seem to rebut his supposed skepticism about imaginative literature, but William Congreve, a particular favorite, may have appealed to Newton on the basis of his plays' tightly rational, clockwork-like construction. Most volumes of imaginative literature from Newton's library date to the later years of his life, when he had a larger home to furnish, and the cataloging of his library, well explained by John Harrison, included some volumes that were owned and read by other members of the household.[13] Some of Newton's library books, as I have described elsewhere, were used for alchemical purposes, like his miniature copy of Ovid's *Metamorphoses*. Newton's copies of Aeschylus date from his years at Trinity College, so they may have been connected to his reading then and not later.[14]

Considered in this context, it becomes newly noteworthy that Hogarth's painting is suffused with fanciful representations contrary to Newton's own bent, especially the Conduitt children's wildly inappropriate choice of text. As in other works by Dryden, *The Indian Emperour* stages heroic tragedy in an exotic location where empires collapse amid lovers who generally do the same. Although Hogarth was not known for turning the tables on those who commissioned work from him, this painting implies that Conduitt was interested in public displays and appropriating Newton's name more than in continuing Newton's legacy through intellectual work. Conduitt was the custodian of Newton's legacy at the Royal Mint and was his would-be biographer, but following Newton's death in 1727, there was a large market for celebratory representations of Newton.[15] This form of cultic commemoration recurs throughout the history of scientific culture, perhaps culminating in the twentieth century's often uninformed fascination with Albert Einstein. For his part, Hogarth was no stranger to the (often unauthorized) reproduction and circulation of his images and endorsed the Engraver's Copyright Act of 1735 in order to protect his own works and those of other engravers.[16]

Hogarth's painting and its memorials to James Thornhill and Newton, as permanent and original commemorations, allow for greater forms of complexity and greater respect for mentorship than Conduitt's world presents to the viewer. Hogarth, son-in-law of Thornhill—for whose portraits Newton himself had sat on multiple occasions—understood the pressures of intergenerational, nonpatrilineal inheritance. The profession of art, in many ways, relies on this kind of mentorship. The Conduitt painting pays homage to Thornhill's portrait of Newton as well, embodied by the empty chair draped in velvet robes like those in which Newton had posed. Successive generations cannot, for Hogarth, rest on the laurels of their great ancestors and instead need to amplify the effects of mentorship by making their own contributions and mentoring future artists. Considered thus, the child actors performing the works of the great Dryden also fail to live up to their subject by simulating a reality they cannot yet feel. Conduitt, who watches the children perform, troubles the legacy of the British state in other ways by misappropriating the work of Britain's former poet laureate. Hogarth admired Dryden, who is cited several times in *The Analysis of Beauty*, especially for his poem written to the seventeenth-century court painter—and another painter of Newton— Sir Godfrey Kneller: "Where light to shades descending, plays, not strives, / Dies by degrees, and by degrees revives."[17] Hogarth and Conduitt were confronted by the legacies of famous men who mentored them but who were not related by blood. Ultimately, the painting reveals that homosocial mentoring should be furthered by its inheritors and not appropriated—mentorship should be "paid forward" rather than cited or, worse, degraded.

Critic Amal Asfour describes *The Indian Emperour*'s connection to Hogarth's own aesthetics, including the painting's "morality of form," which resonates independently of my specific claims here.[18] Asfour writes that "the beauty of [Hogarth's painting] *The Conquest of Mexico* does not consist solely in the pleasures of the eye but in the conjunction of eye and mind—not only in the play of color and form but also in the play of ideas that these provoke."[19] This kind of perpetual interpretable movement is central to Hogarth's art and would become even more mobile in the shift to sequential art, especially the knowledge that what appears to be decorative is often essential. I would extend the implications of Asfour's argument to include this more critical appraisal of the Conduitts. The painting also plays on different forms of representation—painting, performance, imitation, sculpture, and memorial—and aligns with the complexities of nonpatrilineal mentorship that link Hogarth and Conduitt. Likewise, this painting develops the representation of ephemeral and superficial subjects as seen in Hogarth's body of work as a whole.

Because the Conduitts commissioned the painting, their first priority would be their family's representation, which included Newton. The merry and lively scene would have offered endless pleasure to its viewers, as was fit for a conversation painting. In this painting, Newton becomes an icon himself in the form of a bust, an object positioned to enhance Conduitt's reputation. Conduitt used Newton's reputation to further his own public career, and although by all accounts Conduitt was talented and bright, he, like nearly everyone, failed to reach Newton's level of acclaim. Conduitt had succeeded Newton as Master of the Mint and was a distinguished figure in his own right. He is memorialized across from Newton in Westminster Abbey, though there is little doubt that marrying into Newton's family helped secure this permanent honor.

In keeping with Iliffe's analysis of the relationship between Newton and Conduitt when it came to "fanciful" representations, Hogarth here explores the divergence between fancy and reason through art. Newton's bust becomes the performed scene's most significant observer. From the upper left of the painting, portraits of the Conduitts join the bust of Newton as witnesses to the children's performance. Newton's bust sits on the mantel, a three-dimensional object placed next to two-dimensional portraits of his heirs. Even within the picture, the marble bust of Newton has a weight that the canvas Conduitts lack. The bust itself gestures to neoclassical ideals of permanence and would not be out of place next to busts of Greek or Roman mathematicians. The paintings of the Conduitts more closely resemble rococo portraiture, with contemporary hair and clothing. The implication here is that the nearly ephemeral Conduitts have yet to earn their place of permanence as Newton's successors.

Hogarth himself was a fit proposer of an antifanciful Newtonian legacy. As the author of *The Analysis of Beauty,* Hogarth was keenly interested in the intersection of rational principles and art.[20] Hogarth mastered the rococo style, but he disliked its emphasis on frivolous subjects and the upper class: works like *The Rake's Progress* and *Marriage à la Mode* criticize elite decadence using the same levels of exquisite detail as continental paintings that luxuriated in excess. Despite his subjects' rococo adjacency, Hogarth pleaded for artistic principles that were guided by the serpentine "line of beauty" rather than the theatricality and trompe l'oeil that characterized much of continental rococo. Although Hogarth's work shared an investment in detail with rococo, it rarely pursued flights of fancy. As with many works from idiosyncratic eighteenth-century Britain, Hogarth's art engaged in continental conversations on its own terms and in its own time.

Despite long-standing critical investment in the interpretation of ornaments large and small in Hogarth's work, the presence of Newton's bust in the painting has not been considered noteworthy by most Hogarth critics. These critics have understandably focused on the artist's better-known works and his sequential art. When Ronald Paulson mentions this painting, he identifies the relationship between Newton and the Conduitts as well as noting the presence of the bust.[21] Following the work of Paulson, Asfour describes "a moment of familial diversion [as] a moment of compositional diversion" in which a mother telling her daughter to pick up a fallen fan complements the painting's framing of Newton.[22] Asfour argues that this moment reveals "a series of hierarchies and relationships" in the painting.[23] My reading of this familial moment is one of education and duty: taking over generational responsibilities starts with small duties, which are then expanded and amplified so that children take on more responsibility over time. The children on stage are in over their heads; the little girl and her mother are not.

As an artistic representation of Newton, *The Indian Emperour* has many competitors, which may also contribute to its infrequent close study. Newton was a frequent subject of individual portraiture during his life and was painted by Enoch Seeman, Godfrey Kneller, and other important painters. Several busts of Newton appear in other eighteenth-century paintings, even by Hogarth. Hogarth included a bust of Newton in his portrait of Benjamin Hoadly the elder, who had been made a Fellow of the Royal Society at the age of twenty during Newton's presidency. Keynes claims that "why [Hogarth] placed the bust of Newton in this portrait is not known," but at the very least, the audience is meant to compare the sitter and the bust.[24] In this case, there is a more direct reason to include Newton: the sitter knew Newton by acquaintance, and Newton recognized his skills by appointing him a Fellow of the Royal Society. Hogarth himself owned a bust of Newton that had been given to him by the sculptor Louis-François Roubiliac; this is most likely the bust in the painting of Hoadly.[25]

Beyond Hogarth's possession of another bust was his status as the son-in-law of Sir James Thornhill, who had painted three portraits of Newton. Thornhill, best known for painting the dome of St. Paul's Cathedral (1714–17) and the Old Royal Naval College in Greenwich (1707–27), was also court painter to George I and the first English-born artist to be knighted for his art. Hogarth and Conduitt were connected to Thornhill and Newton by marriage rather than blood, so this kind of paternal/mentor relationship without direct kinship hovers over this commission. Among the many surprising things about Conduitt's painting is that Conduitt was *not* painted with a bust of Newton as a three-dimensional sitter; Newton's bust accessorizes Conduitt's opulent drawing room and does not appear to inspire any contemplation at all. Nor was this a family portrait like Hogarth's 1735 *Portrait of a Family* held at the Yale Center for British Art, in which a group of related sitters relaxes and chats in front of a fireplace; the Conduitt portrait is one of far more people, including blood relatives and friends.[26]

Hogarth does more than manage this array of active subjects; he produces his own postmortem portrait of Newton in *The Indian Emperour*. He does not copy the bust of Newton by Michael Rysbrack known as the "Conduitt bust," which would be the obvious choice to include here. Instead of giving us Rysbrack or the Roubiliac bust he owned, however, Hogarth gives us Hogarth: he merges the Rysbrack and the Roubiliac busts, adopting, as Keynes points out, the collared shirt of Rysbrack and the draping of Roubiliac.[27] Rather than accurately reproduce the Conduitt bust in a commission from Conduitt, he offers his own new representation of Newton. This innovation supports my claims about legacy and creation. As a newly created, painted bust by Hogarth rather than a copy of Rysbrack or Roubiliac, its value to the painter as a postmortem "Newton portrait" assigns it greater significance than as an ornament or status symbol. If anything, the portrait bust derails Conduitt's authority by altering his approved representation of Newton.

The placement of the bust further confirms its importance in the painting. Hogarth places the bust of Newton on an austere gray section of the left wall. Underneath it, on the mantel, is a frieze resembling—but not copying—that on Newton's tomb at Westminster Abbey, suggesting other memorials to Newton. A bust of Newton, reminding viewers of his tomb, frames children enacting a play—a juxtaposition of permanence and ephemerality. Hogarth was generally critical of, yet compelled to depict, the ephemeral: ephemeral motifs appear throughout his sequential art, including *The Harlot's Progress*, *The Rake's Progress*, and *Before and After*.[28] In those sequences, ephemeral pleasures—or, at least, ephemeral decisions—yield sorrow and tragedy. This painting links the ephemeral nature of the children's performance, of childhood itself, of the waning empire of Spain whose flourishing was represented in Dryden's play, and of human distractions—exemplified by the fashionable adults socializing on the side. These beautiful ephemera cannot fully distract the viewer's attention from that still, gray channel at the painting's center, where Newton's bust glances outward. The gray and ocher plane that angles through the middle ground of the painting is not simply present but holds the painting's other elements in balance.

Beyond ephemera, images of imitation and inadequacy abound in the painting. Most glaring, of course, is the inappropriateness of children dramatizing the conquest of Mexico—but the simian statue adjacent to the children, whose contrapposto aims the hips at the adult audience and the torso at the children, gestures to other improprieties. Whether it criticizes a commodification of scientific exploration or a British attempt to domesticate the exotic, the inclusion of this monkey gestures to the Conduitts' decadence. Hogarth's famous monkey in plate 2 of *The Harlot's Progress* displays the decadence of exotic pets, the animal natures humans fail to suppress, and of course, imitation or "aping." In the case of *The Harlot's Progress*, Moll Hackabout's imitation of elite status is facilitated only by her youth and novelty, as well as her imitation of a genuine relationship with the men who keep her as a forbidden "pet" of their own. Dryden has also been linked to monkeys before, albeit in an unflattering context: Jacob Huysmans's portrait of John Wilmot, Earl of Rochester (ca. 1675), contains a laureate monkey propped up by books—a dig at those who, like then-laureate Dryden, sought royal favor to endorse their literary merits.[29]

The monkey in this painting, however, is not crowned with laurels. If the Conduitts are merely "aping," this was an especially incendiary comment—

especially with the formerly dubious reputation of Catherine Barton Conduitt, who inherited a suspiciously large estate and many jewels from Charles Montagu, the Earl of Halifax, Newton's friend and patron, for whom Catherine managed a household. When Barton later married Conduitt, this alliance would have been seen as a restoration of Catherine's reputation. Part of the recompense for her rumored involvement as a mistress would have been the knowledge that whoever married Catherine would be Newton's heir apparent.[30]

Hogarth, by contrast, when he painted *The Indian Emperour*, had recently wooed his wife, Jane Thornhill, against her father's wishes: the pair had eloped in 1728, a few years before this was painted. There was a period of uncertainty afterward, but the family would eventually reconcile. Hogarth and Jane did not have biological children but were devoted patrons of the Foundling Hospital, so their commitment to the care of nonbiological dependents extended to future generations. Whatever Hogarth's feelings were about Thornhill's initial resistance to him as a son-in-law, those feelings seem to have abated by the time Hogarth praised his work in *The Analysis of Beauty*: "Let any one take a view of the [ceiling] at Greenwich-hospital, painted by Sir James Thornhil[l], forty years ago, which still remains fresh, strong and clear as if it had been finished but yesterday: and altho' several French writers have so learnedly, and philosophically proved, that the air of this island is too thick, or—too something, for the genius of a painter; yet France in all her palaces can hardly boast of a nobler, more judicious, or richer performance of its kind."[31] The criticism of French style is common in Hogarth's work, but more importantly, the unabashed admiration of Thornhill's painting at Greenwich Hospital reveals the son-in-law's longstanding affection for his late father-in-law.

Hogarth's reverence for his father-in-law extends to his representation of Newton, though Hogarth does not depict all scientific subjects with this level of adulation. For example, in the last frame of *The Four Stages of Cruelty* (1751), the image entitled "The Reward of Cruelty" is set in an anatomy theater, loosely inspired by the dissection frontispiece of Vesalius's *De humani corporis fabrica* (1543). It shows the natural progression of protagonist Tom Nero, who has grown from a child who treated animals cruelly, to an adult who murders, to this end stage: a corpse on the dissection table, ready to receive in death the cuts and stabs he gave in life, with the added cruelty of spectacle. Lookers-on engage in a variety of activities, much like those in Joseph Wright of Derby's *An Experiment on a Bird in the Air-Pump* (1768). Some of the men in attendance take notes, some chat and debate, and a dog like those Nero brutalized sneaks off to eat the criminal's heart, gaining strength from the abuser's flesh. As in Hogarth's other sequential art, moral cause and effect frame the reception of the image; this is done less openly in *The Indian Emperour*, but inheritance—itself a cause-and-effect concept—is a necessary component for understanding it.

The weight of bearing Newton's legacy exerted its own pressures on Conduitt. Conduitt was widely known to be working on his biography of Newton but had not yet written it at the time of this painting. Indeed, this anticipated biography was never published. Conduitt appears, midprocess, in James Thomson's "A Poem Sacred to the Memory of Sir Isaac Newton":

> Conduitt, from thy rural hours we hope;
> As through the pleasing shade where nature pours
> Her every sweet in studious ease you walk,
> The social passions smiling at thy heart
> That glows with all the recollected sage.[32]

Thomson imagines Conduitt as a body infused with Newton's wisdom and friendship, qualities that animate his heart and imbue his daily movements with a gentle Newtonian gravity. Hogarth's Conduitt seems to care more about sociability than being able to carry on Newton's legacy, which may in fact be an impossible task. As the heir to an intellectual giant, one must primarily acknowledge and respect that legacy. Conduitt was not to blame for the difficulty of inheriting Newton's legacy, but the painting implies that the Conduitts were still liable for their own social climbing and frivolity when handed that legacy.

Newton himself was frequently associated with the celestial globe or with copies of his own printed works in portraits; here, the bust of Newton is recognizable independently and identifies the Conduitts as the subject of this painting. The late scientist shifts from being represented as a subject of portraiture, holding objects, to being an object himself. Even though the bust of Newton would appear to be decorative as well as symbolic—balancing the painting's composition or filling a blank space—its placement near the center opposite an empty chair and its austerity complicate this interpretation and shift the balance toward the memorial. The bust is also distinguished from the more ornamental statue and its monkey friend to the rear because of the bust's elevation and Hogarth's use of light to frame it. Including a tomblike space in a domestic scene such as this one repurposes a public depiction of Newton to craft a domestic persona for him. At the same time, Conduitt can now use this domestic scene to create a public persona for his family.

Despite Newton's prominence in the painting, no one is contemplating the memorial: the gazes in the painting are not directed at Newton. However, one point of view directed at Newton is that of the empty chair to the right of the painting, on which a piece of fabric rests. Although the chair balances the composition, it also invites the reader to consider why it is empty and to wonder whose seat it is. Beyond the painting's subtle indictment of performance and the ephemeral, Hogarth presents his own craft as a counter-example to this performance of legacy, by having the memorial to Newton simultaneously serve as a tribute to Thornhill. In creating his own portrait bust of Newton, Hogarth splits painting from sculpture to represent Newton uniquely, in painted form, but not in competition with either busts or portraits of Newton. The empty chair is draped with two-toned velvet, most likely from the wall draperies, that recalls the robes in which Newton had sat for Thornhill during his lifetime, inviting reflection on both Newton's and Thornhill's careers. Thornhill died in 1734, and this painting is said to date from 1732 to 1735, so this tribute aligns with those timelines.

The empty chair is not just anomalous in the context of a sociable painting; it also stands out among representations of chairs. Chairs are customarily present in paintings for the sitter to occupy, not as an empty object of focus. They may be part of a suite of material objects that can "enhance the sense of the sitter's wealth or show his or her graceful interactions with them," as Lin Nulman writes.[33] It would make far more sense to present chairs in this commission in this way, as status symbols. Jacques-Louis David used an empty chair to showcase technique in painting wood grain, perhaps because those who would learn this technique would use it in portraiture.[34] Other writing on domestic interiors in the eighteenth century focuses on the placement of chairs in social seating arrangements.[35] However, the chair in Hogarth's painting is turned away from the viewer, is not overly detailed, and faces the images on the wall rather than the viewer. As such, the chair may be marking the scene as an invite-only one, where the chair is reserved by the piece of drapery, its placement away from the viewer indicating the exclusivity of the scene. The chair directs observers to the wall opposite, including the portraits of the Conduitts, the bust of Newton, the marble frieze, and the wealthy, well-coiffed attendees of the amateur performance.

In Hogarth's other works, empty chairs are generally piled with clutter or are overturned because of carelessness or a struggle. There are overturned chairs in plates 2 and 6 of *Marriage à la Mode* as well as the death scene plate in *The Harlot's Progress*, where a small wooden stool has been knocked over. An upright, empty chair asks the audience to consider who should be sitting in the chair or why no one is sitting in the chair. One interpretation is that Conduitt himself would take up the place opposite Newton, and that would be the likely explanation Conduitt would accept. But given the framing of Conduitt in relation to Newton, Newton's connection to the Hogarth family, and the weight of Newton's bust in the painting, it is likely that Hogarth is less complimentary to the Conduitts and their priorities. The chair, too, is its own memorial to Newton, whose position and works as a natural philosopher and mathematician would remain unsurpassed for many more years. This chair invites spectators to wonder why the person it was saved for has yet to arrive or why no one sits there at this party, thinking about whose legacy helped the Conduitts rise to their station. Its placement at a theatrical performance also invites viewers to consider what is worth watching in the first place.

This painting of children acting, engaging in a performance where they pretend to have relationships other than the ones native to them, extends Hogarth's criticism of ephemeral power and the superficiality of those who inherit it. Some shoes—or chairs—may always be too imposing to fill. Both Hogarth and Conduitt took on the legacies of well-known men who mentored them but were not related by blood. Considered even more broadly, many felt the loss of Newton as an inspiration after his death, especially those of his colleagues in the Royal Society who had benefited from Newton's mentorship. The painting contemplates the way this kind of homosocial mentoring should be honored by its inheritors. Hogarth finds value in the permanence of works that are produced, like the painting, the citation of Thornhill's portrait, and the bust, rather than the ephemeral, superficial performance of a legacy. After Hogarth's death in 1764 following a long illness, his estate was largely controlled not by male inheritors but by the women of his family—his wife, Jane, and his sister Mary. Jane Hogarth worked assiduously to manage his estate, publishing editions of William's engravings and writings as well as pursuing legal action to extend the copyright of his work. These actions embody the legacy Hogarth's painting proposes: a legacy of care, labor, and honor.

Notes

1. Bindman, *World of Art*, 25.
2. Bindman, "Hogarth, William."
3. Rysbrack, Memorial to Newton, Tomb of Sir Isaac Newton, Westminster Abbey.
4. Fara, *Life after Gravity*.
5. Cipriani, Stained-Glass Window Depicting Sir Isaac Newton.
6. Keynes, *Iconography*, 3.
7. Newton, *Opticks*, Query 29.
8. See Miller, *Reading Popular Newtonianism*, chaps. 1 and 2.
9. Wordsworth, *Prelude*, 3.60–64.

10. No authenticated portrait of Hooke survives today.
11. Westfall, *Never at Rest*.
12. Iliffe, *Priest of Nature*, 19–20.
13. Harrison's volume remains useful, but the Newton Project has cataloged additional volumes of Newton's that have been found since its initial publication.
14. Miller, "Masculinity, Space."
15. Feingold, *Newtonian Moment*.
16. The Engravers Act of 1735 became known as the "Hogarth Act" because the artist championed it.
17. Dryden, "To Sir Godfrey Kneller," 69–70.
18. Asfour, "Hogarth's Post-Newtonian," 693.
19. Asfour, 716.
20. Hogarth, *Analysis*.
21. Paulson, *Hogarth*, 175.
22. Asfour, "Hogarth's Post-Newtonian," 696.
23. Asfour, 696.
24. Keynes, *Iconography*, 85.
25. Keynes.
26. Hogarth, *Portrait of a Family*.
27. Keynes, *Iconography*, 81.
28. Paulson, *Hogarth*.
29. Huysmans, John Wilmot, 2nd Earl of Rochester.
30. See Fara, *Life after Gravity*, for more on these years in Newton's life.
31. Hogarth, *Analysis of Beauty*, 119n18.
32. Thomson, *Poem Sacred*, 157–61.
33. Nulman, "Performing Portraits."
34. Crow, *Painters and Public Life*, 252.
35. Lipsedge, *Domestic Space*.

Bibliography

Asfour, Amal. "Hogarth's Post-Newtonian Universe." *Journal of the History of Ideas* 60, no. 4 (October 1999): 693–716.

Bindman, David. "Hogarth, William." *Oxford Dictionary of National Biography*. May 21, 2009. Accessed March 10, 2022. https://www.oxforddnb.com/display/10.1093/ref:odnb/9780198614128.001.0001/odnb-9780198614128-e-13464.

———. *World of Art: Hogarth, 2e*. London: Thames & Hudson, 2022.

Cipriani, Giovanni Battista. Stained-Glass Window Depicting Sir Isaac Newton Being Presented to George III. 1775. Stained glass. Wren Library, Trinity College, Cambridge.

Crow, Thomas E. *Painters and Public Life in Eighteenth-Century Paris*. New Haven, CT: Yale University Press, 1985.

Dryden, John. "To Sir Godfrey Kneller." *The Poems of John Dryden*. Vol. 2. Edited by James Kinsley. Oxford Scholarly Editions Online, May 2013. https://www.oxfordscholarlyeditions.com/display/10.1093/actrade/9780199670222.book.1/actrade-9780199670222-book-1?rskey=mpWNbQ&result=2.

Fara, Patricia. *Life after Gravity: Newton's London Career.* Oxford: Oxford University Press, 2021.

Feingold, Mordechai. 2004. *The Newtonian Moment: Isaac Newton and the Making of Modern Culture.* Oxford: Oxford University Press.

Hogarth, William. *The Analysis of Beauty. Written with a view of fixing the fluctuating Ideas of Taste.* London: J. Reeves, 1753.

———. *Engravings by Hogarth.* Edited by Sean Shesgreen. New York: Dover Fine Art, 1973.

———. *Portrait of a Family.* New Haven, CT: Yale Center for British Art, 1735.

———. *A Scene from "The Indian Emperour," or "The Conquest of Mexico by the Spaniards."* Private Collection, London, 1732–35.

Iliffe, Robert. *Priest of Nature: The Religious Worlds of Isaac Newton.* Oxford: Oxford University Press, 2017.

Keynes, Milo. *The Iconography of Sir Isaac Newton to 1800.* Woodbridge, UK: Boydell & Brewer, 2005.

Lipsedge, Karen. *Domestic Space in Eighteenth-Century British Novels.* London: Palgrave Macmillan, 2012.

Miller, Laura. "Masculinity, Space, and Seventeenth-Century Alchemical Practices." *Gender and Space in British Literature, 1660–1820.* Abingdon: Routledge Press, 2016.

———. *Reading Popular Newtonianism: Print, the "Principia," and the Dissemination of Newtonian Science.* Charlottesville: University of Virginia Press, 2018.

Newton, Isaac. *Opticks: Or, A Treatise of the Reflections, Refractions, Inflexions and Colours of Light. The Second Edition, with Additions.* London: W. and J. Innys, 1718.

Nulman, Lin. "Performing Portraits: Eighteenth-Century Americans at the MFA." *Big, Red, and Shiny.* 2018. Accessed January 23, 2025. https://bigredandshiny.org/16843/performing-portraits-eighteenth-century-americans-at-the-mfa/.

Paulson, Ronald. *Hogarth: His Life, Art, and Times.* 2 vols. New Haven, CT: Yale University Press, 1971.

Rysbrack, Michael. *Tomb of Isaac Newton.* Westminster Abbey, London, 1730.

Thomson, James. *A Poem Sacred to the Memory of Sir Isaac Newton.* London: J. Millan, 1727.

Westfall, Richard S. *Never at Rest.* Cambridge: Cambridge University Press, 1983.

Wordsworth, William. *The Prelude, Or, Growth of a Poet's Mind: An Autobiographical Poem.* London: Edward Moxon, 1850.

"Fully Prov'd by the Plates"
The Unauthorized System of J. T. Desaguliers

AL COPPOLA

In the first week of December 1718, two booksellers advertised the publication of *A System of Experimental Philosophy, Prov'd by Mechanicks* by John Theophilus Desaguliers, Fellow of the Royal Society, the eminent science lecturer, entrepreneur, and popularizer of Newtonian theories.[1] Except that there was no such "System" published—at least according to Desaguliers, who issued a scathing attack on the edition in the papers the very next day.[2] Bezaleel Creake and John Sackfield may have published a book, but it was not *his* book—and in any case, it was no perfected "System" of natural philosophy.

That an author would complain about a bookseller's sharp dealing was hardly a surprise in the media landscape of the early eighteenth century. In the climate fostered by the lapse of the Licensing Act in 1694 and the instantiation of a new system of copyright by the 1710 Statute of Anne, competition among booksellers was as fierce as ever to secure profitable new titles to publish. The new regulations defined copyright as a property owned by authors and sold to booksellers that granted them the right of exclusive publication for a limited period of time, after which the title would lapse into public domain. This contrasted with the old system under the licensing act, which vested in booksellers a perpetual copyright for the books they published.[3] In theory, the new system favored authors and gave them more control over how their books circulated and who could profit from them, but in practice, booksellers were not above exploiting gray areas over who exactly could be considered an "author" of a manuscript and what exactly constituted a work that was eligible for copyright protection in the first place.

In this case, Creake and Sackfield appear to have purchased a copy of the illustrations that Desaguliers had printed himself for the benefit of his lecture attendees,[4] and the booksellers initially asserted their right, with or without Desaguliers's consent, to reprint those plates along with copious lecture notes taken down by Paul Dawson, an attendee at Desaguliers's course and the volume's editor.[5] In making this counterargument, ultimately grounded in eighteenth-century copyright law, the booksellers asserted a remarkable claim: *there is no meaningful difference between the live performance of the experiments, their description in a treatise, or the graphic representation of their matters of fact in the plates.* These experiments, they claim, "are fully prov'd by the plates,"[6] and as such they are able to speak for themselves, whether or not they are accompanied by written descriptions and explanations, whether or not that text is written by Desaguliers, and indeed, whether or not that text is even accurate. In a curious application of legal precedent, the booksellers were claiming that they had merely printed a lawful *imitative adaptation* of a "system" that had already been published in its entirety by virtue of prior public performance and graphic representation.

The booksellers eventually were able to bring Desaguliers around by separately paying him for the copy, and he agreed to add a preface and an errata list to a reissue that appeared in January 1719 in his name with a new title.[7] Creake, Sackfield, and Dawson's disputed *System of Experimental Philosophy* resurfaced about a month later as Desaguliers's *Lectures of Experimental Philosophy,* and the learned world moved on to other matters and fresh scandals. However, what we can reconstruct about the booksellers' rationale for printing the book in the first place gives this episode signal importance for our understanding of not just the marketplace for scientific publishing but the very nature of scientific knowledge in this period. Insofar as the booksellers predicated at least part of their claim on the fact that they bought the illustrations and thus had a right to reprint them, the episode raises important questions about the ontology of scientific illustration and what can count as an authentic and sufficient representation of scientific knowledge. As we will see, the booksellers effectively claimed that the intellectual content of Desaguliers's lectures was fully represented in the illustrations themselves. But at stake in their dispute is an even larger question about what exactly Desaguliers had created and shared with his audience.

Since first coming to London in 1712, Desaguliers had been offering courses of experiments intended to give a coherent account of the new Newtonian natural philosophy. Desaguliers was at pains to distinguish his performances from mere spectacles and amusements, and he emphasized the fact that these were *philosophical* lectures—demonstrations that didn't simply showcase startling or amusing spectacles but instead conveyed matters of fact that gave evidence for and were explained by a larger and more coherent body of knowledge: Newtonian natural philosophy. Desaguliers promoted his work as a "Course in Mechanical and Experimental Philosophy" in newspaper advertisements and in the lecture memoranda he had been printing up for his students since at least 1715,[8] but the booksellers marketed their unauthorized collection of those lectures as a "system." Did Creake and Sackfield oversell what Desaguliers was doing, or did they merely deploy the proper terminology? This episode raises the fascinating questions of what, properly speaking, can constitute a system, and—yet more crucially—who can "own" it. Does a system reside in the specific narrative that delineates it, or is it a concept, a set of relations, a reality that exists independent of any particular description? Moreover, what was the perceived value of marketing Desaguliers's work as, of all things, a "system" in late 1718 as opposed to, say, a collection of "lectures" or a "course" of "philosophy," which is how Desaguliers was wont to identify what he was doing. Indeed, the story of Desaguliers's unauthorized "system" raises critical questions about just where knowledge, authority, and agency reside when we take as the object of our investigation not the innovations and breakthroughs that later ages deemed pivotal but rather the messy, material, and highly performative interlocking complexities that attend the making of science in the long eighteenth century.

Answering this volume's editors' call for investigations into eighteenth-century science studies that take seriously what Rita Felski has called narrative's "makerly mode" and that engage with a wider "range of embodied practices, social routines [and] material artifacts"[9] to trace the full extent of the networks that sustain what Bruno Latour has called "matters of concern,"[10] the story of Desaguliers's unauthorized *System* helps remap the circuits of tacit knowledges, material practices, legal strictures, and regimes of performance that precede and condition the eventual inscription of scientific knowledge.

*   *   *

That the booksellers would jump at the chance to publish Desaguliers's lectures is hardly surprising. The Rev. John Theophilus Desaguliers, Fellow of the Royal Society and that organization's paid curator of experiments from 1716 to 1743, was the preeminent scientific lecturer and entrepreneur in early Georgian England. The son of a Huguenot refugee-turned-Anglican divine, Desaguliers matriculated at Oxford, where he attended John Keill's lectures in natural philosophy. He soon distinguished himself for his grasp of the new science and for his experimental acumen. Eventually, he was tapped to take over

Keill's lectures, and he became a protégé of Isaac Newton, who initially hired him to be his experimental assistant after the death of Francis Hauksbee. In Desaguliers, Newton found a gifted engineer who was also graced with a deep philosophical acumen, and he enlisted him as an ally who might help develop, illustrate, and defend his ideas both in England and abroad.[11] Desaguliers moved to London in 1712, and thanks to his skill in attracting patronage from authorities like Newton, Keill, and Sir Richard Steele, as well as from aristocrats like the Duke of Chandos, Prince Frederick, and Queen Charlotte,[12] Desaguliers set himself up at the very center of what Larry Stewart has called the "rise of public science" in early eighteenth-century Britain, when Newtonianism became ensconced in polite culture and when natural philosophy was enlisted in a wide range of projects for economic and material improvement.[13] As the Royal Society's chief experimentalist and a frequent contributor to the *Philosophical Transactions,* as an experienced engineer who devised or refined useful inventions,[14] as one of the most sought-after performers in London's increasingly crowded field of popular science lecturers, and particularly as the private chaplain and in-house engineering consultant to James Brydges, the Duke of Chandos, an important early capitalist, Desaguliers was perhaps the single most influential natural philosopher of his day if we measure importance not only by the originality of his intellectual contributions but also by the impact that his scientific work had on public affairs.

So it should have raised no eyebrows when the back pages of the *Post Boy* and the *Evening Post* both advertised on December 4, 1718, that a partnership of booksellers was publishing *A System of Experimental Philosophy, Prov'd by Experiments* by no less an authority than John Theophilus Desaguliers. As the booksellers surmised, a definitive "System of Experimental Philosophy" by J. T. Desaguliers would be a valuable property indeed. Desaguliers had been offering demonstration lectures in London ever since 1712, when he made arrangements with Francis Hauksbee's widow to take over his well-known courses. To advertise his lectures, and to give some specimen of what paying auditors could expect for the relatively steep tuition, in 1715 and again in 1717, Desaguliers published descriptive syllabi of his lecture course[15] and sold plates of illustrations to help his auditors master the material.[16] However, at no point had Desaguliers published the actual content of the lectures.

A few considerations would have weighed on him. What might be clear enough in live performance might not have been regular and correct enough for print publication. What's more, there is some question as to whether the content of Desaguliers's lectures was even his in the first place. As Desaguliers would later explain in the "Preface, or Advertisement to the Reader," that he affixed to the authorized reissue of Creake and Sackfield's book in January

1719, the lectures he gave were based on Keill's original lecture memoranda, which was itself a pastiche of experiments extracted out of the major late seventeenth- and early eighteenth-century authorities on mechanics, hydrostatics, optics, and so forth.[17] Setting aside some legitimate questions of authority, originality, and property (I'll return to them soon enough), it is unlikely that Desaguliers would have wanted his lectures published at this point in his career in any form. As can be judged from the fact that Desaguliers had to ask around and surreptitiously cobble together Keill's memoranda from the older students who had been given access to it (and made their own copies)— and this even after he had been asked to step into the master's shoes—we should rightly consider these lecture scripts as a kind of trade mystery, a work product to be jealously guarded against piracy and infringement. It wasn't until 1734 that Desaguliers would voluntarily publish the first of what was to be two volumes of his lectures in *A Course of Experimental Philosophy*. At that time, he was able to assert that he was about to offer his 121st course of experimental philosophy and that of the eleven scientific lecturers then offering courses, he had personally trained nine of them.[18] By that point, he had refined his material to an extent to which he was comfortable putting it into a written record for posterity. But we should also recognize that by that point, his proprietary lectures had served their purpose. After having set on foot so many protégés, it no longer made sense to guard their content.

Publishing his lectures in 1718, however, was not voluntary, and resentments ran so high that the dispute over the edition spilled over into the press, which enables us to reconstruct, to an uncommon extent, the maneuverings of all parties in the imbroglio. Even more critically, we know enough to glean the rationale that lay behind the booksellers' presumptuous actions. Creake and Sackfield began a multipronged advertising campaign to promote their edition of Desaguliers's *System* in the first week of December 1718, and it seems highly likely that these relatively inexperienced booksellers were looking to launch their careers with this high-profile and in-demand publication.[19] They placed initial ads for the book that played up its cultural capital—puffed as "Just Publish'd, Dedicated to Sir Richard Steele"—in two triweekly newspapers that came out on Thursday of that week as well as in a weekly that came out on Saturday.[20] However, based on the follow-up advertisement the booksellers were compelled to take out on Saturday, December 6, Desaguliers apparently did not take this unwelcome news lying down. Two days after their initial announcement, and one day after Desaguliers apparently responded to their scheme in print, the booksellers inserted the following adviso into the *Post Boy* after a reprint of their edition announcement:

> NB: Whereas we have been attack'd after an Ungentleman-like manner, contrary to the Promise of Mr. Desaguliers, by an Advertisement in the S. James's-Post of Friday last, under which are printed these Words, John Theophilis Desaguliers: To prevent Gentlemen from being imposed upon by such-like Advertisements, (which was done by the Persuasion of a TAYLOR) we think ourselves obliged to give Notice to the Publick, that the Book entitled, A System of Experimental Philosophy, &c. advertised by that busy Advertiser, who personates Mr. Desaguliers (and which that ingenious Gentleman will have little Reason to thank him for) contains the Experiments of Mr. Desaguliers (except the Description of the Orrery) which are fully prov'd by the Plates; and that the Copy of the same was purchased by us for a valuable Consideration; and the Plates (far from being surreptitiously obtained, as maliciously as falsely mentioned in the Advertisement) were paid for by us; and that if such-like Advertisements appear any more, the Party shall have the Truth from us, that they are the true Experiments in another manner.
>
> Bezaleel Creake
> J. Sackfield

While the specific issue of the *St. James Post* they cite seems to have been lost to posterity, the editor of the *Post Boy* helpfully reprinted Desaguliers's advertisement directly below Creake and Sackfield's note. Thus, we learn that Desaguliers had initially written,

> To prevent Gentlemen from being imposed upon by an Advertisement of a Book, entitled, A System of Experimental Philosophy, proved by Mechanicks, &c. said to be perform'd by J. T. Desaguliers M.A. F.R.S . . . I think myself obliged to give Notice to the Publick, that the said Book is publish'd without my Knowledge or Consent, that it is full of gross Errors, and very imperfect; that a great Part of it was never mine, and the rest, as well as the Plates, surreptitiously taken from me.
>
> John Theophilus Desaguliers[21]

According to the account that Desaguliers himself gave in the "Preface, or Advertisement to the Reader," that he would eventually affix to this book after the booksellers "made [him] satisfaction and purchased the Copy [from him]," this edition only came into existence because of the actions of a student at one of his lecture courses: "Mr. Dawson (a young Man whom Sir Richard Steel[e] had put under my Care) took a Copy of the Lectures above-mentioned, that they might be of Service to him when he went thro' my courses, and they were afterwards sold and published without my Knowledge."[22] Dawson, for his part, acknowledged the material originated with Desaguliers in his dedication to Steele, where he says his treatise "contain[s] the several Philosophical Experiments shewn by Mr. Desaguliers in his publick Lectures, which [he had] carefully collected."[23] The fact that he identifies Desaguliers as being responsible for "showing" publicly exhibited "experiments" that his treatise then "contains" is fascinating, as it suggests that Dawson understands the scientific knowledge in question—the "system" advertised on the title page—as having a kind of freestanding, detached ontology.

In this regard, Dawson and the booksellers identify and seek to exploit the duality that Clifford Siskin has argued is at the heart of the concept of a system, which he explains can at once refer to an organized and complete relation of concepts as well as to a materially instantiated network of bodies, forces, causes, actions, and so on that has a real existence in the world. It is as if, at a critical period in the emergence of the system as a genre, they understand and perversely seek to profit from the fact that a "system," as opposed to a course or a lecture, is both "a thing in the world *and* . . . a way of constituting the world as a thing."[24] In this line of thinking, Desaguliers might have a proprietary relationship to the experiments he "showed," but that is not the same as owning the knowledge that the experiments "proved"—a "system." As such, the "system" marketed by the booksellers locates its value for the reader in its putative completeness, comprehensiveness, and scalability. That is, this was no partial, essayistic collection of detached lectures offering some insights into limited aspects of natural philosophy but instead an ostensibly comprehensive account of all natural philosophy "prov'd" by the experiments in Newtonian mechanics. When the booksellers assert that they are publishing a "system of experimental philosophy" that "contains the experiments: which are fully prov'd by the plates," they are signaling that, by virtue of it being a system, their book is something that Desaguliers couldn't assert personal ownership of.

In fact, the question of the ownership of this knowledge is vexed in multiple ways. After all, as Desaguliers takes pains to explain in his newspaper notice, "a great Part of it was never mine." I take this to be referring to the fact that the lectures he had been performing were, in a very real sense, not entirely his to begin with, which Desaguliers explains in the preface that he affixed to the January 1719 reissue of that volume. Notably, that imprint replaced the original title page that promoted it as a "system" with a new title page that called the volume simply *Lectures of Experimental Philosophy*.[25] Back at Oxford, Desaguliers relates, Keill had been pressed by his students "to give them something in Writing upon that Subject," so "he wrote a few Papers to serve them as Memorandums." Desaguliers recalled that when he first began to lecture on experimental natural philosophy, "[he] endeavoured to get Dr. Keill's Lectures, (as they were called) which when they were brought [to him, he] found altered according to the Fancy and Number of the Transcribers."[26] Desaguliers corrected and restored Keill's work as well as added new experiments and propositions of his own design that doubled the length of the material. In light of this, it is striking how similar Desaguliers's method was to Dawson's. Dawson, like Desaguliers, had solicited the master's memoranda: Desaguliers says he gave him a copy of his lecture notes "that they might be of service to him when he went thro' [his] Courses." While he didn't set up his own course of lectures like Desaguliers did, Dawson did appropriate them for their own gain by editing them into a treatise that he then sold to some booksellers.

From the bookseller's perspective, they dealt for the manuscript fair and square. Yet we can see them at pains to counter the bad publicity that an author's disavowal of a scientific text might involve—which would have been a particular concern for two young booksellers who were trying to carve out a niche in the market. Things must have been especially delicate insofar as they had invoked the name of Richard Steele, who was patron to both Desaguliers and Dawson. In Desaguliers's preface, he explains that Steele was in fact the very reason that Dawson was his pupil in the first place: he identifies Dawson as "a young man whom Sir Richard Steele had put in [his] care." And Dawson, for his part, lavishly praises Steele in his dedication for being "so remarkably Generous" for "[his] continual Care of [him] from [his] Infant Years"—a curiously intimate declaration that would seem to provide grounds for conjecture that Dawson was Steele's illegitimate child.[27] In some cases, of course, a dispute over unauthorized publication could actually help the sale of a book. In genres such as secret history, erotica, libertine literature, and familiar letters, a bit of scandal was a good thing, and the louder the clamor over its supposed inauthenticity, the more the reading public might assume it had the real thing in its hands. However, the case of scientific publication is very different: a "system" of knowledge might be out there in public lectures for any and all to (pay to) hear, but it would be reasonable to assume that the further one got from the source, the less reliable the text would be. And indeed, this was one of Desaguliers's chief concerns: in return for owning the edition and receiving his own separate payment for the copy, Desaguliers agreed to correct the text's many errors. His errata list covers two pages in eight-point type; there are errors, frequently multiple, on 87 of the book's 201 pages.

But there is another issue at stake in the booksellers' defensiveness. When they assert that the experiments Desaguliers performed in his lectures "are fully prov'd by the Plates" and that the edition contains "the Experiments in another manner," the booksellers are attempting to provide a legal defense for printing a manuscript created by one man (Dawson) based on material created by another man (Desaguliers). They are in fact implying that the book edited by Dawson is a fully legitimate *imitative adaptation* of Desaguliers's literary property. As Simon Stern has explained, while outright piracies of entire books were expressly prohibited by Stationer's Company work rules and statutory law throughout the long eighteenth century, the various legal instruments that were put in place throughout the period—whether we are talking about the 1662 licensing act, the various stopgap measures put in place after its expiry in 1692, the 1710 Statute of Anne, or the 1774 copyright ruling in the House of Lords—nevertheless permitted summaries, adaptations, imitations, parodies, epitomes, anthologies, condensations, indexes, and so forth. All of these forms of imitative paraphrase were considered fair game and a public good insofar as they were believed to keep up a lively traffic of knowledge in the public domain by making new ideas available to as wide a public as possible, particularly for those readers who, for whatever reason, would not or could not read the original text at full length.[28] By saying that the book that they published contains Desaguliers's experiments "in another manner," the booksellers are literally wrapping themselves in copyright law to prevent prosecution for piracy.

Of course, there is a crucial consideration that the booksellers gloss over. The law as it was written and enforced allowed for the imitative replication of any *originally published work*. According to the existing regime of copyright law and Stationers' Company work rules, Creake and Sackfield might be free to paraphrase Desaguliers's lectures so long as he had originally published them. On this point, the booksellers' case is truly fascinating. Since Dawson claimed to transcribe what Desaguliers was publishing in viva voce to his auditors, the closest analog to this case would be the unauthorized publication of a play text. Much like a play, Desaguliers's lectures were performed publicly in voice and action, although most of the "taking" plays in the period appeared in printed editions after a successful run, which of course would directly satisfy the original publication requirement. But what of a play that was never printed yet was performed so often and to such acclaim that someone would go to considerable trouble and expense to assemble and print an unauthorized edition? It appears that eighteenth-century copyright law was just ambiguous enough that one *could* make the case that public performance itself could be considered a kind of original publication. In fact, there was one such dispute later in the period where the actor and playwright Charles Macklin brought legal action against various parties who had published, or had attempted to perform without his permission, his popular 1759 farce *Love à la Mode*.[29]

Macklin jealously guarded his text insofar as he was able to make much more money by performing in the play himself year after year than if he were to print it and take a one-time copyright payment, which would have opened the floodgates for any strolling troupe to stage their own version. This proved to be a very lucrative scenario for Macklin—personal performances of this play are believed to be his chief source of income for a full thirty years until he retired from the stage.[30] However, this state of affairs required constant vigilance and extraordinary secrecy—Macklin was said to have brought the actors' sides to rehearsal "buttoned up . . . in the breast of his great coat" and then collected them from the prompter afterward to take home with him.[31] It also involved frequent legal action. That's precisely what happened when an unauthorized transcription of the first act of the farce was published in the *Court Miscellany* in 1766. The periodical hired the stenographer and Old Bailey court reporter John Gurney to take down the dialogue in shorthand, and it promised to print the second half of the farce by the same means soon after.[32] No second act appeared, however, and by 1770, a lawsuit brought by Macklin against the newspaper publisher finally wound up in Chancery. In *Macklin v. Richardson*, the magazine publishers argued that their case was different from earlier precedents where authors had their manuscripts stolen or copied from them "by reason of the representation of the farce upon the stage, which gave a right to any of the audience to carry away what they could, and make any use of it."[33]

In fact, this argument did not persuade, with the justices flatly determining, "It has been argued to be a publication, by being acted; and therefore the printing is to be no injury to the plaintiff: but that is a mistake. . . . The printing it before the author has, is doing him a great injury."[34] Macklin may have won his suit on these grounds, but the court did also take notice of the fact that it was "not an abridgement, but the work itself," leaving open the question of whether an adaptation or abridgment might have been permissible—that is, whether the *Court Miscellany* would have been in the clear if they had published the play "in another manner," as Creake and Sackfield described their efforts. As Stern tells us, the likelihood of getting prosecuted for a legitimate abridgment was almost zero, and the bar for what might count as an abridgment was set surprisingly low: in the whole of the long eighteenth century, Stern has found only a single case where a defendant "did not successfully defend his publication by characterizing it as an abridgement" when the work in question was even modestly abridged.[35] Creake and Sackfield could have plausibly argued their case on these grounds—perhaps enhanced with the additional pleading that consolidating the concepts conveyed in scientific lectures into a "system" is different from taking down the scripted lines of a play text—but in light of the way in which the courts ruled in *Macklin v. Richardson,* they may very well have failed for the same reason that the magazine did: "Printing it before the author has, is doing him a great injury." Desaguliers never published his experiments in the first place.

Or did he? Here we need to attend closely to the even more radical claim that is implicit in the booksellers' advertisement, where they assert that the experiments in question were "fully prov'd by the plates." After all, it was the illustrations for—and not the text of—the lecture that Desaguliers did in fact publish before the 1718 "system" was ventured. Judging from the copies of these illustrations that are held in the Oxford University Libraries, Desaguliers printed a set of illustrated plates for his lectures in "Mechanicks" (six plates), "Opticks" (five plates), "Hydrostaticks" (three plates), and "Pneumaticks" (six plates), where each set of illustrations was printed facing a page of figure-by-figure "explications."[36] A key element of the booksellers' position is that these plates constituted an original publication of Desaguliers's system—there was not merely its verbal and embodied performance of the lectures but also its self-sufficient and fully comprehensive visual enactment in the diagrams. But how well does that claim hold up? Consider, for example, illustration 14 on the fourth plate of the *System* (see fig. 1), which shows a disembodied human head, floating in a white void space, with a shaft or tube extending out of the mouth and into a rectangular box or basin. All of this is plain enough to see. But what does it *mean?*

In fact, this image cries out for what John Bender and Michael Marrinan, in *The Culture of Diagram,* call correlation. The diagram is the paradigmatic

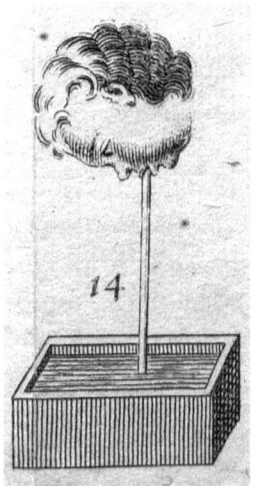

FIGURE 1. Suction illustration from Desaguliers' unauthorized 1719 System. From John Theophilus Desaguliers, *A System of Experimental Philosophy, Explained by Mechanicks* (London, 1719), plate 4, figure 14. (The Library Company of Philadelphia)

Enlightenment representational form, they explain, a kind of "working object," in the sense that Lorraine Daston and Peter Galison theorize, neither a natural object nor a concept or theory but rather a visual object "from which concepts are formed and to which they are applied."[37] According to Bender and Marrinan, "Closer to being things than representations of things . . . diagrams incite a correlation of sensory data with the mental schema of lived experience that emulates the way we explore objects in the world."[38] Not simply "abstractions of reality" or "stripped down versions of the world of experience," a diagram is a thing in itself that calls for a "composite play of imagery and cognition that is the motor-energy of diagram."[39] No single meaning resides within it; instead, a diagram opens itself up to multiple points of view and requires the observer to move back and forth between image, text, and world to produce meaning.

In the case of this "thing in itself," we clearly do not know enough about it to produce a meaningful correlation. Whose head is this? What is it doing? Is that a box of soil, and has the head been planted there, as if on a beanstalk? Are we looking at a stake impaling the head through the mouth? Is the head vomiting or taking nourishment? I think it is safe to say that while this catalog of conjectures is absurd, it is hard to imagine a way of reading this image in isolation that isn't eccentric. The figure appears jumbled together with fourteen other figures on the fourth of ten plates affixed to the back of this collection of lectures, which only compounds its inscrutability. If we look at the source of this image, the original version that appeared in Desaguliers's plates is no more epistemologically secure. Of course, one can immediately see that the original image, which appears as figure 10 in the first plate of "Pneumaticks," is much more finely rendered (see fig. 2).

Indeed, the original set of illustrations, signed by the engraver Sutton Nicholls, exhibits more precision and detail, and they are judiciously laid out with ample space on the page, ensuring that the illustrations for experiments on the same topic from the same lecture appear alongside each other in a logical sequence. By contrast, the copyist who assembled the *System*'s ten plates rendered the images far more crudely, and often confusingly, appearing to have jammed them onto the plates willy-nilly, irrespective of the order in which they are referenced in the text, wherever they might serve to save space and expense. One can see this quite clearly in the fate of the original illustration that appears directly above figure 10, a schematic drawing of a wheel thermometer. In Desaguliers's original plate (fig. 3), one can clearly make out all the critical elements of the mechanism: the tube of mercury with a weight floating on its surface, the string that connects it by pulley to another weight hanging freely, and the needle turning on the pulley that then shows the changes in barometric

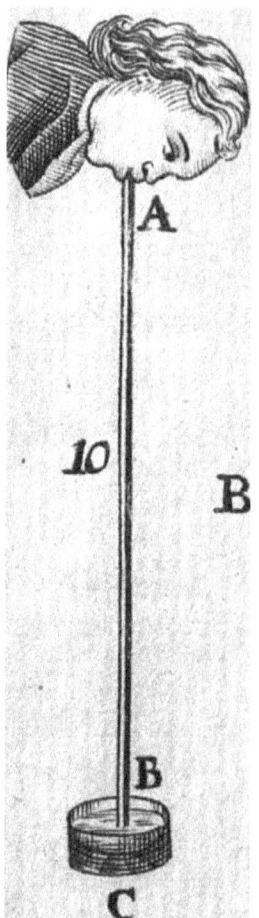

FIGURE 2. Suction illustration from the plates for Desaguliers's lecture courses. From John Theophilus Desaguliers, *Plates of Experimental Philosophy*, "Pneumatics," plate 1, figure 10. (Bodleian Libraries, University of Oxford, Lister L 77, Creative Commons license CC-BY-NC 4.0)

pressure on a dial face. When this image is reproduced in the *System* (fig. 4), not only is it no longer grouped together with all the other pneumatic experiments, but the crudely drawn mechanism is indecipherable.[40]

These mysteries and absurdities surrounding the heads on tubes are scarcely resolved when we read the explanatory texts that are keyed to these images. When one looks at the explication printed with Desaguliers's original plate, we learn that figure 10 "is an Example of Suction; and will shew that Quicksilver can thereby never be rais'd to 29 1/2 inches."[41] If one uses this image as Desaguliers intended, as an aide-mémoire for those who attended his lecture, one could quite plausibly point to this figure and, in an act of correlation, call to mind the experience of watching this experiment and so remember the matter of fact that there was a limit to how high the mercury could be sucked up the tube. However, that knowledge doesn't reside self-sufficiently

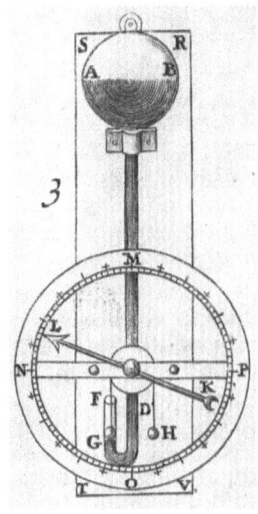

FIGURE 3. Barometer illustration from the plates for Desaguliers's lecture courses. From John Theophilus Desaguliers, *Plates of Experimental Philosophy*, "Pneumatics," plate 1, figure 3. (Bodleian Libraries, University of Oxford, Lister L 77, Creative Commons license CC-BY-NC 4.0)

in the figure. Moreover, when this image is translated to the *System*, when one knows what this figure was intended to represent, it is striking how much less indicative the image is of the core matter of fact. Rough measurements of the width of the head and the length of the tube seem to indicate that the tube as pictured couldn't possibly be larger than twenty-nine inches when considered in proportion to the size of the head. What's more, according to the experimental report that accompanies this figure in the *System*, "Proposition XVIII. *To shew that the Ascent of Fluids in Tubes after Suctions, arises from the Pressure of the Air,*" this figure now supposedly refers not to the suction of mercury but rather to that of water. The reader learns that

> when a Man, by the Muscles of his Breast enlarges the Cavity of the Thorax, then the external Air finding room wherein to expand itself, rushes in at his Mouth into his Lungs; so that if one Orifice of a [T]ube be in his Mouth, and the other immersed in Water, then that Part of the *Superficies* of the Water, which is under the Tube, is free from Pressure; and since the other Parts of the Superficies of the Water are prest by the super-incumbent Weight of the external Air, it must needs be . . . that the Water will ascend up the Tube.[42]

So it is an illustration of an experiment proving the real source of suction in a straw. Of course, it is all so obvious now.

Or is it? In fact, looking at the illustration while hearing the description of the experiment makes it clear how very little this particular diagram actually represents. Where is the expanded chest cavity? In what manner must you

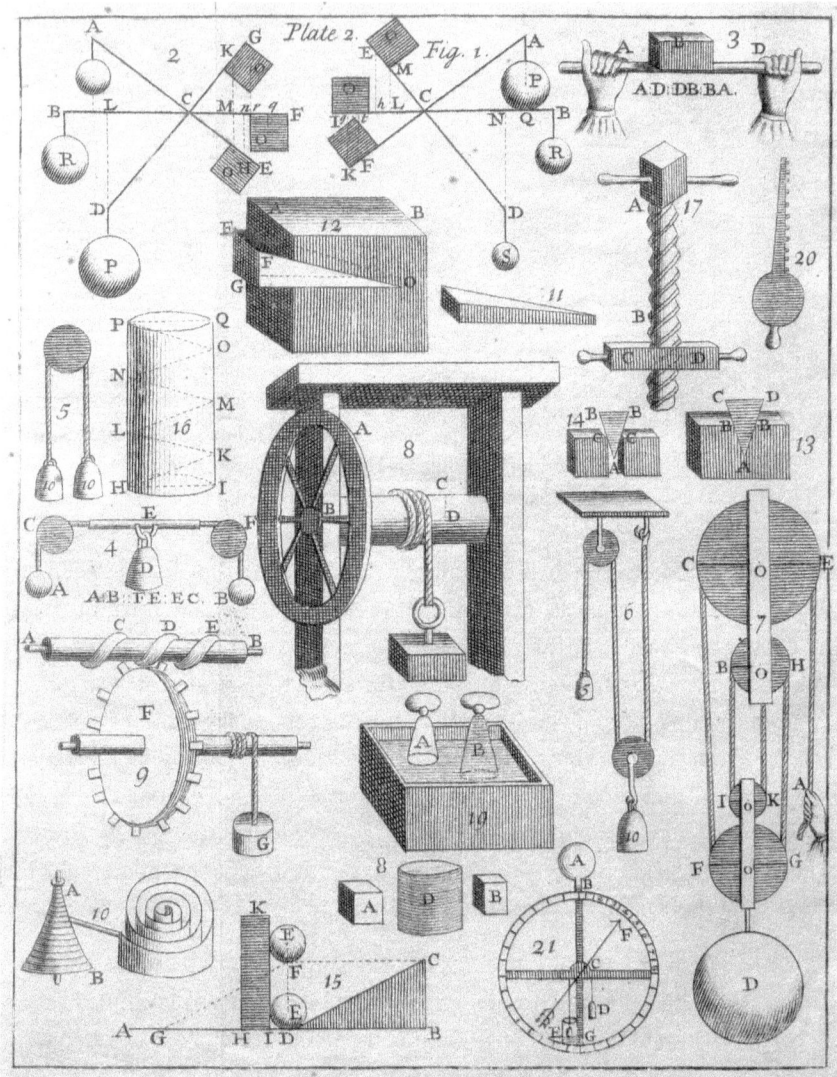

FIGURE 4. Barometer illustration from Desaguliers's unauthorized 1719 *System*. From John Theophilus Desaguliers, *A System of Experimental Philosophy, Explained by Mechanicks* (London, 1719). *Mechanicks* (London, 1719). Plate 2, figure 21 (bottom right), shows the crudely drawn dial barometer. (The Library Company of Philadelphia)

look upon this image in order to actually see the weight of the air bearing down upon the surface of the water? What visual device shows the translation of that surface pressure to the water that ascends the tube? In fine, there is no such device; there are no such keys in this image. The head's fixed downward gaze, which in a teasing moment just now I suggested might be a death's mask, now takes on a more banal, but not any less unsettling, cast. That gaze that bears down onto those opaque lines that join to compose this diagram now strikes me as terrible in its futility. This disembodied head—scowling, transfixed—that invites the viewer to both try the experiment and decipher its diagrammatic representation also serves as an emblem for our inability to apprehend any sort of natural truth self-evidently represented in this diagram. That head is supposed to be a surrogate for the viewer's re-creation of the experiment, but it is more properly a mask for all of us who struggle to correlate what is pictured with what it could plausibly signify. And yet the image of this very head was said to be a perfect and self-sufficient vessel of systematic scientific knowledge in 1719. To hear Creake and Sackfield argue it, this picture truly showed the experiments, albeit in "another manner."

It might be easy to dismiss this claim about the self-sufficiency of the diagrams as a kind of red herring—a patently mercenary dodge on the part of booksellers who find themselves having to defend their right to publish this text at all. But it doesn't explain why they might have believed this to be a plausible argument in the first place. Nor does it address the fact that Desaguliers seems to have felt the need to rebut such a claim in his own preface. There, he said that he had only produced plates for his auditors because they "were at pains to transcribe the lectures," and these illustrations might "save [them] the Trouble of drawing the Figures."[43] But this would seem to patch over a more complex mediation on the value, use, and ontological status of scientific illustration. By the time Desaguliers was prepared to publish his official *Course of Experimental Philosophy* (1734/44), the thing that strikes the reader most forcefully is the sheer volume of footnotes. In this fully realized articulation of his life's work, Desaguliers employs a novel method in which the truths of natural philosophy are presented in a bipartite structure, where his lecture text employs a spectacular instructional method by experiments, even to the point of employing "such ways of demonstrating as are not mathematically true,"[44] while his extensive footnotes labor to correctly prove the propositions by mathematical method. This suggests an increased interest in—but also deep reservations about—the power of visual images to represent natural truth. In a sense, we can interpret Desaguliers's strategy in his final completed masterwork, the *Course of Experimental Philosophy*, as a culmination of his life's work as a *philosophical* performing artist, where striking experiments, described in clear, logical prose and depicted in unambiguous and self-evident images, are arrayed into a logical demonstration of Newtonian laws. These necessary concessions to the needs of nonspecialists are then carefully counterbalanced by a second explanation of those matters of fact in the *Course*'s technical and mathematical footnotes. Fundamentally, it was the market that compelled Desaguliers to furnish illustrations to accompany his demonstration lectures, illustrations that were then taken to be fully representative of the experiments themselves, which in turn were asserted to have formed a completed system. This triggered a kind of epistemic creep that Desaguliers appears to have spent his entire career attempting to negotiate and properly harness. Dawson, Creake, and Sackfield—and the readership that they intended to serve—may have desired completed "systems" and truths "fully prov'd by the Plates," but Desaguliers was too well acquainted with their limits to consent to allow his work to circulate under such false and exaggerated circumstances.

In closing, I want to make one more observation about the larger epistemic implications of the publication of Desaguliers's unauthorized *System*. We are accustomed, thanks to Simon Schaffer and Steven Shapin, to think about the production of scientific knowledge through a regime of witnessing where the performance and reperformance of experimental trials were held to produce the facts of empirical knowledge, both in person and virtually through the mediation of print publication.[45] Witnesses, both physically present and virtual, were needed to observe and affirm that the trials so exhibited/described/depicted did actually produce the matters of fact in question. In this regime of what Shapin and Schaffer pointedly call "modest witnessing," the experimenter modestly made no positive assertions about what was and wasn't a matter of fact outside of the demonstration of an experiment or the relation of an experience, much less what cause might explain those phenomena. Instead, it was left to the judgment of the witnesses, and this modesty also required the experimentalist to assume the role of a disinterested gentleman who disavows any financial interest in the outcome of the experimental trials. Rather, in the circle of modest witnessing that gathered around the performance of matters of fact, auditors and readers who sought entry into that space were required to ascribe to that self-same modesty and gentlemanly disinterest so that experimental trials, freely performed for qualified auditors and/or published in a prose style sufficiently detailed, furnished their witnesses with all the requisite information, and nothing more, to enable them to make informed judgments.

It is easy to see how the case of Desaguliers's unauthorized *System* highlights the pressure that model came under in the expanding marketplace of scientific knowledge in eighteenth-century England. In this regard, what I am arguing here complements what Aaron R. Hanlon describes elsewhere in this collection, where the full sweep of the "literary technology" of modest witnessing must embrace both the controlled and authorized representations of qualified practitioners and the remediations and contestations of those representations in other nonelite and nonspecialist discourses.[46] The case study I have been developing in this essay demonstrates that when experimentalism moved out of the closed community of gentlemen that controlled access to spaces like the meetings of the Royal Society, the experimental performances that were exhibited, and the forms of knowledge they could be understood to take, interacted in volatile ways with the forms of property those experiments were obliged to become. For it was the logic of print culture, as interpreted by a pair of booksellers who sought to creatively apply it in a bid for financial gain, that urged on the epistemic creep that made it conceivable to assert that Desaguliers's illustrations were self-evident and that a series of philosophical lectures could be marketed as a "system." A system, after all, was understood to be a comprehensive network of relations that didn't merely describe the workings of the natural world but actually *was* the workings of the natural world and so was fungible as a kind of property that could belong to no one and could not be subject to copyright, not even the celebrated experimentalist and Fellow of the Royal Society who was so famous for lecturing about it.

Notes

1. John T. Desaguliers, *A System of Experimental Philosophy, Prov'd by Mechanicks* (London: Printed for B. Creake at the Bible and Ink-Bottle in Jermyn-Street St. James, J. Sackfield in Lincolns-Inn-Square, and Sold by W. Mears at the Lamb without Temple-Bar, 1719). The title page suggests that William Mears was only contracted to resell this edition, which was owned and brought out by Bezaleel Creake and John Sackfield, who performed the role of booksellers. The pair placed advertisements announcing that the book was "This Day is Publish'd" in the December 2–4, 1718, editions of the thrice-weekly *Post Boy* and *Evening Post* as well as in the November 29–December 6, 1718, edition of the *Weekly Packet*.
2. *Post Boy*, December 4–6, 1718.
3. Simon Stern, "Copyright, Originality and the Public Domain in Eighteenth-Century England," in *Originality and Intellectual Property in the French and English Enlightenment*, ed. Reginald McGinnis (London: Routledge, 2008), 69–101. Cf. also Adrian Johns, *Piracy: The Intellectual Property Wars from Gutenberg to Gates* (Chicago: University of Chicago Press, 2009), chap. 6.
4. A copy of the illustrations in question is housed in the Christ Church Library, University of Oxford, bound with twelve other items in a volume with the shelfmark A 156: J. T. Desaguliers and Sutton Nicholls, engraver, *Mechanicks. An explication of the first plate* (London?, 1719?). A second copy, formerly

owned by the naturalist Martin Lister, is held in the Bodleian Library, shelfmark Lister L 77, which bears the handwritten title *Plates of Experimental Philosophy*. This is the title I will prefer to use in this essay, as the library catalog's title only refers to the first set of plates; there are sets of plates for other sciences, like optics and hydrostatics, also included in the volume. In both of the Oxford Libraries copies, Desaguliers's twenty illustrations, engraved by Sutton Nicholls, appear without a printed title page in a series of twenty bifolium, with each plate appearing on the verso of the first leaf along with a printed description of its contents on the facing recto of the second leaf.

5. An abbreviated account of the unauthorized publication of Desaguliers's lectures appears in Adrian Johns, *The Nature of the Book: Print and Knowledge in the Making* (Chicago: Chicago University Press, 1998), 180–82, where it serves as an example of how natural philosophers were "particularly vulnerable to the practices of piracy because of the importance of printing in establishing their knowledge and techniques." Johns doesn't consider the extent to which *not* appearing in print—or appearing in print in only limited and highly controlled ways—was central to Desaguliers's professional and scientific objectives at this time. This episode is also mentioned in Audrey T. Carpenter, *John Theophilus Desaguliers: A Natural Philosopher, Engineer and Freemason in Newtonian England* (London: Continuum, 2011), 34–35. On the subject of early modern scientists' wariness about print publication, see Nicole Howard, *Loath to Print: The Reluctant Scientific Author, 1500–1750* (Baltimore: Johns Hopkins University Press, 2022), which does not discuss this episode.
6. *Post Boy*, December 4–6, 1718.
7. John T. Desaguliers, *Lectures of Experimental Philosophy* (London: Printed for W. Mears, at the Lamb without Temple-Bar; B. Creake, at the Bible in Jermyn-Street; and J. Sackfield, in Lincolns-Inn-Square, 1719).
8. John T. Desaguliers, *A Course of Mechanical and Experimental Philosophy; Whereby Any One, Altho' Unskill'd in Mathematical Sciences, May Be Able to Understand All Those Phænomena of Nature, Which Have Been Discover'd by Geometrical Principles, or Accounted For by Experiments; and Mathematicians May Be Diverted in Seeing Those Machines Us'd, and Physical Operations Perform'd, Concerning Which They Have Read. Given by John Theophilus Desaguliers, M.A. F.R.S. at His House* (London, 1715?).
9. Rita Felski, *The Limits of Critique* (Chicago: University of Chicago Press, 2015), 191.
10. Bruno Latour, "Why Has Critique Run out of Steam? From Matters of Fact to Matters of Concern," *Critical Inquiry* 30, no. 2 (2004): 225–48.
11. Patricia Fara, "Desaguliers, John Theophilus (1683–1744), Natural Philosopher and Engineer," in *Oxford Dictionary of National Biography*, September 23, 2004.
12. The *Weekly Journal or British Gazetteer*, September 14, 1717, includes a notice that Desaguliers was performing his lectures for the royal family at Hampton

Court Palace, and a sermon he preached before the king was published at this time. John T. Desaguliers, *A Sermon Preach'd before the King, at Hampton-Court; on Sunday, Sept. 29th, 1717* (London, 1717). In 1737, Desaguliers would initiate Prince Frederick into the Masonic lodge he helped found.

13. Larry Stewart, *The Rise of Public Science: Rhetoric, Technology, and Natural Philosophy in Newtonian Britain, 1660–1750* (Cambridge: Cambridge University Press, 1992), remains the best source for details on Desaguliers's remarkable career. See also Fara, "Desaguliers"; and Carpenter, *John Theophilus Desaguliers*.

14. Jacques Oznam, *Treatise of Fortification . . . Done into English, and Amended in Several Places by J. T. Desaguliers, Hart Hall, Oxon* (Oxford, 1711); Nicholas Gauger, *Fires Improv'd: Being a new method of building chimneys, so as to prevent their smoking . . . Made English and Improved by J. T. Desaguliers, M.A. F.R.S.* (London, 1715); Edmé Mariotte, *The Motion of Water, and Other Fluids. Being a Treatise of Hydrostaticks . . . Translated into English. Together with a little treatise of the same author, giving practical rules for fountains, or Jets d' Eau. By J.T. Desaguliers, M. A. F. R. S. Chaplain to the Right Honourable James Earl of Caernarvon. By whom are added, Several Annotations for Explaining the doubtful Places* (London, 1718).

15. The earliest record we have of what Desaguliers was presenting in his lectures is a brief four-page syllabus: John T. Desaguliers, *A Catalogue of the Experiments in Mr. Desaguliers's Course* (London, 1713?). This catalog indicates that subscriptions for the twenty-one-lecture course were a substantial two guineas. Virtually the same catalog was reprinted as John T. Desaguliers, *A Course of Mechanical and Experimental Philosophy* (London, 1715?), which now promoted his lectures as a course in "philosophy." By 1717, these promotional pamphlets, now being published in both English and French, swelled to eighty pages and included a handful of schematic diagrams interspersed in the main text. Desaguliers explained in his preface to the reader that these volumes contained the "minutes" of his lectures for use as "memorandums" for those who had completed the course and wished to recall the chief points: John T. Desaguliers, *Physico-mechanical Lectures. Or, an account of what is explain'd and demonstrated in the course of mechanical and experimental philosophy, given by J. T. Desaguliers, M. A. F. R. S. Wherein the Principles of Mechanics, Hydrostatics and Optics, are Demonstrated and Explain'd by a Great Number of Experiments. Design'd for the Use of All Such as Have Seen, or May See Courses of Experimental Philosophy* (London, 1717), A2; John T. Desaguliers, *Leçons physico-mechaniques, ou abrege du cours de philosophie, mechanique & experimentale de Jean Theophile Desaguliers, Maitre es Arts, & Membre de la Societe Royale* (London, 1717).

16. While there were small diagrams inserted into the 1715 publications, by 1718, Desaguliers had printed up a set of large illustrated plates for the use of his students when they went through his course. Cf. Desaguliers,

"Mechanicks"; and Desaguliers, *Plates of Experimental Philosophy*. The twenty plates in that series form the basis for the ten plates that appear in the copies of *A System of Experimental Philosophy* (1719) and *Lectures of Experimental Philosophy* (1719).

17. Desaguliers, *Lectures of Experimental Philosophy*.
18. John T. Desaguliers, *A Course of Experimental Philosophy*, 2 vols. (London, 1734 and 1744), 1:x.
19. Beginning in 1716, Bezaleel Creake's name appears as a bookseller on just fifteen titles through the end of 1719 aside from the various editions associated with the Desaguliers project. John Sackfield was somewhat more established, having been named as publisher on twenty-one titles between 1716 and 1719 exclusive of Desaguliers's texts. Sackfield and Creake start to appear as publishers together on a number of titles in 1719, most notably a similar work of natural philosophy, a translation of Herman Boerhaave's *Institutiones medicae* (Leiden, 1709) entitled *A Method of Studying Physick*. In the initial canceled edition of Desaguliers's *System*, the title page also notes that the book was to be "sold by" W. Mears. Mears, by contrast, had already been well established in the trade as a bookseller, with 112 titles to his credit in English Short Title Catalog dating back to 1707. When Desaguliers's text is reissued as *Lectures of Experimental Philosophy*, Mears is promoted to a "Printed for" credit and appears first in the list of booksellers, before Creake and Sackfield, suggesting that he played a pivotal role (perhaps providing the requisite capital) in satisfying Desaguliers's claims, making him payment, and gaining his participation in the publication.
20. *Post Boy*, December 2–4, 1718; *Evening Post*, December 2–4, 1718; *Weekly Packet*, November 29–December 6, 1718.
21. *Post Boy*, December 4–6, 1718.
22. Desaguliers, *Lectures of Experimental Philosophy*.
23. Desaguliers, *Lectures of Experimental Philosophy*.
24. Clifford Siskin, *System: The Shaping of Modern Knowledge* (Cambridge, MA: MIT Press, 2016), 24. On this subtle but critical distinction, Siskin offers the helpful example of the telephone system, which is "something both conceptual—a way to organize communications—and concrete—a physical wired or wireless network that can 'go down'" (20).
25. Desaguliers, *Lectures of Experimental Philosophy*. In the *Evening Post*, January 13–15, 1719, the booksellers advertised that the now renamed book was "Just published, corrected by the Author," adding that it was "All Carefully Examined and Corrected by Mr. Desaguliers." Advertisements for this edition appear through January and February in multiple periodicals.
26. Desaguliers, preface, *Lectures of Experimental Philosophy*.
27. Desaguliers, preface, and Dawson, dedication, *Lectures of Experimental Philosophy*. After being promoted to captain in the Coldstream Guards and coming to London as Commander Lord Cutt's personal secretary, around 1697 Steele

fathered a child by Jacob Tonson's unmarried daughter. Neither George A. Aitken's *Life of Richard Steele* (2 vol., London: W. Isbister, 1889) nor Calhoun Winton's 2004 entry on Steele in the *Oxford Dictionary of National Biography* mentions a second child out of wedlock. Catherine Dawson, identified as a widow in St. Martin's in the Fields, is a major creditor throughout Steele's life, and his letters betray some anxiety and frustration at being in her debt. Seven months after marrying his second wife, Mary Scurlock, Steele writes to Scurlock in April 1708, "I have sent Dawson thirty pounds, and will not rest till I have enough to discharge Her" (Aitken, vol. 1, p. 206). In 1714, she is owed £500 in a list of creditors drawn up around the time Steele gains the license for Drury Lane theater. While he would reassure his wife in a letter of February 1716/17 that he was working hard to pay down her debt—"every day, I do something towards this and next week shall pay off Madam Dawson"— Dawson appears as one of Steele's most important and urgent creditors when he retires to Wales and draws up a scheme to pay down his debts by signing over his proceeds from his share of the playhouse. Out of the £400 he had in hand to satisfy creditors before the contractual payments began, Dawson received £100, the largest sum among just six disbursements made at that time (Aitken, vol. 2, pp. 49, 12, 298–300). If Paul Dawson is her son by Steele, assuming he was born sometime between 1697 and 1705 when he married his first wife, Margaret Stretch, a widow with a Barbados fortune, Dawson would have been between thirteen and twenty-one when Desaguliers's *System* was published in December 1718. It might be relevant that Dawson's public declaration of Steele's "continual Care of [him] from [his] Infant Years" would have appeared in the same month as, and possibly only after, Steele's wife's death. It seems unlikely that Dawson would have embarked on this controversial plan to publish Desaguliers's lectures entirely on his own initiative. Did Steele himself set this scheme in motion as a means of partly satisfying his debt to Catherine by generating income for Paul through the monetizing of the intellectual property of a figure who, as another one of his clients, would have been beholden to him? It would be reckless to assert this based on the evidence at hand, but it seems foolish to ignore the possibility. While this would seem to have been done without Desaguliers's knowledge or consent, he did eventually receive payment from the booksellers—who in this case paid twice for the same copy—and by December 1719, Desaguliers would be offering his lectures in a high-profile course at the Censorium—that is, "At Sir Richard Steeles great Room in Villers-Street, York-Buildings"—while a second edition of the *Lectures* went to press (*Daily Post*, November 23, 1719; *Evening Post*, December 1–3, 1719). Cf. Calhoun Winton, "Steele, Sir Richard (bap. 1672, d. 1729), Writer and Politician," *Oxford Dictionary of National Biography*, September 23, 2004.

28. Stern, "Copyright," 71–76.
29. Stern, 78.

30. Richard R. Findlay, "Macklin's Acting Version of *Love à la Mode*," *Philological Quarterly* 45, no. 4 (1966): 751.
31. Findlay, 751, quoting John O'Keefe, *Reflections* (London, 1826), 2:315–16.
32. W. Matthews, "The Piracies of Macklin's *Love à la Mode*," *Review of English Studies* 10, no. 39 (1934): 315.
33. Charles Ambler, *Reports of Cases Argued and Determined in the High Court of Chancery* (London, 1790), 696.
34. Ambler, *Reports of . . . Chancery*, 696.
35. Four of these cases failed because the texts in question, like the *Court Miscellany*'s farce, made no effort whatsoever to abridge the original text, while the fifth case we do not know the issue of: a rogue publisher brought out a new edition of *Paradise Lost*, which added but 28 notes to the original publisher's 1,500. There, publication was enjoined pending a hearing on "whether the alterations make it a new work,'" which so far as we know never happened. Stern, "Copyright," 78.
36. Desaguliers, *Mechanicks*; Desaguliers, *Plates of Experimental Philosophy*.
37. John Bender and Michael Marrinan, *The Culture of Diagram* (Palo Alto, CA: Stanford University Press, 2010), 33; Lorraine Daston and Peter Galison, *Objectivity* (New York: Zone Books, 2007), 19.
38. Bender and Marrinan, *Culture of Diagram*, 21.
39. Bender and Marrinan, 19, 23.
40. Desaguliers, *Plates of Experimental Philosophy*, "Plate 1 Pneumatics," figure 3; Desaguliers, *System of Experimental Philosophy*, plate 2, figure 21. The head and tube image, "Plate 1 Pneumatics," figure 10 in *Plates*, appears on plate 4, figure 14, of the *System*.
41. Desaguliers, *Plates of Experimental Philosophy*, 15.
42. Desaguliers, *System of Experimental Philosophy*, 199–200.
43. Desaguliers, *Lectures of Experimental Philosophy*.
44. Desaguliers, *Course of Experimental Philosophy*, ix.
45. Steven Shapin and Simon Schaffer, *Leviathan and the Air-Pump: Hobbes, Boyle, and the Experimental Life* (Princeton: Princeton University Press, 1985).
46. See Aaron R. Hanlon, "Literary Technologies of the Sextant in Eighteenth-Century Britain," in this volume.

PART III

Embodiment

Plants, Principles, Strata

John Woodward's Improbable Corpuscles

HELEN THOMPSON

Transmuted Plants and Empirical Science

In the midst of his anti-alchemical manifesto *The Sceptical Chymist* (1661), Robert Boyle endorses the following account of vegetable growth: "The main Body of the Plant consisted of Transmuted Water."[1] This is a surprising statement to encounter in Boyle's polemic against alchemists' imprecise, even fanciful representational practices, which are licensed, Boyle complains, by the "unreasonable Liberty they give themselves of playing with Names at pleasure."[2] On the topic of vegetation, *Sceptical Chymist* cites the chymical innovator Joan Baptista Van Helmont's famous willow tree experiment—a young tree and oven-dried soil are watered and weighed for five years; the tree gains hundreds of pounds while the soil loses "about two ounces"[3]—whose proof that water comprises the tree's sole source of nourishment implicates not the contemporary referent H_2O but "water" as a vibrantly, even divinely impregnated, composite and cosmogenic body. To stress his approval of Helmont's experiment (which Boyle replicates with "Squash, which is an Indian kind of Pompion"), Boyle invokes the grown willow tree as "so notable a Quantity of Transmuted Water."[4] Clearly, Boyle's indictment of alchemy's unreasonable liberty of naming does not extend to the stuff of vegetables. Plants that occupy *Sceptical Chymist* as transmuted water signal the split status of transmutation in Restoration-era science and beyond: the official disrepute of gold-making or *chrysopoeia* belies the broader experimental and explanatory salience of what might be called the transmutation concept, a matter-based logic of radical ontological change.[5] One of transmutation's epitomes, in this latter sense, was plant germination and growth; another was human digestion.

In 1699, the geologist, fossil collector, antiquarian, Gresham College professor of physick, and member of the Royal Society and College of Physicians John Woodward (1665 or 1668–1728) published experimental results in the Royal Society's *Philosophical Transactions* that disproved the transmutation of water into plant. Woodward intuited both that plants transpire—taking in, then emitting water—and that they are nourished by particles of soil conveyed in that fluid vehicle. In what follows, I explore the controversy elicited by Woodward's efforts to mobilize particles as agents of geological and medical change in *An Essay toward a Natural History of the Earth* (1695) and *The State of Physick* (1718). For Woodward and his critics, inference about imperceptible particulate causes was coextensive with debate over observed data. Rather than disputing the centrality of micromatter to geological and physiological causation, Woodward's critics lambasted Woodward's often implausible representations of it. To anticipate what the Woodward pamphlet wars tell us about the inference of imperceptible causes and their representation in empirical natural philosophy, I recap the mirror-knowledge paradigm claimed for Restoration-era science by Steven Shapin and Simon Schaffer and then Pamela H. Smith's divergent account of embodied or artisanal epistemology before returning to Woodward's micromaterial argument that plants transpire.

Evoking the descriptive intricacy of Boyle's writing style, Shapin and Schaffer remark, "It assured the reader that . . . he was in fact being faithful to reality. Complex and circumstantial accounts were to be taken as undistorted mirrors of complex experimental outcomes."[6] As the authors note, it is Richard Rorty who theorizes this epistemological figure for "undistorted" mimesis, ascribed to René Descartes and John Locke, whose guarantee of authentic observational knowledge pivots upon the capacity of "the mind as a great mirror, containing various representations—some accurate, some not."[7] Of course, Thomas Sprat's *History of the Royal Society* (1667) affords the much-cited metaphor that the *History*'s own disclosure of chymical trade secrets proves unable to vindicate: "The mind of Man is a Glass, which is able to represent to it self, all the Works of Nature."[8] Such "mirror-imagery," also evoked by Rorty as "the original sin of epistemology,"[9] defines hegemonic "knowledge as a set of immaterial representations" or "accuracy of representation."[10] Transposed into the history of experimental science, Rorty's genealogy of mimetic mirror-knowledge compels Shapin and Schaffer to oppose Thomas Hobbes, for whom epistemological certainty is "a function of convention," and Boyle, an exemplar of "the intuitions of the empiricist . . . [who] regards the man-made component of knowledge as a distortion of the mind's mirroring of reality."[11] For those who adopt mirror-knowledge as the schema for knowledge of the kind promoted in the *Philosophical Transactions,* a constructivist critique thereby comes baked in. Shapin and Schaffer's constructivist twist on Rorty's elision of empiricism and mirror-knowledge entails exposure of the social (or, unmentioned by Shapin and Schaffer, extractive or imperialist) determinants of truth occluded by the illusion of its mirroring. Yet Shapin and Schaffer themselves repress the failure of Royal Society science to uphold the mimetic mirror scheme, whether through aggressive ideological fealty, overtly structuring extractive and mercantile commitments, or, most germane here, explanatory recourse to micromatter that eludes mimetic reflection. Shapin and Schaffer's constructivist matter of fact thus enshrines a tendentiously literalist mimetic schema of scientific understanding and experimental insight. As Boyle's account of vegetable growth already suggests, the mirror model does not capture the representational self-reflexivity of scientific debate, which explicitly grappled with—that is, was not in constructivist denial of—inductive and figural complexities precipitated by the causal powers of micromatter.

On a historically and methodologically divergent track, Pamela H. Smith traces the development of early modern European experimental and artisanal practice through the sixteenth-century alchemist and doctor Paracelsus. This corporeal epistemology, achieved through sustained "bodily experience of the particulars of matter,"[12] explicitly rejects visual idealization of the encounter with nature: "Paracelsus defined *scientia* . . . as the divine power in natural things that the physician must 'overhear' and with which he must achieve union. . . . Observation with the eyes was not enough, for 'seeing is only like a peasant sees when he looks at a psalter. He sees only the letters, and cannot say anything more about it.'"[13] Rather than mimetic mirroring, denial of social determination, and referential transparency, Smith's Paracelsian derivation of embodied knowledge posits representation as indissolubly physical testimony to the knower-maker's dynamic struggle with matter: "This imitation of nature produces an effect—a work of art—that displays the artisan's knowledge of nature and in itself constitutes a kind of knowledge. The background to all these claims was the conviction that knowledge is active and knowing is doing."[14] While Woodward's formidable skill as a collector and analyst of geological samples (or, as Sean Silver puts it, a "rock hound"[15]) defines Woodward as an embodied knower in Smith's sense, my argument in this essay, advanced against the mirror paradigm, concerns the battle of Woodward and his critics over micromatter that cannot, in Paracelsus's words, be seen with the eyes.[16] As the fierceness of their conflict will show, this is no denial of the representational constituents of experimental truth but, on the contrary, open antagonism over the logic and stakes of referential claims for material causes situated far outside the domain of Rorty's literalist mimesis.

Despite Francis Bacon's avowal that "every natural action is transacted by means of the smallest particles, or at least by things too small to make an impression on the senses,"[17] many present-day appraisals of natural philosophy continue to take mimetic mirroring as the outer limit of empirical representation and causation. To gesture beyond the mirror scheme, I employ the term *corpuscular insight* to link perceived phenomena and the micromatter that brings them about. The provoking concern in the Woodward controversy, again, is not whether geological or medical effects are transacted, in Bacon's words, by imperceptible particles but how their discursive *representation* may be more or less plausible. Woodward's deployments of particles trigger the question, If probabilistic standards apply to imperceptible corpuscular causes, then what, if not mimetic resemblance, constitutes such probability?

Woodward's "Some Thoughts and Experiments concerning Vegetation" debunks Boyle's contention that plants grow from transmuted water by querying the transparency of the liquid they take in:

> *Earth* is an *opake Body;* but it may be so far dissolved, reduced to so extreme small Particles, and these so *diffused* through the *watery Mass,* as not sensibly to impede *vision,* or render the Water much the less *diaphanous.* Silver is an *Opake,* and indeed a very *dense Body;* and yet, if perfectly dissolved in *Sp.* of *Nitre* . . . it does not *darken* the *Menstruum,* or render it less *pellucid* than before. . . . So that were there *Water* any where found so *pure,* that the quickest Eye could discover in it no *terrestrial intermixture; that* would be [so] far short of a *Proof,* that in reality there was *none.*[18]

Woodward cites an *experimentum crucis* well known to Boyle: enabled by the discovery that mineral acids like "*Sp. of Nitre*" separate silver into minuscule parts or corpuscles, the experimenter proved the metallic integrity of these *minima* when a base added to the acid solution precipitated out the original silver.[19] Woodward displays a sophisticated awareness of the repercussions for observational method when he evokes the dissolution of solid bodies like silver and earth into the pellucidity of apparently "*pure*" liquid. Because "the quickest Eye" cannot discern suspended particles, perfectly clear fluid constitutes no "*Proof*" of their absence. Woodward's own plant experiment, which commenced in 1691, elegantly supports his claim for water's indetectable particulate intermixture: he filled identical covered vials with equal volumes of water from various sources, inserted cuttings (mostly sprigs of mint), watered as needed, and measured the weights of vegetable growth and fluid consumed. The disproportionately small "Encrease of the Plant to the Expence of the Water"—at its starkest, a ratio of 1 to 1,052 grams in the case of *"Hyde-Park Water"* infused with *"Garden Mould"*[20]—led to the sole Woodwardian hypothesis deemed true today: *"The much greatest part of the Fluid Mass that is thus drawn off and convey'd into the Plants, does not settle or abide there: but passes through the Pores of them, and exhales up into the Atmosphere."*[21]

By showing that imperceptible particles of soil nourish his mint, Woodward refutes Boyle's claim that plants consist of transmuted water. Advancing an account of water's micromaterial texture from which panspermic powers are banished, Woodward argues that it is structured as a matrix of geometrically spherical *minima* whose empty spaces bear *"terrestrial Matter"*: *"perfectly dissolved, and reduced to single Corpuscles,"* waterborne soil particles thus "enter the *Tubules* and Vessels of Plants."[22] However, the asterisk Woodward attaches to the *reductio* experiment he recapitulates above (dissolved silver *"does not darken* the *Menstruum,* or render it less *pellucid* than before*"*) vexes his attempt to mechanize the nutritive transfer between earth and plant. Stipulating that "the *Silver* be pure *and absolutely* refin'd,"[23] this asterisked caveat cannot be applicable to earth, especially earth recruited to nourish a host of vegetative species, as Woodward is acutely aware: "It is not possible to imagine how *one, uniform, homogeneous Matter,* having its *Principles* or *Original Parts* all of the same Substance . . . should ever constitute Bodies so egregiously *unlike,* in all those respects as *Vegetables of different kinds* are."[24] Here Woodward articulates a conundrum he tries to resolve with microrefinements of the claim that *"Each* [kind] . . . takes forth *that peculiar Matter* that is *proper* for *its own Nourishment."*[25] Woodward thus delimits the tautology or *mis-en-abîme* that will also define his efforts to reduce geological phenomena to mechanistic *minima* and motion: If terrestrial matter enters the plant in the form of "single *Corpuscles"* or particles, how is each earthy *minimum* simultaneously distinguished as a particular plant's *"peculiar Matter"*?

The promise of micromaterial structure or texture to deliver a nonessential rationale for the empirical distinctiveness of things—that is, "to reduce the qualitative multiplicity of the phenomenal world to a tightly delimited set of structural factors"[26]—sits in tension with the insinuation of innate identity that populates Woodward's arsenal of corpuscular principles with the peculiar matter of every known kind of plant. This tension is amplified by Woodward's adoption of the alchemical keyword *"Principles"* (*"or Original Parts"*): central to Boyle's indictment of chymists' referential sloppiness, this signifier designates the two or three chymical substances (mercury, sulfur, sometimes salt) from which all metals are made. Despite his familiarity with *Sceptical Chymist,*

Woodward is far from observing its call to linguistic rigor: Woodward's principles smuggle *"Substance"* (every plant's peculiar matter) into a mechanistic scheme he touts for its power to replicate planetary cosmogony in *An Essay toward a Natural History of the Earth* and to reduce the etiology of human illness to bilious failures of digestion in *The State of Physick*.

Writing of the generative role of "seminal tinctures" in the Paracelsian physician Petrus Severinus's *Idea Medicinae Philosophicae* (1571), Jole Shackelford asserts their congruity with the seemingly discrepant agency of mechanism: "'Mechanical' in Severinus' terminology is less related to the modern concept of machinery than it is to the person who repairs that machinery This 'mechanic' is an active, vital agent who carries out a process."[27] In reference to Woodward's insistence on the solely geometric qualities of corpuscles, I do not employ the words mechanism and mechanical in Shackelford's Paracelsian sense, even though its significance to Restoration-era science persists through Boyle's defense of plant transmutation and beyond. In a footnote explaining early modern conceptions of the genetic agency that Woodward's particles lack, Shackelford affirms the hermeneutic utility of anachronism: "One is tempted, for the sake of illustration, to liken *semina* and tinctures to DNA and RNA: The *semina* possess the knowledge necessary for their expression as a body, whereas the tinctures may transfer that knowledge to another body."[28] By contrast, Woodward's commitment to a sharply impoverished repertoire of corpuscular-scale knower-makers deprives him of any analog for "DNA and RNA" except the substance preemptively incarnated by his principles, which defer the agency of biological and geological making to corpuscular parts already made of the bodies they will make. Woodward's *"Original Parts"* are not *minima* restricted to three chymical types and abetted by the supplemental powers of *semina* and tinctures but minuscule samples of every species of body in the world.

Rather than the corpuscular paradigm or the transmutation concept, it is Woodward's reliance on species-specific principles in *Natural History of the Earth* and *State of Physick* that elicits the core scientific objection posed by the physicians John Arbuthnot, John Freind, and Richard Mead and the Scriblerian trio Arbuthnot, John Gay, and Alexander Pope, among others. Their escalating pamphlet war, accompanied by dramatic satires of Woodward, testifies to the explanatory force of corpuscles whose ambivalent capacity to entrench and transmit substance Woodward flouts, his opponents allege, to the point of improbable absurdity.[29]

Woodward's *Reductio* Flood: Deluge, Dissolution, and Specific Gravity

Woodward relies on the silver experiment, known as *reductio in pristinum statum,* to theorize the empirical nuance of transpiration: it is corpuscles of earth hidden in clear water that really nourish plants. Woodward's recourse to the *reductio* proof anticipates its centrality to his major geological intervention, a corpuscular account of the biblical flood and its stratigraphic aftermath.[30] The chymical dissolution of obdurate bodies into imperceptible solution inspired Woodward, a gifted geologist, to elucidate the puzzling anomaly of fossil shells, which were often found far from natural sources of water, interlarded with or composed of stone, metals, or minerals, or instantiating unknown kinds. Denying that such fossils are *lusus naturae* (ludic excrescences of nature) or organic mineral growths and countering Thomas Burnet's thesis, in *Sacred Theory of the Earth* (1681), that the terraqueous globe was reshaped by the flood, Woodward claimed in his *Essay toward a Natural History of the Earth . . . With an Account of the Universal Deluge* that a subterranean oceanic abyss inundated the earth's surface, dissolving stony bodies into particulate suspension:

> I shall have shewn that at the time of the Deluge (when these Shells were brought out upon the Earth, and reposed therein in the manner we now find them) Stone, and all other solid Minerals, lost their solidity: and that the sever'd Particles thereof, together with those of the Earth, Chalk, and the rest, as also Shells, and all other Animal and Vegetable Bodies, were taken up into, and sustained in, the Water: that at length all these subsided again promiscuously, and without any other order than that of the different specifick Gravity of the several Bodies in this confused Mass.[31]

Despite the perplexities entailed by this hypothesis, we can note its fidelity to the arc of the *reductio* proof. Woodward's floodwater claims the capacity to dissolve "Stone, and all other solid Minerals," as well as "Earth, Chalk, and the rest" into "sever'd Particles," thereby mimicking the original biblical chaos with a diluvial menstruum through which diffused bits of matter float "promiscuously." His deluge accounts for primordial artifacts "brought out upon the Earth," mixed with stony matter, and deposited far from their antediluvian home; it also explains the apparition of novel fossils, Woodward claims, exhumed from the sea's depths.

However, the projected closure of the *reductio* reaction—encapsulated by the meaning of *reductio* as the return of suspended silver to its metallic state—is thwarted by difficulties unique to Woodward's scenario. Whereas floating silver particles reconvene as visible metal at the bottom of a flask, Woodward must explain how the dissolved earth's "confused Mass" falls back into the exact same geological order as before (shells may migrate, but "the Terrestrial Parts of the Globe, Metalls, Minerals, Marble . . . settled down again in or near the same Place from which they were before taken up"[32]). Still faithful to a mechanistic articulation of the corpuscle, Woodward denies the agency of any seminal "plastick Vertue concerned in shaping" the diffused bits,[33] marshaling only two forces to reassemble the postdiluvial earth in biblical time: the "specifick Gravity" characterizing different kinds of bodies and "meerly the Configurations of the Particles . . . and the simple Motion of the Water."[34] *Natural History* heavily exploits the former. After the flood renders every rock and mineral's "constituent Corpuscles all disjoyned, their Cohaesion perfectly ceasing," the separated *minima* "precipitated and subsided towards the bottom . . . according to the Laws of Gravity." These constituent parts "formed the *Strata* of Stone, of Marble, of Cole, of Earth, and the rest . . . meerly by the disparity of the Matter, of which each consisted, as to Gravity . . . That for this reason, the Shells of those Cockles, Escalops, Perewinkles [sic], and the rest, which have a greater degree of Gravity, were enclosed and lodged in the *Strata* of Stone, Marble, and the heavier kinds of Terrestrial Matter: the lighter Shells not sinking down till afterwards, and so falling amongst the lighter Matter."[35] Woodward predicates the organized "Subsidence" of the flood's disjoined corpuscles on their different weights;[36] particles that "have a greater degree of Gravity" determine the settling of geological layers as well as the presence of intercalated shell species in consonantly weighted strata.

But by ascribing this result to the "disparity of the Matter" alone, Woodward appears to ignore a proviso advanced by Royal Society member William Molyneux in a *Philosophical Transactions* entry Woodward cites in *Vegetation* to support his own theory of aqueous intermixture. Minuscule parts of heavy bodies, like iron, may not sink in lighter bodies, like water, because they are impeded, Molyneux writes, by "the Natural Congruity of the Parts of a Liquor, whereby they desire, as't were, to unite and keep together, just as we see two Drops of Water on a Dry Board . . . do jump and Coalesce."[37] (Here Molyneux intuits the Earth-sustaining quiddity of water's hydrogen bonds.) Despite "each Particle (how small soever) of Iron being heavier then a like Particle of *Aqua-Fortis*," the dissolved metal "does not fall to the Bottom."[38] Though the mechanist Woodward does not countenance the agency of intraparticulate "desire" in either *Vegetation* or *Natural History*, Molyneux's insight exposes a deep contradiction between Woodward's two texts: while plants transpire because water conveys soil granules upward in a microspherical matrix,

Woodward's postdiluvial *menstruum,* sieve-like, lets corpuscles of every kind of matter sink through it.

Such micromaterial incoherence irked the Scottish mathematician, doctor, and satirist John Arbuthnot. Woodward's *Earth,* as Joseph Levine affirms, "brought him fame and celebrity," testimony to the mainstream currency of corpuscular explanation and its close pressure on observational data.[39] (Woodward insists that his account of stratigraphic settling is "guided wholly by *Matter of Fact.*"[40]) Arbuthnot's *Examination of Dr. Woodward's Account of the Deluge* (1697) does not dismiss but rather interrogates the consistency of the corpuscular insight that ascribes empirically perceived phenomena like Molyneux's jumping drops of water to the mechanical, seminal, or desiring agency of imperceptible particles. Concerning the uneven effects of Woodward's diluvial dissolution, Arbuthnot is quick to ask, "What dissolv'd the Fossils? and at the same time spared the Animal and Vegetable Substances?"[41] Italicizing Woodward's words, Arbuthnot extends this query:

> The next Miracle is, the Dissolution of all Solids (except Vegetable and Animal Substances) into their *constituent Parts;* of this the Dr. says he will assign a plain Physical Reason. I must beg his pardon if I think it cannot be very plain. I will not trouble my self any more with guessing, but this I know, if any Man besides the Doctor should have pretended to such a Secret, it would have found the same Credit as the Philosophers Stone . . .
>
> If the Cohesion of all Solids perfectly ceas'd, and their constituent Corpuscles were disjoin'd, why were not those of Sand, Gravel, and Earth so too? for they are little Solids, and have their constituent Parts as well as the larger; and what dissolves the one will dissolve the other.[42]

What agent of separation or *spagyria*—to foreground the likeness of dissolution and chymical analysis—dissolves rocks while leaving shells, plants, and animals intact? *Earth* gestures toward, but does not provide, a response to this question: "I shall assign a plain and Physical Reason, taken meerly from the *Cause* of the Solidity of these Mineral Bodies; which I shew to be quite *different* from *that,* whereunto Vegetables and Animals owe the Cohæsion of their parts."[43] In response, Arbuthnot satirizes an occulted "Reason" whose bankruptcy is foreordained by its likeness to "the Philosophers Stone." But if he takes the stone as an easy trope for science nobody will credit, Arbuthnot does not dismiss Woodward as a charlatan on par with chrysopoeiac pretenders. Instead, Arbuthnot elucidates his critique along the lines of the structure-substance tension that Woodward's flood so flagrantly exacerbates. In this instance, Woodward conserves the texture of "Sand, Gravel, and Earth" while bigger rocks dissolve; Woodward thus, Arbuthnot argues, grants irreducible substance to solids on the basis of size. Arbuthnot argues on mechanico-corpuscular grounds: because small rocks "have their constituent Parts" too, no menstruum that dissolves sandstone can leave a grain of sand intact. He jokes about the species-scale incoherence triggered by arbitrarily corrosive floodwater—"I cannot see why, amongst other Stowage, *Noah* ought not to have had a Green-House and Fish-pond"[44]—but Arbuthnot seriously engages the logic of Woodward's scenario at the corpuscular level.

In Arbuthnot's view, Woodward conserves substance under cover of the reductionist impetus of severed parts. For example, Woodward insists that the flood reconstitutes earthly geology "by dissolving it: by reducing all the Matter of it to its first constituent Principles."[45] As in *Vegetation,* the phrase "first constituent Principles" connotes separation into irreducible units. But as Arbuthnot cautions, "it is extremely hard to find the Specifick Gravity of Bodies reduced into small Particles, such as Sand and Earth" because small particles surrender their empirical likeness to the bigger things they make up: "If the Solids were reduced into their Elements, which I think the Dr. means by *their constituent Parts,* it is hard to tell what Specifick Gravity those would have: it is probable [*sic*] not the same with their Composita, which require a Mixture of heterogeneous Matter to make them up."[46] Arbuthnot's use of the word "probable" explicitly extends his critique of Woodward's flood-*reductio* theory into the domain of representation: probability, coeval with the emergent narrative standard of *vraisemblance,* or "truth to nature,"[47] activates this incipient verisimilar mode not in service of mimetic mirroring but rather to appraise Woodward's corpuscular insight. This truth to nature does not recur to analogy—whereby empirical things operate as mimetic surrogates for tiny bodies—but rather hinges on Arbuthnot's awareness that particulate *minima* relinquish the specific gravity of the larger solid they once composed. The crux of Arbuthnot's competing intuition hinges on a particle's break from the phenomenal features of its macroscopic body and subsequent claim to discrepant—if still empirically intuited—attributes like an iron corpuscle's buoyancy in the midst of desiring particles of water or a silver corpuscle's invisibility in a clear bath of nitric acid. Arbuthnot flags his refusal of Woodward's claim that the same specific gravity qualifies both rocks and their severed corpuscles by renaming the latter entities "Elements": while these are not elements in any modern sense, they deny the qualitative sameness of composite solids and constituent parts. Elements, the *minima* into which antediluvian bodies dissolve, do not act like the solids they make up. The violation of probability of which Arbuthnot accuses Woodward does not presume undistorted transcription of minute particulars or mimetic simulation subtended by analogy, because Arbuthnot asserts the nonresemblance of microparts and the macroscopic things they comprise.[48]

Arbuthnot argues for the role of "surface" area and tension in determining the "Celerity" with which different forms of matter sink:

> For example, a Stone of an hundred weight will fall sooner to the bottom, than the Powder of the same Stone. . . . The Consequence of this will be, that the Parts of Animals, which were the greater Solids, could never be buried in Matter of the same Specifick Gravity with themselves: Yea, throw an Oyster-shell into Water, and at the same time the constituent Parts of the heaviest Metals, the Shell will fall soonest to the bottom. . . . By the same Rule, the larger Grains of Sand would fall lower than the imperceptible constituent Parts of other Fossils.[49]

Because empirically quantified specific gravity does not dictate how quickly a body's "imperceptible constituent Parts" subside, shells will fall faster than heavier rocks if the latter have been dissolved into "Powder." Arbuthnot rejects Woodward's explanatory premise with disarming humor: "God forbid I should limit Omnipotency, but as to the Second Causes, I must remain an Infidel till the Doctor's larger Work appears."[50] As one literary-historical corollary of scientific debate that turns on corpuscular insight, Arbuthnot's satire can be read not as generic Scriblerian antagonism to micromaterial inquiry but as debate over a representation of imperceptible causes to which rules of probability should apply.

Perverse Geology: *Three Hours after Marriage*

Coauthored by Gay, Arbuthnot, and Pope, the Scriblerian drama *Three Hours after Marriage* (1717) caricatures Woodward as Doctor Fossile, an incompetently Latinate virtuoso, would-be chrysopoeiac alchemist, and queer cuckold.[51] *Three Hours after Marriage* borrows stock tropes from Thomas Shadwell's *The Virtuoso* (1676), which characterizes the natural philosopher Sir Nicholas Gimcrack as "the finest speculative gentleman in the whole world"—most notably iterating Gimcrack's pretense to be "much skill'd in Rosicrucian learning. I am one of the *vere adepti.*"[52] Shadwell's play is far less concerned with the epistemic distortions unleashed by micromatter than with Gimcrack's fixation on vermin, as the virtuoso's uncle Snarl intones: "My nephew [is] . . . good for nothing but useless experiments upon flies, maggots, eels in vinegar, and the blue upon plums, which he finds to be living creatures."[53] In only one instance, during a dialogue between Gimcrack and his nieces' libertine suitors, does *The Virtuoso* address empirically intuited access to the microworld, which extends the critique imminent in Gimcrack's animation of "the blue upon plums":

> SIR NICHOLAS. . . . But now he talks of eels, I'll show you millions in a saucer of vinegar. They resemble other eels save in their motion, which in others is sideways, but in them upwards and downwards thus, and very slow.
> LONGVIL. We have heard of these, sir, often.
> SIR NICHOLAS. Another difference is these have sharp stings in their tails. By the way, the sharpest vinegar is most full of 'em.
> BRUCE. Then certainly the sharpness or biting of vinegar proceeds from those stings striking upon the tongue.
> SIR NICHOLAS. I see you are a most admirable observer. It must needs be so.[54]

In *The Virtuoso*'s central reference to microcausation, Bruce provides a mock rationale for vinegar's sourness: its "sharpness or biting" is effected by tiny biting eels. The butt of this satire is both Gimcrack's own lax standard of probability—"It must needs be so"—and the animist devolution of biting taste into infinitesimally small biting creatures. In Shadwell's mockery of materialist improbability, perceived qualities refer downward to teensy actants conjured to enact them directly upon the receptive sense organ. (Shadwell writes in oddly prescient anticipation of Isaac Newton's *Opticks* [1704], according to which acidic taste transpires when attractive matter rips corpuscles off the tongue.)

Whereas *The Virtuoso* spoofs microcreatures whose explanatory salience is exhausted by their reenactment of phenomenal experience, *Three Hours* satirizes Woodward's diluvial geology. Fossile's farcically literary niece Phoebe Clinket composes a play within a play entitled, in an iteration of Woodward's *Natural History*, *The Universal Deluge*, which opens, "All the Fields beneath are over-flowed, there are seen Cattle and Men swimming. The Tops of Steeples rise above the Flood." Clinket's listener asks, repeating Arbuthnot, "If Stones were dissolved, as a late Philosopher hath proved, how could Steeples stand?"[55] This iteration of Arbuthnot's *Examination* taps the comedic potential of corpuscular illogic: micromaterial implausibility translates into a theatrical punchline. At least some in *Three Hours*'s audience would appreciate the standard of corpuscular probability that dictates Fossile's characterization, surely penned by Arbuthnot, as "the Man that has the Raree-Show of Oyster-shells and Pebblestones."[56] Spectacularly incoherent corpuscular insight claims the alternative absurdity of a street player's "Raree-Show."

Alchemy exemplifies natural philosophical ineptitude in neither Gimcrack's nor Fossile's drama. Gimcrack's consummate demonstration of the inutility of his labors occurs when he claims to "swim most exquisitely on land" because "I content myself with the speculative part of swimming."[57] (Waterless swimming is too speculative a practice to stand in for intractably messy alchemical pursuit.) But while Gimcrack has a wife and mistress and partakes in the sex comedy that parallels his lampooning, *Three Hours*'s major extrapolation of virtuosic fecklessness is marked by Fossile's lack of heteronormative acumen. The anterior sexual experience of his bride-to-be, Townley, is redundantly underscored; the play's running jokes concern both Fossile's reluctance to initiate consummation—"Now I think of it, my Dear; *Venus,* which is in the first Degree of *Capricorn,* does not culminate till Ten; an Hour, if Astrology is not fallible, successful in Generation"[58]—and his failure to detect Townley's departure from the sexual passivity normatively entailed upon new wives. Fossile professes total incomprehension of the sex/gender system whose empirical signs he must decode: "Why are there no external Symptoms of Defloration, nor any Pathognomick of the Loss of Virginity but a big Belly? Why, has not Lewdness it's Tokens like the Plague?"[59] When Fossile is duped by Townley's lover Plotwell, disguised as the itinerant Polish chymist "Doctor *Cornelius Lubomirski,*" the absurdities of Woodward's geological obsession collude with the appeal of an obviously spurious chymical litmus:

> PLOTWELL. Be dere any secret in the Hydrology, Zoology, Minerology, Hydraulicks, Acausticks [sic], Pneumaticks, Logarithmatechny, dat you do want de Explanation of?
> FOSSILE. This is all out of my Way. Do you know of any Hermaphrodites, monstrous Twins, Antidiluvian Shells, Bones, and Vegetables?
> PLOTWELL. Vat tink you of an Antidiluvian Knife, Spoon, and Fork, with the Mark of *Tubal Cain* in *Hebrew,* dug out of de Mine of *Babylon?*
> FOSSILE. Of what Dimensions, I pray, Sir?
> . . .
> PLOTWELL. . . . me have prepare a certain Liquor which discover whether a Woman be a Virgin or no.
> FOSSILE. A curious Discovery! have you any of it still?[60]

In his account of the flood, Woodward cites *"Tubal-Cain, an Instructer of every Artificer in Brass and Iron"* to assert the identity of pre- and postdiluvial metal.[61] The fake collector's eponymously stamped "Antidiluvian Knife, Spoon, and Fork" present geo-biblical antiquarianism as a site of evidentiary gullibility continuous with Fossile's inexpertise on the marital marketplace. Lubomirski's "certain Liquor" serves most emphatically to suture observational and heteropatriarchal credulity, because Fossile employs the potion to vindicate Townley's chastity: "O thou spotless Innocence! I cannot refrain Tears of Joy."[62]

Three Hours asserts this endpoint as its protagonist's most discrediting claim to experimental knowledge—"I have prov'd thee Virtuous"—through the play's end, when Townley is exposed as bigamous and Fossile, glad to acquire an heir without "the Act it self,"⁶³ adopts her natural child. Fossile's adjunct offspring amplifies the artificiality of a kinship line already foreshadowed by "a discourse begun during Woodward's lifetime," as Will Burgess suggests, that claimed his scrupulously preserved specimens "as a surrogate for children in light of Woodward's 'well-known' homosexuality."⁶⁴

Three Hours channels its portrayal of Woodward's queerness, which was explicitly sketched by some pamphleteers, into Fossile's credence in a virginity indicator that harks back to Paracelsian spiritual medicine.⁶⁵ The play's mockery of this remnant of embodied artisanal epistemology is thereby continuous with its ridicule of Fossile *as* a deviant body. *Three Hours* targets the perverse epistemology of the dabbler in biblical artifacts—"an Antediluvian Trowel, unquestionably the Tool of one of the *Babel* Masons!"—and sterile monstrosities, overdetermined as Fossile's failure to reproduce.⁶⁶ (The nongenerativity of geo-biblical time is underscored by Fossile's self-characterization as "almost at my grand Climacterick," when Woodward would have been around fifty at the play's production.⁶⁷) Corpuscular illogic is consonant with the spectacle of Fossile's incapacity to intuit patriarchal probability, coded for dramatic consumption as the gender trouble inherent in a prospective bride named Townley.

Woodward's Stratigraphic Stomach

Marking another chymical reformer's approval of the transmutation concept, the Dutch medical and chemical professor Herman Boerhaave (1668–1738) limited radical micromaterial transformation to six classes of "vegetable matter."⁶⁸ Boerhaave defends the irreducibility of metallic corpuscles but admits fermentative transmutational processes that could still brew beer.⁶⁹ In *The State of Physick: and of Diseases* (1718), Boerhaave's correspondent Woodward repeats the prohibition and its fermentative—or digestive—exception: "Brought to any Test, to any the highest Rack and Torture: committed to the most intense Fire, to the strongest Menstruums, they [metal corpuscles] come ever forth without any Essential Change. Which may serve to shew how vain, delusive, and ill grounded the Essayes for Transmutation of Metals have been.... But, in an Affair of vastly greater Moment, the Life and Health of Man, the Doctrine of Transmutation obtains, to this Day, as generaly [sic] as ever."⁷⁰ Woodward's contemporaries did not dispute his transmutative rationale for digestion. Arbuthnot's *Essay concerning the Nature of Ailments* (1731)—which notes its own author's reliance on *"the most learned and industrious* Boerhaave"—confirms that "Nature is at a great deal of Labour to transmute Vegetable into Animal Substances."[71] The canonical standing of the transmutation concept as a warrant for nutritive transfer is registered by Samuel Johnson's *Dictionary* (1755), which cites Arbuthnot's *Ailments* under the heading *transmutable:* "The Fluids and Solids of an Animal Body, are easily transmutable into one another."[72] (Here Arbuthnot evokes a body's imminent uptake of the blood-borne nutritive liquor chyle.)

Instead, Woodward's antagonists attacked *State of Physick*'s starkly reductionist etiology of human illness, which attributes disease to "vitiated Biliose Salts" produced in the stomach;[73] they also assailed Woodward's claim that a distended belly presses "the Artery behind it," restricting blood flow to the brain and affecting both "the Passions" and "the Affair of Cogitation."[74] In the first instance, Woodward returns again to the *reductio* reaction—now catalyzed by "hot, sharp, corrosive" digestive acid[75]—to motivate a scene of stratigraphic subsidence akin to that of his deluge:

> The morbid Principles that are lightest, and float at Top [*sic*], in the Stomach, are recent, less putrid, and offensive: and those, which stagnate at the Bottom, are more stale, vitious, gross, ponderous, and noxious. As the others, above, are pass'd off, these, finaly riseing in their Turn, act a Part, and produce Symptoms, as much more troublesome than those others, as they are more vitious. They, who attend to the Operation of Vomits, rightly manag'd, cannot be Strangers to this Order of Things in the Stomach: or ignorant that the most dangerous, active, and potent Principles there lye commonly deepest.[76]

According to Woodward, gravity separates lighter digestive content from matter that stagnates below: he claims that a stuffed stomach extends "below the Pylorus, or Pass into the Guts," to suggest that putrefying principles may occupy the lower belly since the date of "the Birth of the Patient."[77] The stomach's stratigraphic "Order of Things" reciprocally vindicates Woodward's signature therapeutic modality, the vomit. True to Woodward's mechanism, no extraneous *semina* of disease operate here: the efficacity of the "rightly manag'd" vomit turns on its extractive, even archaeological, expulsion of the strictly internal pathology of "morbid Principles that had layn long lurkeing in the Body."[78]

Woodward employs his favored term "Principles" for the cause of illness into which stagnated food transmutes. He thus violates Boyle's effort to delimit what entities "are more worthy to be call'd by the Name of a Principle (which ought to be pure and homogeneous)."[79] Critics of *State of Physick* flag the explanatory nullity of Woodward's faux-reductionist word use: "Have you any clearer Conception of the Bile than you had before? . . . No doubt when any of the Constituent Parts exceed their just Proportion, Inconveniences will happen; but has not the other Juices of the body the same Title to do Mischief as the *Bile,* and for the same Reason?"[80] But the pamphlet war fomented by *State of Physick* most derisively targeted Woodward's argument for the mechanical impact of a swollen stomach—its obstruction of arterial blood to the brain—on cognition. Amplifying the immediacy of the brain-belly link, Woodward denied the existence of the subtly material nervous medium known as animal spirits to insist on "the strict Intercourse and Reciprocation betwixt the Stomach and Brain."[81] His vomits thus remove not only stagnating salts but their affiliated mental content: "As these Salts are the Instruments that concurr to the Produceing of Cogitation, being thus multiplyed, they render the Thoughts more intense. . . . As those Salts cause these Disorders, so the Removal of the Salts, particularly by Vomit, puts an End to the Disorders."[82] Woodward borrows Boerhaave's neologism "Instruments" to dignify the mechanism perpetrated by multiplied "biliose" salts, but this semantic edit did not cushion the affront of his vomitive materialization of thought.[83] The first pamphleteer to react was the Royal College of Physicians doctor John Freind:

> A Lady of Quality was troubled with melancholy Dreams. . . . You judg'd very rightly . . . that the entire Set of those wrong Ideas arose wholey from this biliose Principle flowing continually out of the Stomach. . . . [Y]our Affairs not permitting you to attend in Person, whereby too large an Egress was permitted to the Cogitative Principles of the Bile . . . she brought up finaly all her Religion; and had from that time no biliose Matter left to assist her in the Affair of *Faith.* A deplorable Instance of the artless Management in the Administration of a Vomit.[84]

As materialist reduction exhaustively realized by Jonathan Swift's *Tale of a Tub* (1704), Woodward's therapeutic "Removal of the Salts" dictates Freind's inevitable next step: the "Lady of Quality" pukes up not salts but religion. "Cogitative Principles of the Bile" skewer the metaphysical folly and threadbare explanatory salience of Woodwardian principles with the lady's expulsion of not just nightmares but an underlying stratum of faith.

Woodward's insistence on the intercourse of mind and belly mobilized his most memorably satirized foray into dietetics, a jeremiad against luxurious foodstuffs. The flood, he argues, belatedly realized Adam's divine punishment by thinning the planet's topsoil, ordaining for Christians an extractive regime of agrarian cultivation.[85] In *State of Physick,* a fall into easy eating reoccurs with the "continued Course of Gluttony" enabled by "new Modes of Cookery" and "the late great Multiplication of Pastry-Cooks in the City": these engender "Vice and Immorality: Irreligion, Impiety: Passion, Animosity, Contention . . . Stupidity: Poverty: Discontent: Sickness."[86] Cementing Woodward's most durable literary-historical persona, Scriblerians spoofed the surfeit of pastry cooks lamented by "Don Bilioso de l'Estomac" until, at least, Henry Fielding's reference in *Shamela* (1741) to "the luscious Temptations of Puddings and Custards, exciting the Brute (as Dr. *Woodward* calls it) to rebel."[87] The satiric drama *Three Hours* spoofs the identity of food and mental content that compels Woodward's hostility to "new Modes" of luxurious baking:

> TOWNLEY. So the Wit of one's Posterity is determin'd by the Choice of one's Cook.
> FOSSILE. Right. You may observe how *French* Cooks, with their high *Ragousts,* have contaminated our plain *English* Understandings. Our Supper to Night is extracted from the best Authors.[88]

Lauren Kassell's archival recovery of early modern alchemical and spiritual medicine recasts Woodward's diatribe against pastry cooks as one endpoint of a medieval and Paracelsian justification of disease, according to which "before the fall of *Adam* all things were good, all things came unto him. . . . But afterward part of it was joyned to poison, part of it so fast lockt up, that without great sweate of browes he should not eat of it."[89] By contrast, Woodward's antagonists appear—at least rhetorically—to defend the ascendant hegemony posited by Shapin and Schaffer, which asserts the dominance of a disembodied ideal of philosophico-scientific knowledge.[90] Indeed, the "artisanal understanding of the material world" reconstructed by Smith was repressed, she argues, in Woodward's age: "Enlightenment science would come to elide its artisanal origins . . . [and] the bodily epistemology that had been articulated by artisans."[91] The Woodward controversies demonstrate the partial nature of this elision: "artisanal origins" whose practiced grasp of matter's powers exceeds empirical delimitation linger into a representational dispute in which truth claims summon divergently intuited corpuscular causes. This engagement with the representational pressure of micromaterial causation recalls Smith's claim for the maker-knower's productive contact with matter's capacities: "The final result for which the artisan strove was an imitation of nature much more profound than the reflection of nature in a mirror; beyond verisimilitude, the artisan sought a knowledge of materials and an ability to produce."[92] The Woodward pamphlet wars affirm the ongoing epistemological currency of representational claims on corpuscular materiality. A representational mode not reducible to mimetic reflection continues to animate Arbuthnot's claim for the improbability of a menstruum in which dissolved *minima* do not sink with the same celerity as the rocks they compose.

Woodward's rote caricature as a disciple of "that great Foreigner *Bombast ab Ohenheim,* in his Philosophical Treatise *de Generatione Stultorum*" belies the role of corpuscular insight in eighteenth-century scientific epistemology.[93] The Scriblerian mock biography *Memoirs of the Extraordinary Life, Works, and Discoveries of Martinus Scriblerus* (1741) lists among Martin's medical discoveries "a Menstruum to dissolve the [kidney] Stone, made of Dr. Woodward's Universal Deluge-water."[94] Yet again, Woodward's flood dissolvent is held to a standard of probability that locates corpuscular causation at the crux of both experimental knowledge and its representation.

Notes

1. Robert Boyle, *Sceptical Chymist,* in *Works of Robert Boyle,* ed. Michael Hunter and Edward B. Davis (London: Pickering & Chatto, 1999), 2:256.
2. Ibid., 291.
3. On the willow-tree experiment, see Walter Pagel, *Joan Baptista Van Helmont: Reformer of Science and Medicine* (Cambridge: Cambridge University Press, 1982), 49–57; citation from Helmont (trans. Pagel), 53.
4. Boyle, *Sceptical Chymist,* 255, 257.
5. However, Boyle was a committed, practicing alchemist. See Lawrence M. Principe, *The Aspiring Adept: Robert Boyle and His Alchemical Quest* (Princeton: Princeton University Press, 1998).
6. Steven Shapin and Simon Schaffer, *Leviathan and the Air-Pump: Hobbes, Boyle, and the Experimental Life* (Princeton: Princeton University Press, 1985), 64.
7. Richard Rorty, *Philosophy and the Mirror of Nature* (1979), introduction by Michael Williams, afterword by David Bromwich (Princeton: Princeton University Press, 2009), 12.
8. Thomas Sprat, *The History of the Royal Society* (London: Printed by T. R. for J. Martyn, 1667), 97.
9. Rorty, *Mirror of Nature,* 60n32.
10. Ibid., 93 and 113. Rorty claims that Locke is a Platonic idealist: "In constructing both a Lockean idea and a Platonic Form we go through exactly the same process—we simply lift off a single property from something (the property of being red, or painful, or good) and then treat it as if it itself were a subject of predication, and perhaps also a locus of causal efficacy" (ibid., 32). Locke's *Essay concerning Human Understanding* (1690) argues precisely the opposite:

that language falsely reifies intractably relational qualities produced in us, as Locke puts it, by bodily micromatter stimulated by worldly micromatter. On Locke's qualities, see my *Fictional Matter: Empiricism, Corpuscles, and the Novel* (Philadelphia: University of Pennsylvania Press, 2017).
11. Shapin and Schaffer, *Leviathan and the Air-Pump,* both p. 150.
12. Pamela H. Smith, *The Body of the Artisan: Art and Experience in the Scientific Revolution* (Chicago: University of Chicago Press, 2004), 110.
13. Ibid., 87.
14. Ibid., 149.
15. See Sean Silver, *The Mind Is a Collection: Case Studies in Eighteenth-Century Thought* (Philadelphia: University of Pennsylvania Press, 2015), 74, on Woodward's geological cabinet as "a design and way of ordering the world" (75) and Woodward as a collector who "thinks through the materials of his concern" (76).
16. W. G. Burgess evokes Woodward as "an embodied interpreter . . . who must see, taste, handle, weigh, break, burn, dissolve, and scrutinize his specimens." Burgess, "Instead of Children: Legacy and Embodied Interpretation in the Woodwardian Museum," *Studies in Philology* 118, no. 4 (Fall 2021): 772.
17. Francis Bacon, *The New Organon* (1620), ed. Lisa Jardine and Michael Silverthorne (Cambridge: Cambridge University Press, 2000), 107 (bk. 2, aphorism 6).
18. John Woodward, "Some Thoughts and Experiments concerning Vegetation," *Philosophical Transactions* (January 1, 1699), 195, accessed February 2017, https://royalsocietypublishing.org/doi/10.1098/rstl.1699.0040.
19. See William R. Newman, *Atoms and Alchemy: Chymistry and the Experimental Origins of the Scientific Revolution* (Chicago: University of Chicago Press, 2006), 23–25 and chap. 4: "The reduction of dissolved metals into their original or 'pristine' state (*reductio in pristinum statum*) became a sort of crucial experiment," 24.
20. Woodward, "Some Thoughts," 205. I round off the second figure.
21. Ibid., 208.
22. Ibid., 209 and 211.
23. Ibid., 195.
24. Ibid., 215.
25. Ibid., 216.
26. Newman, *Atoms and Alchemy,* 176.
27. Jole Shackelford, *A Philosophical Path for Paracelsian Medicine: The Ideas, Intellectual Context, and Influence of Petrus Severinus (1540/2–1602)* (Copenhagen: Museum Tusculanum Press, 2004), 178.
28. Ibid., 177n85.
29. For Woodward's side of this altercation, see [Richard Steele,] *The Antidote. Number II. In a Letter to the Free-Thinker* (London: J. Roberts, 1719), 15–18;

reprinted in Joseph M. Levine, *Dr. Woodward's Shield: History, Science, and Satire in Augustan England* (Berkeley: University of California Press, 1977), 16–17. For discussion of the pamphlet wars, see Alexander Wragge-Morley, "Medicine, Connoisseurship, and the Animal Body," *History of Science* 60, no. 4 (2022): 481–99; Burgess, "Instead of Children," 765–86; Sophie Vasset, "Medical Laughter and Medical Polemics: The Woodward-Mead Quarrel and Medical Satire," *XVII–XVIII* [en ligne], 70 (2018), consulté le 06 avril 2021, https://doi.org/10.4000/1718.514; Levine, *Woodward's Shield*, chaps. 1–7; and Lester M. Beattie, *John Arbuthnot: Mathematician and Satirist* (New York: Russell and Russell, 1935).
30. On theories of diluvial dissolution by John Ray and William Whiston, see Robert Markley, *Fallen Languages: Crises of Representation in Newtonian England, 1660–1740* (Ithaca: Cornell University Press, 1993), 117–24 and 204–7.
31. John Woodward, *An Essay toward a Natural History of the Earth* (London: R. Wilkin, 1695), 29.
32. Ibid., 261.
33. Ibid., 191.
34. Ibid.
35. Ibid., 74, 76.
36. Ibid., 75.
37. William Molyneux, "A Discourse on this Problem; Why Bodies Dissolved in Menstrua Specifically Lighter Than Themselves, Swim Therein," *Philosophical Transactions* 16, no. 181 (January 1, 1687): 92, accessed April 14, 2021, https://royalsocietypublishing.org/doi/10.1098/rstl.1686.0015.
38. Ibid., 88.
39. Levine, *Woodward's Shield*, 35.
40. Woodward, *Natural History of the Earth* (1695), 2.
41. John Arbuthnot, *An Examination of Dr. Woodward's Account of the Deluge* (London: C. Bateman, 1697), 7.
42. Ibid., 10.
43. Woodward, *Natural History of the Earth* (1695), 108.
44. Arbuthnot, *Examination*, 12–13.
45. Woodward, *Natural History of the Earth* (1695), 89.
46. Arbuthnot, *Examination*, 19, 20.
47. For an excellent genealogy of *vraisemblance* as it emerged through French heroic romance, see Ros Ballaster, *Seductive Forms: Women's Amatory Fiction from 1684 to 1740* (Oxford: Clarendon Press, 1992). Citation on p. 43.
48. The foreclosure of corpuscular causation from empirical science asserted by Shapin and Schaffer has resulted in dubious claims for the hypertrophied role of observational detail in natural philosophy. J. Paul Hunter characterizes natural philosophy as the paratactic accumulation of particulars with

no conceptual paradigm: "Not only did modern science have no clear program of usefulness and no plan for sorting between practical and impractical methods, but it offered no authoritative conclusions and depended simply on the subjective application of individual observation and experience." Hunter, "Robert Boyle and the Epistemology of the Novel," *Eighteenth-Century Fiction* 2, no. 4 (1990): 289.
49. Arbuthnot, *Examination*, 22–23. Arbuthnot uses the word *fossil* in the sense of something "dug out of the earth." See Samuel Johnson's *A Dictionary of the English Language* (London: W. Strahan et al., 1755), which cites both Woodward and Arbuthnot.
50. Arbuthnot, *Examination*, 18.
51. In his advertisement for the play, John Gay acknowledges "the Assistance" of "two of my Friends." John Gay, [John Arbuthnot, Alexander Pope,] *Three Hours after Marriage. A Comedy, as It Is Acted at the Theatre Royal* (London: Bernard Lintot, 1717), [sig. A2r]. See *A Complete Key To the New Farce, Call'd "Three Hours after Marriage"* (London: E. Berrington, 1717), 4, for this identification and much scurrilous detail. Al Coppola's *The Theater of Experiment: Staging Natural Philosophy in Eighteenth-Century Britain* (Oxford: Oxford University Press, 2016) argues that "*Three Hours after Marriage* is decidedly old-fashioned in its adherence to Jonsonian humors comedy"; the play "adopts a mode of satiric comedy whose epistemology is precisely what it finds so objectionable about the humorists the play would target" (108, 109).
52. Thomas Shadwell, *The Virtuoso* (1676), ed. Marjorie Hope Nicholson and David Stuart Rodes (Lincoln: University of Nebraska Press, 1966), 18, 106.
53. Ibid., 101.
54. Ibid., 101–2.
55. Gay, Arbuthnot, and Pope, *Three Hours after Marriage*, 21.
56. Ibid., 71.
57. Shadwell, *Virtuoso*, 46–47. See Coppola, *Theater of Experiment*, 44–46, on this scene and the theatricality of science.
58. Gay, Arbuthnot, and Pope, *Three Hours after Marriage*, 3.
59. Ibid., 31.
60. Ibid., 41–42.
61. Woodward, *Natural History of the Earth* (1695), 259.
62. Gay, Arbuthnot, and Pope, *Three Hours after Marriage*, 50.
63. Ibid., 55, 80.
64. Burgess, "Instead of Children," 767. Burgess's cited text comes from Silver, *Mind Is a Collection*, 83.
65. On Woodward's sexuality, see *An Account of the Sickness and Death of Dr. W—DW—RD* (London: J. Morphew, 1719), a mock obituary: "It will be expected I should say something for the satisfaction of the *Ladies*, who will

be inquisitive of *what Sex* he dyed: The Account of his Direction will inform them in that Particular; and altho' from the Softness of his Voice something may have been suggested to his Disadvantage in their Esteem, yet I know not whether that Constitution is not more eligible, that inclines one to the *Goût* of *Italy* and *Spain,* and gives a Man a stronger relish for the more *manly* Pleasures of those *warmer Climates*" (8). See also *Harlequin-Hydaspes: or, The Greshamite. A Mock-Opera* (London: J. Roberts, 1719): "All the World knows by the *Doctor's* Voice and beardless Chin, of what Sex he is" (26). On Paracelsian medicine and virginity, see note 89.

66. Gay, Arbuthnot, and Pope, *Three Hours after Marriage,* 61.
67. Ibid., 3.
68. John C. Powers, *Inventing Chemistry: Herman Boerhaave and the Reform of the Chemical Arts* (Chicago: University of Chicago Press, 2012), 86.
69. See ibid.: "The influential Joan Baptista van Helmont had denoted the term 'ferment' to refer to any instance where a vital agent—a *semen*—acted to transmute one chemical species to another. . . . In this context, fermentation was the foundational process of all true chemical change" (85).
70. John Woodward, *The State of Physick: and of Diseases* (London: T. Horne, 1718), 36.
71. John Arbuthnot, *An Essay Concerning the Nature of Ailments* (Dublin: S. Powell, 1731) sig. a2r, 100.
72. Citing ibid., 19; Johnson, *Dictionary,* s.v. "Transmutable."
73. Woodward, *State of Physick,* 23.
74. Ibid., 4, 4, 5.
75. Ibid., 66.
76. Ibid., 110.
77. Ibid., 98, 234.
78. Ibid., 235.
79. Boyle, *Sceptical Chymist,* 307.
80. [Richard Mead,] *A Serious Conference between Scaramouch and Harlequin,* by Momophilus Carthusiensis (London: J. Roberts, 1719), 18.
81. John Woodward, *Natural History of the Earth* (London: T. Edlin, 1726), 79. Of the pioneering brain anatomist Thomas Willis and animal spirits, Woodward writes, "He takes them merely on Trust; without ever going about to advance one Argument that there are realy *Animal Spirits,* or any such Thing as a *Nervous Fluid,* in the Body." Woodward, *State of Physick,* 167. Woodward subsequently amplified his indictment of "Fictions, introduced into the Philosophy of the last Age . . . the *Materia subtilis* of the *Cartesians.*" Woodward, *Natural History of the Earth* (1726), 29.
82. Woodward, *State of Physick,* 16.
83. On Boerhaave's "instrument theory," see Powers, *Inventing Chemistry,* 72–83. Wragge-Morley suggests, "What Woodward had in effect done, therefore, was to argue that the mind was in some sense controlled by the animal body

rather than the immaterial soul." Wragge-Morley, "Medicine, Connoisseurship," 493.
84. [John Freind,] *A Letter to the Learned Dr. Woodward. By Dr. Byfielde* (London: James Bettenham, 1719), 13–14.
85. What Tobias Menely might call Woodward's "ecological unconscious" structures his vision of extractive improvement in tandem with his corpuscularianism. See Menely, *Climate and the Making of Worlds: Toward a Geohistorical Poetics* (Chicago: University of Chicago Press, 2021).
86. Woodward, *State of Physick*, 142, 194, 196, 196.
87. Henry Fielding, *Joseph Andrews* and *Shamela*, ed. Douglas Brookes-Davies (Oxford: Oxford University Press, 1999), 307. See [Richard Mead,] *The Life and Adventures of Don Bilioso de L'Estomac* (London: J. Bettenham, 1719).
88. Gay, Arbuthnot, and Pope, *Three Hours after Marriage*, 70.
89. Lauren Kassell, citing I. W., *A Coppie of a Letter . . . by a Learned Physician*, in "'The Food of Angels': Simon Forman's Alchemical Medicine," in *Secrets of Nature: Astrology and Alchemy in Early Modern Europe*, ed. William R. Newman and Anthony Grafton (Cambridge, MA: MIT Press, 2001), 363. Kassell also mentions the capacity of the "astrologer-physician" (372) to determine whether a woman is a virgin.
90. Steven Shapin, "The Philosopher and the Chicken: On the Dietetics of Disembodied Knowledge," in *Science Incarnate: Historical Embodiments of Natural Knowledge*, ed. Christopher Lawrence and Steven Shapin (Chicago: University of Chicago Press, 1998). See Wragge-Morley's "Medicine, Connoisseurship" for the argument that mind-body dualism was not rigorously upheld even by Woodward's opponents.
91. Smith, *Body of the Artisan*, 25, 236.
92. Ibid., 98.
93. [Mead,] *Serious Conference*, 24. Markley trenchantly observes of a post-Newtonian ideology of mathematicized science, "Paracelsus symbolizes the would-be scientist who surrenders to the chaos of a feminized nature." *Fallen Languages*, 202.
94. [John Arbuthnot, Alexander Pope, Jonathan Swift, John Gay, Thomas Parnell, and Robert Harley,] *Memoirs of the Extraordinary Life, Works, and Discoveries of Martinus Scriblerus*, ed. Charles Kerby-Miller (New York: Oxford University Press, 1988), 168.

"Mice in a Barn" or "Every Little Miss"?
Figurative Imagination and Demographic Narratives of the Long Eighteenth Century

LISA FORMAN CODY

In 1723, Bernard de Mandeville compared an increasing population to "the never-failing Nursery of Fleets and Armies." In the 1730s, Richard Cantillon wrote that population grows "like mice in a barn." In 1755, Benjamin Franklin warned that numbers in America could spread like fennel or even the "polypus," an aquatic animal that could endlessly regenerate itself. In 1798, Thomas Robert Malthus described population, among other things, as "famished wolves in search of prey." In 1807, William Hazlitt argued that population behaved like "every little Miss," a young lady in search of a husband.[1]

What do all these endless metaphors mean? And do they even matter? After all, the fundamental causes of growth and decline have long been starkly plain to authors for centuries: no community can grow beyond what its resources will support, and populations decline with death and rise with birth.[2] Though these variables seem quite straightforward, even the authors of the sparest prose or arithmetic narration could not help but interject a vivid analogy or two. And some authors, like the seventeenth-century political arithmetician William Petty and later Rev. Malthus, scarcely wrote a paragraph without rhetorical devices. Why did these authors lean on figurative language when underlying demographic features were understood?

Scholars across disciplines have shown how rhetorical tropes are not mere embellishments but essential to conveying meaning and making arguments. Poets and playwrights played with language to tell their stories but, as Jess Keiser and Frank Boyle observe in this volume, so did natural philosophers and policy makers.[3] They used metaphors, analogies, and narration as representational and cognitive tools to help make abstractions concrete, new theories comprehensible, and political policies seemingly natural.[4] As

Willard Quine observed, "Along the philosophical fringes of science we may find reasons to question basic conceptual structures and to grope for ways to refashion them. Old idioms are bound to fail us here, and only metaphor can begin to limn the new order."[5] This may help explain why Petty's and Malthus's texts were rhetorically superabundant. The two shared little: they belonged to entirely different times and intellectual regimes; one valorized population growth, the other wrung his hands. But both forced their readers to imagine population in vivid and very new ways through figurative comparisons and narrative devices. They needed their metaphors.

Even when political economists were not making "new" claims, they analogized because the subject of population was abstract and intangible. "Population" was (and is) amorphous, difficult to imagine because it is something to which we all belong but cannot literally see in its entirety while retaining sight of its individual members, including ourselves.[6] Characterizing it as "mice" or "every little Miss" inspires an imaginable thing that has specific traits and qualities, giving population form and argument. When population is described as a rock, we can't imagine it changing in size, but when population is compared to rabbits, we envision its rapid growth.[7] Cultures treat apt metaphors as so natural and true that they no longer seem like figures of speech. These obviously true metaphors turn into clichés, a presumed truth that justifies deeper presumptions and even policy decisions. The flippant simile that some people "breed like rabbits" has rationalized forced sterilizations as well as the decision to let a subcontinent starve.[8]

Some metaphors become clichés, but many do not. Some die, and some no longer make sense. Verbal analogies provoke stories and images, but the features that come to mind may not be what the author from another culture or time intended. We can imagine Cantillon's mice, but are our associations aligned with his? We might be tempted to lump his mice with rabbits and other rapidly reproducing pests, all easy breeders signifying reproductive growth. But in this case, his analogy may have made associations that are no longer its primary signification. Metaphorical fields mutate, losing meaning or acquiring new ones, particularly as authors push their comparisons to make potent political or moral arguments, which give new meaning to the metaphor itself.[9] We might think that comparing human reproduction to that of any animal—foxes, mice, rabbits—means essentially the same thing across time. But this is not always true because the referent-object can vary or mutate in its meaning. Not only does the history of population theory and its variables have an intellectual history, but so too do the metaphorical fields that theorists invoked across time.

This essay examines how population authors used metaphors not as embellishments but as arguments. Figurative language could be meaningful and revealing, not only colorful and seemingly random. Metaphors were often used at points where authors most needed to make the abstract concrete, the debatable incontrovertible, the new convincing, and the political most compelling. In some cases, we continue to share the salient meaning of an analogy or image they invoked. But in some cases, we do not. It is here that unpacking the historical meaning of metaphorical fields can reveal authors saying something different from what we today presume they are saying.

Breeding Buildings, Infertile Foxes

Without regular censuses, early modern British theorists had to triangulate data to estimate population size and describe demographic patterns.[10] They worked by extrapolation and analogy, transforming numbers of houses, churches, chimneys, imports, roads, and other physical things into human headcounts.[11] These things served as "multipliers," presumably stable representatives that could be arithmetically formulated in relation to other quantities.[12] For example, the London merchant and early actuarial statistician John Graunt explained how an inanimate object could serve as a multiplier to estimate population size: "I took the Map of *London* . . . drawn by a Scale of Yards. Now I ghessed that in 100 Yards square there might be about 54 Families."[13]

This process raised questions on at least two fronts. First, observers debated the numeric value of the "thing": How many people were in a family or used a single chimney? Second, theorists did not share meaning when relying on the built environment for their multipliers and metaphors. Some believed that enclosure, decaying old buildings, and the construction of great houses provided evidence of a shrinking population—"the subject of this sinne is houses and lands," lamented Robert Powell.[14] But others looked at the exact same things as signs of growth. The financial speculator and London builder Nicholas Barbon extolled "New Buildings" as "instrumental to the preserving and increasing of the number of Subjects . . . for Houses are Hives for the People to breed and swarm in, without which they cannot increase. . . . There is that peculiar advantage that ought to be ascribed to the Builder, that he provideth the place of Birth for all other Acts, as well as Man."[15]

Buildings facilitated growth, both economic and demographic. When John Graunt fastidiously analyzed decades' worth of the weekly London Bills of Mortality, he found high rates of infant and childhood mortality, which meant that London's "teeming women" had relatively less impact on growing the city than migration to the city. After the 1666 plague, the city fully repeopled itself within two years not through "a supply by Procreations [but] . . . by new Affluxes to *London* out of the Country."[16] In average and healthy years, birth could not offset death. In years of high adult mortality, pregnant women did not offset demographic catastrophe: "The more sickly the years are, the less fecund, or fruitfull of Children they also be."[17] Graunt's views were shared by both contemporaries who did not treat fertility as positive and modern demographers who have shown "no natural increase in the number of London-born children surviving was possible in its hazardous disease environment."[18]

London was deadly. Its mortality rates were high, including those of infants and children. The city could not replace itself and seemingly relied on migration from other places to grow its population. Early modern theorists interpreted London's demographic challenges as arising from the differences between urban and agricultural work. Urban labor was cerebral, rural labor bodily. Graunt explained, "The minds of men in *London* are more thoughtfull and full of business then in the Country, where their work is *corporal* Labour, and Exercizes. All of which promote Breedings, whereas *Anxieties* of the minde hinder it."[19] Mercantilists valued the land and its labor, but some doubted whether urban culture was productive or healthy. Cities' reliance on intellectual professions and costly consumerism required disproportionate support from the countryside. This imbalance was expressed through different depictions of masculinity. Rural life was virile and robust; urbanity was effeminate and impotent.[20] As the Reverend John Brown would later complain about dissipated, consumer-oriented modern culture, "No increase of Numbers in the inferior Ranks can possibly make amends for this internal and capital Defect. Such a Nation can, at best, only resemble a large *Body*, actuated (yet hardly actuated) by an incapable, a vain, a dastardly, and effeminate *Soul*."[21]

Early modern authors relied on corporeal metaphors and personification to depict both the nation and the concept of population. The medically trained Nicholas Barbon chose a healthy image inspired by William Harvey's discovery of the circulation of the blood: "The Metropolis is the heart of a Nation, through which the Trade and Commodities of it circulate, like the blood through the heart, which by its motion, giveth life and growth to the rest of the Body."[22] Barbon disagreed with the "gentlemen"—that is, Graunt—who criticized London's expansion, reporting that they "use for Argument a *simile* from the Rickets, fancying the City to be the Head of the Nation, and that it will grow too big for the Body."[23] He refuted their characterization, but such a sickly image showed the supposed danger of imbalance. One author worried that "London . . . is perhaps a Head too big for the Body, and possibly too strong: That this Head grows three times as fast as the Body unto which it belongs."[24]

Some corporeal analogies invoked female generation and nurture—for instance, when land was compared to a womb or the breasts.[25] But even these symbols of potential growth were cast as limited and limiting by mercantilists. Thomas Harrington described the earth as a "dug" (or breast) that was sucked dry by places like the Netherlands, which had "weaned" itself of its own land's resources; consequently, Amsterdam "thereby become[s], as it were, one city that sucks all the world."[26]

Population theorists described actual women as failing to reproduce even when they were sexually active. In fact, the female sexual body could even cause *decrease*. Graunt explained that "the Adulteries and fornications, supposed more frequent in *London* than elsewhere, do certainly hinder Breeding. For a Woman, admitting ten Men, is so far from having ten times as many Children, that she hath none at all."[27] What might seem biologically illogical to us was perfectly common knowledge in the early modern period. As seventeenth-century medical authors explained, the "luxuriant, and the whorish crew" of cities are barren "because by frequent coition their bodies become empty of seed" or "by reason of their frequent cohabitation with men, whereby the neck of the *Matrix* is made so slippery, that it cannot retain the mans [sic] seed."[28] Medical authors followed Hippocrates, who had explained that "the often use of the act of Copulation makes the Womb slippery, and hinders conception."[29]

Graunt compared the paradox of urban debauchery to the reproductive patterns seen in polygamous cultures, where "the Increase of Man-kind would be but like that of *Foxes* at best."[30] We might think that both foxes and polygamous households with their extra women could reproduce large numbers, but this is not what contemporaries believed. Foxes were notoriously poor reproducers *because* they were so sexually active: "If you castrate none . . . it is highly probable, that, every of the twenty *Males* copulating with every of the twenty *Females*, there will be little or no conception in any of them. . . . And this I take to be the truest Reason, why *Foxes, Wolves,* and other *Vermin Animals,* that are not gelt, increase not faster than *Sheep*, when as so many thousands of these are daily Butchered."[31] Sexual activity consequently *prevented* populations from expanding. Licit marital relations produced children, but sex outside marriage did not grow the population but shrunk it. London's fornication failed to offset its fatalities.

Working Mice, Racial Weeds

In the eighteenth century, authors began comparing population to small and social creatures like bees and mice instead of macrocosmic bodies or mercantilist piles of gold.[32] The Irish-French banker and profiteer Richard Cantillon said that "Men multiply like Mice in a barn if they have unlimited Means of Subsistence."[33] Though mice had long had sexual connotations and were observed to reproduce very quickly, which might have foreshadowed Malthusian equations between sexual drive and resources, this was not earlier observers' only characterization of mice.[34] As much as they bred quickly, they also were known for their incessant activity. Barns and fields might have had too many destructive, greedy mice, but nobody accused them of laziness. They were viewed as laborious creatures who were only limited by the availability of work—the size of the "barn."[35]

Theorists had long recognized the value of labor, lauding the work of the countryside. Most forms of labor were characterized as potentially generative and expansive. For instance, as the preeminent seventeenth-century political arithmetician William Petty explained, one trade, like watchmaking, spurred others, like the making of springs and cases, and even bakers required the work of millers and oven-makers; or, as he stated, "manufactures will beget one another."[36] Barbon had compared buildings to hives not because they served as shelter but because they were factories. The French philosopher Montesquieu explained a nation grows when it "sends out colonies and whole nations, like swarms of bees, to look for other places to live." Comparing colonization to swarms of bees setting up new hives—instead of packs of predators, which we might consider a more apt metaphor today—emphasized labor and production. The bee was so synonymous with work that its very definition was in its labor, and as a laboring creature, it was an idiom for a hard worker. As Samuel Johnson defined it, a bee was both "the animal that makes honey, remarkable for its industry and art," and "an industrious and careful person."[37]

Though bees and mice multiplied—the latter irritatingly so—the primary attributes attached to these creatures (whether male or female) were productivity and consumption, not sexual reproduction. In fact, some eighteenth-century authors echoed the classical view that bees and mice multiplied "spontaneously" rather than sexually at all. (One author dismissed the belief that mice were "generated by putrefaction, or drop out of the clouds," reminding us that far into the eighteenth century, sexual reproduction was not yet considered a universal trait of animals.)[38]

To be sure, authors recognized that the number of sexually active women was necessary to estimate the extent of possible population growth. But

Cantillon and others did not see them as the prime movers of growth because their ability to reproduce was predicated on men's labor; they noted that a woman would not "become a mother" unless her partner was willing to run the risk.[39] What mattered more than reproduction was production, so much so that the reason the Catholic clergy were costly was not because they failed to reproduce but because they did not engage in productive work. Cantillon was not alone in arguing that the problem with Catholic priests was not their celibacy but their "idleness."[40]

By way of counterfactual comparison, animals known to multiply rapidly were *not* invoked by population authors until the 1790s.[41] In the 1734 play *The Lady's Revenge*, two servants in love speculate what will happen if they have sex and are discovered by their employer: "And then farewell all Hopes of this noble Settlement. We live in a Garret, breed like tame Rabbits, wear out the Cloaths we got in Service, and having no Money to buy more, stink in coarse Rags, and mutually curse each other, to the melodious Concert of half a Dozen squaling Brats about our Ears."[42] Rabbits reproduced in large numbers—and so might foolish servants—but the national population did not seem to grow at this pace. Britons were depleted by nearly constant warfare with France, plus there were not enough of them to colonize the globe. If anything, mid-century populationists eager for troops and sailors would have liked Britons to multiply like rabbits. But as proponent of population growth Jonas Hanway remarks, "Is it in nature for women to breed like cats or rabbets? They have usually one child at a birth, at the distance of at least ten months. Can they nurse a dozen at a time?"[43]

Though many population authors remarked upon non-Europeans, they rarely treated anybody other than the French as Britain's main demographic competitor. And few people anywhere really seemed to multiply very quickly or abundantly. This changed with Benjamin Franklin's 1755 *Observations concerning the Increase of Mankind*, which was concerned with not only the French but also everybody else. By replacing the traditional contest with France with one between the "black," "tawny," and "swarthy" and the white English, he biologized economic and demographic competition.[44] His metaphors were naturalistic, which helped turn both rapid reproduction and race into facts.

The North American colonists doubled themselves every twenty years because they had eight children per family rather than the imagined English four. Franklin explained that Americans' ability to have (theoretically) twice as many children as Britons was due to the vast American lands and opportunities that gave all men willing to work the means to marry much earlier than their European cousins.[45] More babies expanded numbers, of course, but this growth was about work. He explained, "The great increase of Offspring . . . is not always owing to the greater fecundity of Nature" but to "industrious education."[46] Echoing the typical Protestant indictment of lazy Spanish Catholics, Franklin described Spain as sparse, "owing to national pride and idleness."[47] Propensity for labor was implicitly Protestant. Franklin was thus not too worried about the idle Spanish, but he lamented London's unwillingness to prevent Dutch and German immigrants from coming to the English colonies—these Protestants were hard workers too. This is classic mercantilism so far, but Franklin turned fellow European Protestants into a much worse threat by racializing them.

The word *race* had been used for centuries to categorize different regions, linguistic groups, or nations; it was not primarily a biological signifier until the 1680s in French.[48] By the 1750s, *race* still was used to speak of nations, but as it increasingly signified phenotypical and color associations, it provided its own metaphorical field of lightness and darkness, which Franklin inventively exploited. As a colonist writing not just to London policy makers but also to his fellow settlers, including slave owners, he turned immigration into a racial battle between white Anglo-Americans and everybody else, portrayed as predatory "races" who were less "white." Franklin listed everybody who was "black," "tawny," and "swarthy." These nonwhites were Africans, Asians, Italians, and Spaniards but also "*Russians* and *Swedes* . . . as are the *Germans* also."[49] Russians, Swedes, Germans, and the Dutch "*Palatine Boors*" were not white but *Black*.[50] Only the "*Saxons*" and English "make the principal Body of White People on the Face of the Earth." Franklin told his readers that he "wish[ed] their [white] Numbers were increased."[51]

Franklin explained the struggle between the English and others through a botanical metaphor. If the earth were "vacant of other Plants, it might be gradually sowed and overspread with one Kind only; as, for instance, with Fennel."[52] The nefarious power of the comparison might be lost on those who do not garden. Fennel is invasive and hardy, able to choke and overrun nearly every other plant nearby.[53] Franklin proposed that the English could take over North America like fennel, but so too could any other hardworking "black" race if the metropole would not check non-Anglo immigration.

What of "black" Africans here? Franklin objected to the mass importation of Africans because they were inferior to "the lovely White and Red" but also because slaves themselves were "not so generally prolific . . . being work'd too hard, and ill fed, their Constitutions are broken, and the Deaths among them are more than the Births."[54] Franklin did not claim that Africans were naturally unable to reproduce, but that the inherently abusive nature of slavery itself prevented successful pregnancies. Slavery was bad in additional ways, including its impact upon Europeans: "Slaves also pejorate the Families that use them; the white Children become proud, disgusted with Labour, and being educated in Idleness, are rendered unfit to get a Living by Industry."[55] For Franklin, who was already eponymous with the virtues of hard work, slavery prevented industry and demographic growth among both Africans and whites.

Franklin used "blackness" to categorize all Anglo-Saxon competitors, from Africans to other Europeans. His characterization of African inferiority and its debasing effect on English virtue developed a metaphorical field of "blackness." To prove the danger of Swedes, Russians, Germans, and Dutch in Pennsylvania and the mid-Atlantic colonies, Franklin turned them into a blot on the North American landscape. He placed Native Americans somewhere in between white and Black as "red" people who posed little threat to English agriculture and manufacture as hunters in the woods. He asked, "And while we are . . . *Scouring* our Planet, by clearing *America* of Woods, and so making this Side of our Globe reflect a brighter Light to the Eyes of Inhabitants in *Mars* or *Venus*, why should we in the Sight of Superior Beings, darken its People? Why increase the Sons of *Africa*, by Planting them in *America*, where we have so fair an Opportunity, by excluding all Blacks and Tawneys, of increasing the lovely White and Red?"[56] "Planting" was a deliberate act, the gardener's choice. Transplants in foreign soil ran risks—like fennel, they might choke out all other plants. Why not breed "lovely" whites instead of darker races?[57]

If London were to continue ignoring the hordes of "black" Europeans taking opportunities away from Anglo colonists, Franklin warned of what happens in nature by invoking the "polypus," a generic name for a category of zoophytes, or ocean invertebrates (including some jellyfish and corals), that had captured scientific attention in the 1720s onward for its ability to regenerate and multiply itself into as many pieces as it was cut.[58] He argued, "A Nation well regulated is like a Polypus; take away a Limb, its Place is soon supply'd; cut it in two, and each deficient Part shall speedily grow out of the Part remaining." With "Room and Subsistence enough, as you may . . . make ten Polypes out of one, you may of one make ten Nations . . . or rather, increase a Nation tenfold in Numbers and Strength."[59] Franklin suggested two ways that population might regenerate itself: through proper regulation or through expansion. North America offered the possibility of both. London could protect "white" interests in North America and enjoy the riches gained by its growing population. But whatever London did or did not do, settler populations would expand given the continent's vast resources. In 1755, Franklin implicitly warned what Britain's choice was: accede to "white" Americans' interests or face the consequences that they would grow in numbers and be ungovernable. Franklin's "political polypus" was so notorious that when later English authors used the metaphor to explain any form of population growth, they were considered unpatriotic.[60]

Multiplying Numbers, Fertile Imagination

By the 1760s, authors began treating sex as the most powerful variable in the population equation.[61] William Temple explained, "The desire of union between the sexes, is so strongly implanted in mankind by the wise Author of nature, that a man may with as much reason expect to see the laws of vegetation suspended, as marriages to stop among the bulk of people."[62] Lust was so powerful, however, that neither reason nor suffering could stop it. Adam Smith noted that "Poverty, though it no doubt discourages, does not always prevent marriage." In fact, poverty could have the ironic effect of "seem[ing] even to be favourable to generation. A half-starved Highland woman frequently bears more than twenty children, while a pampered fine lady is often incapable of bearing any, and is generally exhausted by two or three." The tragic result was naturalized through a botanical metaphor: "The tender plant is produced, but in so cold a soil and so severe a climate, soon withers and dies."[63] Sir James Steuart moved analogically from biology to physics, turning the battle between desire and resources into a predictable law: "The generative faculty resembles a spring loaded with a weight, which always exerts itself in proportion to the diminution of resistance"; when there is plentiful food, the spring will uncoil, and "generation will carry numbers as high as possible," but when supplies grow thin, "the force of it becomes less than nothing."[64]

Malthus went even further. He portrayed the tension between sex and resources as a battle between life and death. As a fundamental truth, he moved from metaphor to mathematics: "The passion between the sexes has appeared in every age to be so nearly the same that it may always be considered, in algebraic language, as a given quantity."[65] Math is a symbolic language too, but one that in Malthus's age was commonly understood as representing eternal truths.[66] In fact, as Arthur Walzer observed, Malthus deliberately echoed Sir Isaac Newton's *Principia*, first, by entitling his work *Principles*; second, by making his argument a mathematic proof; and third, by presenting population as immutable a law as any law of calculus.[67] Unlike a metaphor that relies on readers conjuring up "mice," "fennel," "vegetation," "springs," and so on in their own minds, Malthus's readers understood that a mathematical equation or algebraic figure was universally agreed upon—everybody knew what any number meant. When the Cambridge mathematics graduate turned sex and food into equations, he was proclaiming no need for metaphor.

In fact, Malthus's work was redolent with rhetoric, and here his critics attacked.[68] Many claimed that his arguments were built on false analogies and stylistic bravura, not facts. Some disagreed with his metaphorical fields. Like many critics, the pseudonymous Simon Gray accused Malthus of placing "Man . . . not on the same footing with animals but with vegetables."[69] While they resisted his particular analogies because they supposedly conjoined unrelated or wrongly related things (e.g., humans and vegetables), critics were not rejecting the ontological power of metaphor. Gray and others instead proposed different metaphorical fields because they recognized figuration was cognitively and politically persuasive.

Malthus used synecdoche, allegory, and narrative descriptions too—for example, representing the consequences of overpopulation through a pitiable description of poor, starving peasant boys so hungry that they "'are very rarely seen with any appearance of calves to their legs.'"[70] He inverted the traditional valorization of the countryside and repudiated faith in rural labor as replenishing the nation. Critics recognized his storytelling power, pointing out that he could have just as easily chosen positive allegories. The Reverend James Law, who promoted granting acreage to all Britons, argued that had Malthus used another "personification,—the situation of that far-fetched mighty personage, John Bull," he would have seen that England's population was naturally productive, responsible (and beef eating).[71]

Other critics accused Malthus of slyly creating a drama between the individual and the whole by relying far too heavily on personification.[72] The liberal journalist William Hazlitt explained that Malthus had turned individuals into "vegetables" that were controlled by an anthropomorphic monster: "Population was in fact the great devil, the untamed Beelzebub that was only kept chained down by vice and misery."[73] Malthus had turned population into a living, breathing antagonist: "Our author has been hurried into an unfounded assumption by having his imagination heated with a *personification*. He has given to the principle of population a personal existence, conceiving of it as a sort of infant Hercules, as one of that terrific giant brood, which you can only master by strangling it in its cradle; forgetting that the antagonist principle which he has made its direct counterpoise, still grows with its growth and strengthens with its strength, being in fact its own offspring."[74] Hazlitt's invocation of infanticide was satirical, but it also underscored what he considered Malthus's murderous political views that the poor laws should be abolished.[75]

Like nearly all of Malthus's critics, Hazlitt recognized the power to persuade through rhetoric. But tropes were not truths. They were highly personal, concocted in the writer's imagination. Hazlitt suggested not everybody was uncontrollably driven by desire. In fact, it was Malthus who was obsessed with sex. Hazlitt teased, "I should suppose that Mr. Malthus to be a man of warm constitution and amourous complexion. . . . But the women are the devil . . . the smiles of a fair lady are to him irresistible; the glimpse of a petticoat throws him into a flame." Malthus meant for math to be irrefutable and true, but Hazlitt mocked how he "gravely reduces the strength of the passion to a mathematical certainty," which "is sure to have the women on his side."[76] In other words, Malthus had projected his own repressed desire onto the entire field of political economy, mistaking his fear of personal self-control for a universal truth.

Instead of using himself for his metaphorical field, according to Hazlitt, Malthus should focus on women's reproductive decisions if female fertility was the determinant for population growth. Look then at how Englishwomen behave: "Almost every little Miss, who has had the advantage of a boarding school education, or been properly tutored by her mamma, whose hair is not of absolute flame-colour . . . waits patiently year after year, looks about her, rejects or trifles with half a dozen lovers . . . till she is at last smitten with a handsome house, a couple of footmen in livery, or a black-servant, or a coach with two sleek geldings, with which she is more taken with than her man."[77] Like Cantillon and earlier theorists who predicated women's reproduction on men's ability to support them, Hazlitt argued that human population was controlled by women's marital decision-making. The Reverend Malthus presumed that all people could scarcely control their desire—this was both an early modern view and one reinforced by Christian teaching. Early modern theorists like Graunt and Petty depicted both sexes as fornicators and emphasized women's lustfulness. By Malthus's age, however, contemporaries were articulating new ideas about human sexuality, which had thoroughly mixed results. As many historians have long noticed, sexual desire was increasingly bifurcated along gender lines, with all men associated with dangerous passion and bourgeois women with sexual modesty, even apathy.[78] Hazlitt satirized these gendered transformations depicting the typical woman as both acquisitive and asexual, less interested in "her man" than his things. The situation was not as dire as Malthus feared, Hazlitt insisted. Perhaps philosophers, population theorists, and policy makers might be able to regulate population just fine. After all, they should "be able to manage these matters as decently and cleverly as the silliest women can do at present!"[79]

Hazlitt's response to Malthus was cheerful and cheeky, a happy collapsing of population with middle-class manners. But that's not how Britain's early nineteenth-century poor law bureaucrats and political figures viewed the relationship between sex, gender, and population. Nor is it how nineteenth-century Britons viewed lower-class women, who were sometimes depicted as sexually voracious and dangerously powerful.[80] Malthus himself sympathized with single mothers' burdens, but later commentators characterized these women far more negatively, explaining that they were "nine times out ten, less the seduced than the seducer."[81] When asked about out-of-wedlock pregnancies, parish officers highlighted the danger of sex among the poor. Unmarried mothers were "pests."[82] No longer described as productive busy bees, the poor—especially bastard bearers—were now cast as vermin. This transformation in population theory returned humans to comparison with lower life-forms, now laden with classism, sexism, and the apparent truth of mathematics. Population had become a cornerstone of political economy and political policy with proof in the census from 1801 forward. Bureaucrats argued that the *math*—like "nine out of ten" female seducers and the fact of the nation's rapidly expanding numbers in the census—defined what population was. By the 1820s, demography's metaphorical field was entirely mixed, the comparisons shaped and impelled by political ideology.[83]

Notes

I am grateful to the anonymous outside readers and David Alff, Carla Bittel, Bill Forman, Dan Livesay, Lindsay O'Neill, Tawny Paul, Danielle Spratt, Amy Woodson-Boulton, and especially Erika Rappaport for their insights and helpful suggestions.

1. Bernard de Mandeville, *The Fable of the Bees*, 2nd ed. (London: Edmund Parker, 1723), 328; Richard Cantillon, *Essai sur la nature du commerce en general* ("London" [Paris]: "Fletcher Gyles," 1755), 110; Cantillon died in 1734, but manuscripts of his work circulated among economic theorists, who may have had the *Essay* posthumously published; Richard van den Berg, *Richard Cantillon's Essay on the Nature of Trade in General: A Variorum Edition* (London: Routledge, 2015), 5; [Benjamin Franklin], *Observations on the Late and Present Conduct of the French. . . . to Which Is Added, Wrote by Another Hand: Observations concerning the Increase of Mankind, Peopling of Countries, &c.* (Boston: S. Kneeland, 1755), 13; [Thomas Robert Malthus], *An Essay on the Principle of Population* (London: J. Johnson, 1798), 45; [William Hazlitt], *A Reply to the "Essay on Population"* (London: Longman, 1807), 56.
2. James Bonar, *Theories of Population from Raleigh to Arthur Young* (London: Allen & Unwin, 1931).

3. See Keiser's "Science for the Birds: Figurative Language and The History of the Royal Society" and Boyle's "Jane Barker and Virgin Anatomy" in this volume.
4. The literature here is enormous, but some relevant texts here include Arthur E. Walzer, "Logic and Rhetoric in Malthus's *Essay on the Principle of Population, 1798*," *Quarterly Journal of Speech* 73, no. 1 (February 1987): 1–17; Emily Martin, "The Egg and the Sperm: How Science Has Constructed a Romance," *Signs* 16, no. 3 (Spring 1991): 485–501; Anna Wierzbicka, "What Is a *Life Form*? Conceptual Issues in Ethnobiology," *Journal of Linguistic Anthropology* 2, no. 1 (June 1992): 3–29; Mary Poovey, *Making a Social Body: British Cultural Formation, 1830–1864* (Chicago: University of Chicago Press, 1995); Catherine Gallagher, *The Body Economic: Life, Death, and Sensation in Political Economy and the Victorian Novel* (Princeton: Princeton University Press, 2006); Devin Griffiths, *The Age of Analogy: Science and Literature between the Darwins* (Baltimore: Johns Hopkins University Press, 2016); Tita Chico, *The Experimental Imagination: Literary Knowledge and Science in the British Enlightenment* (Stanford: Stanford University Press, 2018); Emily Steinlight, *Populating the Novel: Literary Form and the Politics of Surplus Life* (Ithaca: Cornell University Press, 2018); Andrea Charise, *The Aesthetics of Senescence: Aging, Population, and the Nineteenth-Century British Novel* (Albany: State University of New York Press, 2020); Charlotte Sussman, *Peopling the World: Representing Human Mobility from Milton to Malthus* (Philadelphia: University of Pennsylvania Press, 2020); and Robert Mitchell, *Infectious Liberty: Biopolitics between Romanticism and Liberalism* (New York: Fordham University Press, 2021).
5. W. V. Quine, "A Postscript on Metaphor," *Critical Inquiry* 5, no. 1 (Autumn 1978): 161; Hanna Pulaczewska, *Aspects of Metaphor in Physics* (Berlin: De Gruyter, 1997).
6. On the power of metaphor in early modern astronomy texts, including those of Bernard Le Bovier de Fontanelle, see Frédérique Aït-Touati, *Fictions of the Cosmos: Science and Literature in the Seventeenth Century* (Chicago: University of Chicago Press, 2011), 79–88.
7. George Lakoff, *Women, Fire, and Dangerous Things: What Categories Reveal about the Mind* (Chicago: University of Chicago Press, 1987); George Lakoff and Mark Johnson, *Metaphors We Live By* (Chicago: University of Chicago Press, 1980); George Lakoff and Mark Turner, *More Than Cool Reason: A Field Guide to Poetic Metaphor* (Chicago: University of Chicago Press, 1989).
8. Madhusree Mukerjee, *Churchill's Secret War: The British Empire and the Ravaging of India during World War II* (New York: Basic Books, 2010), 205; Elena R. Gutiérrez, *Fertile Matters: The Politics of Mexican-Origin Women's Reproduction* (Austin: University of Texas Press, 2008), 11, 52. The rabbits metaphor is so recognizably distasteful now that it in turn has become part of a trope to characterize racism or classism: for example, Gutiérrez's preface

begins, "[This book] is an exploration of the ways we have come to think about the reproduction of women of Mexican origin in the United States. In particular, I look closely at one of the most popular and longstanding public stereotypes that portray Mexican American and Mexican women as 'hyperfertile baby machines' who 'breed like rabbits'" (xi).

9. Brad Pasanek, *Metaphors of Mind: An Eighteenth-Century Dictionary* (Baltimore: Johns Hopkins University Press, 2015).

10. Though seventeenth-century English-language speakers did not mention it, other early modern states tabulated human numbers more directly. Younghoon Rhee, "A Comparative Historical Study of the Census Registers of Early Choson Korea and Ming China," *International Journal of Asian Studies* 2, no. 1 (January 2005): 25–55; Gunnar Thorvaldsen, "An International Perspective on Scandinavia's Historical Censuses," *Scandinavian Journal of History* 32, no. 2 (September 2007): 237–57. Adam Smith noted this material, but Malthus would question its trustworthiness.

11. William Petty, *Another Essay in Political Arithmetick . . . , 1683* (London: Printed for H. H. for Mark Pardoe, 1683); De Souligné, *London Bigger Than Old Rome* (London: Printed by A. S., 1701); De Souligné, *A Comparison between Old Rome in Its Glory . . . and London . . .* , 2nd ed. (London: John Nutt, 1706).

12. Andrea A. Rusnock, *Vital Accounts: Quantifying Health and Population in Eighteenth-Century England and France* (Cambridge: Cambridge University Press, 2002).

13. John Graunt, *Natural and Political Observations . . . Made upon the Bills of Mortality* (London: Printed for Tho[mas] Roycroft, 1662), 61.

14. Robert Powell, *Depopulation Arraigned, Convicted and Condemned* (London: R[ichard] B[adger], 1636), 65.

15. Nicholas Barbon, *An Apology for the Builder; or a Discourse Shewing the Cause and Effects of the Increase of Building* (London: Cave Pullen, 1689), 27, 31. On Barbon and his metaphors, see David Alff, *The Wreckage of Intentions: Projects in British Culture, 1660–1730* (Philadelphia: University of Pennsylvania Press, 2017), 54–55; on Barbon and contemporary debates more broadly, see Joyce Oldham Appleby, *Economic Thought and Ideology in Seventeenth-Century England* (Princeton: Princeton University Press, 1978).

16. Graunt, *Natural*, 38.

17. Graunt, 40.

18. Gill Newton and Richard Smith, "Convergence or Divergence? Mortality in London, Its Suburbs and Its Hinterland between 1550 and 1700," *Annales de démographie historique*, no. 126 (2013/12): 48. Though contemporary and modern demographic interpretations aligned for the seventeenth century, they do not for the eighteenth century. Births increased (due to the drop in the

age of female marriage from the early eighteenth century forward), but contemporaries did not see this. In the eighteenth century, they debated whether the population was growing; after the 1801 census, growth was demonstrated, but according to Cem Behar, few authors addressed fertility even into the nineteenth century. Behar, "Malthus and the Development of Demographic Analysis," *Population Studies* 41 (1987): 276.
19. Graunt, *Natural*, 46.
20. Raymond Williams, *The Country and the City* (New York: Oxford University Press, 1973).
21. [John Brown], *An Estimate of the Manners and Principles of the Times,* 7th ed. (London: L. Davis and C. Reymers, 1758), 191. Robert Wallace's treatise also argued that luxury caused depopulation: *A Dissertation on the Numbers of Mankind in Antient and Modern Times* (Edinburgh: G. Hamilton and J. Balfour, 1753).
22. Barbon, *Apology,* 30. Matthew Wren described cities as "both the Heart and the Head": *Monarchy Asserted* (Oxford: W. Hall for F. Bowman, 1659), 152.
23. Barbon, *Apology,* 2–3; Alff, *Wreckage,* 55. Graunt first used the phrase in his epistle, *Natural,* n.p.
24. Anon., *A Computation of the Increase of London* (London: s.n., 1719), 6–7.
25. Beginning with Charles Cotton's *Erotopolis: The Present State of Betty-Land* (London: Tho[mas] Fox, 1684), the *"Merryland"* genre of pornography conflated landscape and sexual bodies; Karen Harvey, *Reading Sex in the Eighteenth Century: Bodies and Gender in English Erotic Culture* (Cambridge: Cambridge University Press, 2004).
26. Quoted in Bonar, *Theories,* 57.
27. Graunt, *Natural,* 46.
28. Anon., *The Compleat Doctoress, or A Choice Treatise of All Diseases Insident to Women* (London: Edward Farnham, 1656), 133.
29. Nicholas Culpeper, *A Directory for Midwives* (London: J. and A. Churchill, 1701), 23.
30. Graunt, *Natural,* 51. The suspicion that polygamy failed to increase population was nearly ubiquitous during the long eighteenth century; see Alfred Owen Aldridge, "Population and Polygamy in Eighteenth-Century Thought," *Journal of the History of Medicine* 4, no. 2 (Spring 1949): 129–48. For a synopsis of why polygamous cultures can in fact have high overall fertility rates, see Bruno Schoumaker, "Across the World, Is Men's Fertility Different from That of Women?," *Population & Societies* no. 548 (October 2017): 1–5.
31. Graunt, *Natural,* 48; Beryl Rowland, *Animals with Human Faces: A Guide to Animal Symbolism* (Knoxville: University of Tennessee Press, 1973), 161–67.
32. Ted McCormick, *William Petty and the Ambitions of Political Arithmetic* (Oxford: Oxford University Press, 2009); Ted McCormick, "Alchemy in the

Political Arithmetic of Sir William Petty," *Studies in History & Philosophy of Science* 37, no. 2 (2006): 290–307; Graunt, *Natural*, 96–98.
33. "Les Hommes se multiplient comme des Souris dans une grange, s'ils ont le moïen de subsister sans limitation. . . ." Cantillon, *Essai*, 110.
34. Tudor-Stuart author Edward Topsell said the mouse was "in general most libidinous" and cited sexualized references going back to Xenophon. Edward Topsell, *The Historie of Foure-footed Beastes* (London: William Iaggard, 1607), 20; Gordon Williams, *A Dictionary of Sexual Language and Imagery in Shakespearean and Stuart Literature* (London: Athlone Press, 1994). On the multiple traits of mice, see Lucinda Cole, *Imperfect Creatures: Vermin, Literature, and the Sciences of Life, 1600–1740* (Ann Arbor: University of Michigan Press, 2016), 28–29, 39, 42–46.
35. On mouse slang, see Rowland, *Animals*, 127–29; and Rowland, "Forgotten Metaphor in Three Popular Children's Rhymes," *Southern Folklore Quarterly* 31 (1967): 12–19.
36. Petty, *Another Essay*, 37.
37. Samuel Johnson, "bee," *A Dictionary of the English Language* (London, 1755), 1:n.p.; Dror Wahrman, *The Making of the Modern Self* (New Haven: Yale University Press, 2004), 3–6. For additional examples of metaphorical bees helping make economic arguments, see Mandeville, "The Grumbling Hive," in *Fable*, 1–22; and Jonathan Swift, who portrayed bees as hardworking and responsive to nature—like the ancients—while spiders and the moderns are critical and destructive in *A Tale of a Tub. . . . to Which Is Added, an Account of a Battle between the Antient and Modern Books in St. James's Library*, 4th ed. (Dublin: s.n., 1705), 136–40.
38. Richard Brookes, *A New and Accurate System of Natural History* (London: J. Newbery, 1763), 301. On the long-held beliefs that mice resulted from spontaneous generation (rather than sexual intercourse), see Cole, *Imperfect Creatures*, 28–29. See Clara Pinto-Correia, *The Ovary of Eve: Egg and Sperm and Preformation* (Chicago: University of Chicago Press, 1997), 340n5, for late eighteenth-century claims that Egyptian mice sprang from the mud of the Nile.
39. "Une Fille prend soin de ne pas devenir Mere, si elle n'est mariée; elle ne se peut marier si elle ne trouve un Homme qui veuille en courir les risques." (A young woman is careful not to become a mother unless she is married; she cannot marry unless she can find a man who is willing to take the risk.) Cantillon, *Essai*, 105. Translation mine.
40. "Le célibat des Gens d'église n'est pas si désavantageux qu'on le croit vulgairement; . . . mais leur fainéantise est très nuisible." (The celibacy of clerics is not as disadvantageous as commonly believed; . . . but their indolence is very damaging.) Cantillon, 126. Translation mine.

41. The first population discussion I have found comparing rapid human reproduction and rabbits is an article in the *Whitehall Evening Post*, April 26–28, 1798: "Nothing but salt is wanted to fill the isles and Northern coasts of Scotland with as numerous swarms of men of hardy and well-behaved fishers and sailors.... Give but the Highlanders salt for their fish and their [oatmeal], and they will breed like antelopes and rabbits."
42. William Popple, *The Lady's Revenge, or The Rover Reclaim'd* (London: J. Brindley, 1734); on servants' and other dependent laborers' vulnerability, see Patricia Fumerton, *Unsettled: The Culture of Mobility and the Working Poor in Early Modern England* (Chicago: University of Chicago Press, 2006); for employers' range of reactions to married and pregnant servants, see Tim Meldrum, *Domestic Service and Gender, 1660–1750: Life and Work in the London Household* (Harlow: Routledge, 2000), 114–17.
43. Jonas Hanway, *Letters on the Importance of the Rising Generation for the Laboring Part of Our Fellow-Subjects* (London: A. Millar, 1767), 1:11.
44. [Franklin], *Observations*, 14.
45. For the endless bounty that North America supposedly could provide, see Robert Markley, "'Land enough in the World': Locke's Golden Age and the Infinite Extension of 'Use,'" *South Atlantic Quarterly* 98, no. 4 (Fall 1999): 817–37; and Sussman, *Peopling the World*, 182–87.
46. [Franklin], *Observations*, 10. Franklin's work was quickly and widely disseminated in Britain, for example, in *Gentleman's Magazine* 25 (November 1755): 483–85; Alan Houston, *Benjamin Franklin and the Politics of Improvement* (New Haven: Yale University Press, 2008), 106–46.
47. [Franklin], *Observations*, 12.
48. Pierre H. Boulle, "François Bernier and the Origins of the Modern Concept of Race," in *The Color of Liberty: Histories of Race in France*, ed. Sue Peabody and Tyler Stovall (Durham: University of North Carolina Press, 2003), 11–27.
49. [Franklin], *Observations*, 14.
50. [Franklin], 13.
51. [Franklin], 14.
52. [Franklin], 12.
53. In the 1750s, satirists played with scientific and reproductive possibilities, conflating plants and humans; they provided an interesting example of botanical reproduction serving not as a metaphor for the zoological but rather as a literal, if speculative, comparison. See Danielle Spratt, "Surrogacy and Empire in *The Man-Plant* and Eighteenth-Century Vernacular Medical Texts," in *The Routledge Companion to Humanism and Literature*, ed. Michael Bryson (New York: Routledge, 2022), 229–47.
54. [Franklin], *Observations*, 15, 8.
55. [Franklin], 8.

56. [Franklin], 14–15.
57. Arguments about the slave trade, slavery, and abolition were filtered through the category of population, and these conversations were rich with analogies and figurative language, no matter which side; as one example of a metaphorically dense defense of slavery, see John Hippisley, *Essays: I. On the Populousness of Africa* (London: T. Lownds, 1764). North American colonists feared that Africans were reproducing at more rapid rates than northern Europeans; Katherine Paugh, *The Politics of Reproduction: Race, Medicine, and Fertility in the Age of Abolition* (Oxford: Oxford University Press, 2017), 24–30. Indeed, North American slaves' birth rates in the eighteenth and nineteenth centuries were high (compared to those of European settlers and Caribbean slaves); for an explanation based on the gendered labor requirements of sugar production versus tobacco and cotton cultivation, see Michael Tadman, "The Demographic Cost of Sugar: Debates on Slave Societies and Natural Increase in the Americas," *American Historical Review* 105, no. 5 (December 2000): 1534–75.
58. [Franklin], *Observations*, 13; Susannah Gibson, "On Being an Animal, or, the Eighteenth-Century Zoophyte Controversy in Britain," *History of Science* 50, no. 4 (2012): 453–76.
59. [Franklin], *Observations*, 13.
60. "G.W.," *Gentleman's Magazine* 52 (August 1782): 373; G.W. was critiquing John Howlett, *An Examination of Dr. Price's Essay on the Population of England and Wales* (Maidstone: For the author, 1781), 22. Howlett defended his patriotism (and the usefulness of metaphor) in *Gentleman's Magazine* 52 (October 1782): 473–75.
61. On the rising interest in reproduction among male theorists and doctors by the 1760s, see Lisa Forman Cody, *Birthing the Nation: Sex, Science, and the Conception of Eighteenth-Century Britons* (Oxford: Oxford University Press, 2005).
62. [William Temple], *A Vindication of Commerce and the Arts* (London: J. Nourse, 1758), 15.
63. Adam Smith, *An Inquiry into the Nature and Causes of the Wealth of Nations*, 2 vols. (London: W. Strahan and T. Cadell, 1776), 1:96–97.
64. Sir James Steuart, *The Principles of Political Oeconomy* (1767; Chicago: University of Chicago Press, 1966), 32.
65. [Malthus], *Essay*, 128.
66. George Lakoff and Rafael E. Nuñez, *Where Mathematics Comes From: How the Embodied Mind Brings Mathematics into Being* (New York: Basic Books, 2001).
67. Walzer, "Logic and Rhetoric," 4–12.
68. *British Critic* 17 (1801): 278–82; James Grahame, *An Inquiry into the Principle of Population* (Edinburgh: James Ballantyne and Col., 1816), 31–34, 58; George Ensor, *An Inquiry concerning the Population of Nations: Containing a Refutation of Mr. Malthus's Essay on Population* (London: E. Wilson, 1818), 80–81, 126, 132.

69. [Simon Gray], George Purves, pseud., *Gray versus Malthus: The Principles of Population and Production Investigated* (London: Longman, 1818), 127, 132.
70. [Malthus], *Principles*, 73, quoted in Walzer, "Logic and Rhetoric," 14–15.
71. James Thomas Law, *The Poor Man's Garden. . . . with Remarks, Addressed to Mr. Malthus . . .* , 3rd ed. (London: Gilbert and Rivington, 1830), 19.
72. James P. Huzel, *The Popularization of Malthus in Early Nineteenth-Century England: Martineau, Cobbett, and the Pauper Press* (Aldershot: Ashgate, 2006).
73. [Hazlitt], *Reply*, 56.
74. [Hazlitt], 90.
75. Hazlitt's jibes, of course, echoed those of Jonathan Swift in *A Modest Proposal Preventing the Children of Poor People* (Dublin: S. Harding, 1729); for the debates about improvement through biological and social reproduction, see Jenny Davidson, *Breeding: A Partial History of the Eighteenth Century* (New York: Columbia University Press, 2009), esp. 81–83, 183–86.
76. [Hazlitt], *Reply*, 140, 141.
77. [Hazlitt], 52–53.
78. Nancy F. Cott, "Passionlessness: An Interpretation of Victorian Sexual Ideology, 1790–1850," *Signs: Journal of Women in Culture & Society* 4, no. 2 (Fall 1978): 219–36; Marjorie Levine-Clark, *Beyond the Reproductive Body: The Politics of Women's Health and Work in Early Victorian England* (Columbus: Ohio State University Press, 2004); Faramerz Dabhoiwala, *The Origins of Sex: A History of the First Sexual Revolution* (Oxford: Oxford University Press, 2012).
79. [Hazlitt], *Reply*, 52–53. There's some irony in Hazlitt accusing Malthus of sexual obsession, as he was himself troubled by sexual yearnings and a series of unhappy marriages (and divorces). He was mocked by conservatives for his supposed libertine behavior and for his "emotionally raw, at times vulgar and embarrassing," 1823 work on a failed sexual affair. Jonathan Bate, "Hazlitt, William (1778–1830), Writer and Painter," *Oxford Dictionary of National Biography*, September 17, 2015, https://www.oxforddnb.com.
80. Anna Clark, *The Struggle for the Breeches: Gender and the Making of the British Working Class* (Berkeley: University of California Press, 1997), 44; Linda Nead, *Victorian Babylon: People, Streets, and Images in Nineteenth-Century London* (New Haven: Yale University Press, 2000).
81. John Easterly, "Appendix C: Communications," in *Report from His Majesty's Commissioners for Inquiring into the Administration and Practical Operation of the Poor Laws* (London: n.p. 1834; reprint in *Irish University Press Series of British Parliamentary Papers* [Shannon: Irish University Press, 1970]), 17:399–400c, 409c.
82. Edward Tregaskis, "Appendix C: Communications," in *Report*, 17:399–400.
83. Lisa Forman Cody, "The Politics of Illegitimacy in an Age of Reform: Women, Reproduction, and Political Economy in England's New Poor Law of 1834," *Journal of Women's History* 11, no. 4 (Winter 2000): 131–56; Marjorie

Levine-Clark, "Engendering Relief: Women, Ablebodiedness, and the New Poor Law in Early Victorian England," *Journal of Women's History* 11, no. 4 (Winter 2000): 107–30; Anna Clark, "The New Poor Law and the Breadwinner Wage: Contrasting Assumptions," *Journal of Social History* 34, no. 2 (Winter 2000): 261–81. For the broader developing contexts, see David Philips, "Three 'Moral Entrepreneurs' and the Creation of a 'Criminal Class' in England, c. 1790s–1840s," *Crime, Histoire & Sociétés* 7, no. 1 (2003): 79–107; and Oz Frankel, *States of Inquiry: Social Investigation and Print Culture in Nineteenth-Century Britain and the United States* (Baltimore: Johns Hopkins University Press, 2006).

Jane Barker and Virgin Anatomy

FRANK BOYLE

To support her claim that women and men have identical intellectual capacities, Mary Astell, in her *Defence of the Female Sex* (1696), calls on the latest evidence from physicians and anatomists: "Neither can it [any defect in women] be in the Body, (if I may credit the Report of learned Physicians) for there is no difference in the Organization of those Parts, which have any relation to, or influence, over the Minds; but the Brain, and all other Parts (which I am not Anatomist enough to name) are contriv'd as well for the plentiful conveyance of [animal] Spirits, which are held to be the immediate Instruments of Sensation, in Women, as Men."[1] No writer is better positioned to help us understand the provenance and implications of Astell's new-scientific claims than Jane Barker. Born in 1652 in Northamptonshire, Barker first published poems in late 1687 and her first of several novels in 1713. She died at nearly eighty years of age in 1732. Among her contributions to the writing of the period were "patchwork" novels, narratives created as conversations between women where various episodes correspond to the patches added to the quilts being created by the women's hands. In the most autobiographical of these works, *A Patch-Work Screen for the Ladies* (1723), Barker incorporates a number of poems from her 1688 *Poetical Recreations,* including several discussing the death of her beloved brother Edward in 1675.[2] She credits Edward, who studied for a medical career at Oxford, with sharing his education with her, including fostering her skill in "simpling," or herbal pharmaceuticals, as well as her interest in anatomy. Projecting herself into the career left absent by her brother, Barker draws on the new scientific culture and evidence to figure what we now call the "modest witness"—an agent of scientific objectivity—as a category where a chaste and curious mind, not biological sexual characteristics, qualify one for a role.[3]

Among the many tragedies suffered by Galesia, Barker's literary persona in the novels, are fraught romances with false suitors who wound and disillusion Galesia emotionally but fail to stain her virtue.[4] Learning from these sordid affairs, Galesia embraces the unmarried life, as did Barker herself. The *Patch-Work Screen for the Ladies* embroiders the biographical into the narrative by including Barker's poem from the 1688 collection, "A Virgin Life." Galesia's patchwork companion responds to the poem by remarking on her "submissive and resign'd" mind (*GT*, 140), but Kathryn King, noting that the poem has Barker defending herself from "Mens almost Omnipotent Amours" (*PR*, 12), sees Barker drawing "out the ambiguities in relations of domination and submission" and modeling "the exemplary female life in terms of socially responsible celibacy."[5] Even those who read her embrace of celibacy in relation to her conversion to Catholicism see complexities: on one hand, there is "bitterness" in her acceptance of her circumstances, and on the other, she figures convents as places "of political and cultural resistance, allowing women an alternative to marriage and motherhood."[6] Recently, Alice Tweedy McGrath, building on the work of Susan Lanser and Paula Backscheider, has read Barker's "patchwork aesthetic as a methodological and formal model of queer failure," a form glancingly linked to *Tristram Shandy*, which will not submit to heteronormative accountability.[7] In her own words, Barker claims the virgin life allows her to attend to her "whole lives business," to be as "good a Subject as the stoutest Man," and "to serve her God, her Neighbour, and her Friends" (*GT*, 140).

Beyond her patchwork aesthetics and her "challenge to the heteropatriarchy," Barker and her writing are often discussed in relation to medicine and the new science.[8] There is evidence that she practiced as a physician in London in the 1680s and may have written prescriptions.[9] Moreover, she wrote with pride about her medical skills and knowledge, most notably in a poem that in the 1688 collection is called "A Farewell to Poetry with a Long Digression on Anatomy." A *Magic School Bus*–like trip through the human body in which the speaker is guided by leading anatomists, this poem was revised sometime in the following decade, and then more than thirty-five years after its initial publication, it was reframed and extensively elaborated on in *Patch-Work Screen for the Ladies*. Jess Keiser has read this poem as the exemplar of what he terms "nervous fiction," a hybrid mode in which scientific and poetic discourses are not only linked but generative of each other, while Rachel Mann ties the understanding of science in the poem to Barker's extensive experience as a coterie poet, where narratives emerge from a patchwork of crucial social exchanges and vetting whether among scientists or poets.[10] I propose in this essay to link Barker's thinking about sex and celibacy to the new science and to knowledge production. Astell points to the evidence that both male and female brains are "contriv'd . . . for the plentiful conveyance of Spirits," but Barker figures her own chaste female brain engaged, along with modest male scientists, in a literally visceral search for "New Lights" to illumine "Nature's obscurity" (PR, 99).

My approach to getting at Barker's thinking about both science and single life will be to recover as much as possible the narrative of the *Poetical Recreations* from the author's subsequent revisions and reworking. My rationale for this is that Barker's committed scientific study—in all her accounts—revolves around her brother and his death in 1675 when she was in her twenties and extending perhaps into her thirties. The revisions of the Magdalen Manuscript date from after her conversion to Catholicism and are believed to have been complete in 1704 when she was in her fifties, while the *Patch-Work Screen for the Ladies*, which includes a version of both "A Virgin Life" and the anatomy poem, appears when she is in her seventies.[11] Noticing that *Poetical Recreations* figures the voice of a talented and educated young female poet at the center of an overwhelming male coterie argues for reading that voice apart from—at least as a first step—the voice of the older talented and educated writer who has repositioned herself religiously, politically, and as an artist whose coterie is now a community of patchworking women.[12]

An Invitation to Celibacy

Suffused with grief for her brother, Barker's volume begins with the young female poet's passionate invitation to a group of admiring young men to join her in chaste intercourse within and about teeming nature. It proceeds to a clarion statement of personal identity focused on living life as a virgin; foregrounds the speaker's education, medical knowledge, and poetic skills; includes witty advice to men on artistic and intimate matters; and ends with a rejection of poetry in favor of the study of anatomy.[13] Far from incongruous or random, these seemingly disparate elements of the collection reveal a speaker advocating for something like what Benjamin Kahan (working in a later period and without reference to science or empiricism) has termed "celibacy as a sexuality," one that is understood as "an organization of pleasure rather than a failure, renunciation, or even ascesis of pleasure."[14] In her brother, in his scientifically minded colleagues and friends, in the university men of her own coterie, and in her heroes among the famous men of the new science, Barker figures not only an opening for a woman like her whom patriarchal norms would ordinarily exclude but also a path forward for both men and women to escape a heteronormative world she relentlessly portrays as deadening to human engagement and pleasure. This essay reads *Poetical Recreations* as Barker performing dramatic exchanges—her normative nubile value for a life of extended intellectual pleasure and her recognized poetic talent in favor of joining the project of empiricism via anatomy—in service of a manifesto-like invitation to her scientifically minded peers to, as Donna Haraway has put it, "queer the modest witness."[15] In Barker's terms, this means embracing a way to sustained pleasure and "recreations" in collective research that obviates the misconceived need for fleeting sexual pleasures and stultifying human procreation.

Poetical Recreations opens with a preface by the publisher, Barker's friend Benjamin Crayle, followed by six poems praising Barker's verse by those "that know how to judge of Poetry." These encomiums are extravagant, making Barker the successor to the "Great Orinda" (Katherine Philips, 1631–64), comparing her to Johnson and Shakespeare, and tracing a parallel, via metaphors of womb and birth, between her "Almighty Pen" and the "Almighty" who created the "Infant World" out of chaos.[16] Her own verse begins appropriately enough with "An Invitation to My Friends at Cambridge," a poem that invites her male coterie to join her in her rural retreat where nature's and her own "innocence" overlap to create a prelapsarian space of distracting "joy" (*PR*, 1). Much has been appropriately made of this poem's extended reflection on the tree of knowledge and "Our Maker's Laws" that sanction excluding women while rewarding men with "Luxurious Banquets" of education.[17] But Barker is not (or not only) opening her volume with what was a recognized injustice within her coterie.[18] Instead, Barker opens a space similar in striking ways to that of Andrew Marvell's "The Garden," where society, defined largely in terms of heterosexual pursuits and norms, can be escaped by a mind freed "To a green thought in a green shade." While Marvell's speaker locates his "happy garden-state" in the time when "man there walked without a mate" (ll. 57–58), Barker's speaker figures a chaste pastoral retreat where she, an innocent Eve, invites men to "participate" in the pleasures that are found beyond "this World's Gallantry," a prelapsarian space where "we"—not the isolated male or female self—"have full inlargement of the Mind" (*PR*, 2).

Barker's rural retreat contrasts human heterosexual games with the pervasive sex that drives nature:

> For here's no *pride* but in the *Sun's* bright *beams*;
> Nor *murmuring*, but in the Crystal *streams*
> No *avarice* is here, but in the *Bees*,

> Nor is *Ambition* found but in the *Trees*.
> No *Wantonness* but in the frisking *Lambs*,
> Nor *Luxury* but when they suck their *Dams*.
> Nor are there here Contrivances of States;
> Only the Birds contrive to please their Mates;
> Each minute they alternately improve
> A thousand harmless ways their artless love. (PR, 2)

Ambition, wantonness, luxury, murmuring, frisking, sucking, and *contriving* all have their innocent analogs in the natural world but here emphasize that this female invitation to men—this invitation to collaborate—is outside and beyond the economy of human romance:

> 'Tis such a pleasing solitude as yet
> Romance ne're found, where happy Lovers met:
> Yea such a kind of solitude it is,
> Not much unlike to that of Paradise,
> Where all things do their choicest good dispence.
> And I too here am plac'd in innocence. (PR, 3)

Milton's *Paradise Lost* is a tragic love story with heteronormative sex before, during (as consummation of), and after the Fall. Barker reads Eden as a place of solitude where sex is not for, or at least between, humans. She here figures herself an unmarried Eve enthralled by fecund nature and welcoming of young men whose "Vertues," enhanced by their educations, "are bright and fair" (PR, 4).

Despite the early prominence of the poem "A Virgin's Life" in the 1688 collection, Barker is engaged throughout the volume in various conversations about heterosexual norms. These include a number of poems where the speaker describes her own, always unhappy, romances, each of which reinforces her sense of the delusions and false promises that invariably attend these activities. More surprising are poems in which the female speaker advises her male friends about their marriage prospects. In a poem that directly precedes "A Virgin Life," Barker addresses an "adopted Brother, on the nigh approach of his Nuptials," with this image:

> Then let not Marriage thee in danger draw,
> Unless thou'rt bit with Love's *Tarantula*;
> A Frenzy which no Physick can reclaim,
> But Crosses, crying Children, scolding Dame . . . (PR, 11)

While Barker is graphic, if hardly original, in comparing love to a lethal venom, notice that it is love's antidote she is actually warning about. Married life is defined by "Anger's Calenture"—that is, a disease that makes one seek suicide. The choice her "adopted" brother is just about to make will "spoil" his "good humour" and "soil" his "wit and friendship" (PR, 11). If poets have often been lauded for insights beyond direct experience, what is important to notice here is less the almost comic horror of married life and more the loss of enduring pleasures marriage invariably effectuates:

> From Married Men wit's Current never flows,
> But grave and dull, as standing Pond, he grows;
> Whilst th' other like a gentle stream do's play ... (PR, 12)

The playful pleasures and fecund energy of flowing nature are contrasted with the "grave," stultifying stasis of marriage. Even when Barker is not opining about marriage, as in a poem in which she writes to praise a friend's verse, her governing metaphors for suffering are the costs incumbent to sexual passion:

> For those that needs will taste of *Parents* joys,
> Must too indure the plague of Cradle-noise. (PR, 42)

Moralists have always reveled in the link between illicit sex and disease, but the venereal condition Barker decries is the link between married sex and procreation. While her famous "A Virgin Life" poem—which we will consider directly—says little about sex, the volume in which it appeared repeatedly figures sex as, in itself, desirable or at least ardent but also an activity that binds men and women alike to a zombie-like existence in which wit, humor, curiosity, and the capacity for pleasures of every sort, including eventually that of sex itself, are siphoned away by the conditions of married life.

The version of "A Virgin Life" in the 1688 volume is significantly different from what appears in the *Patch-Work Screen* published thirty-five years later. Instead of concluding with a life in service of "her God, her Neighbour, and her Friends" (GT, 140), the service in the earlier poem is to her god, but the poem ends with the virgin life allowing her to "enjoy her Books, and Friends" (PR, 13). Moreover, while the later version emphasizes community, the early poem pictures the virgin poet in

> Her Closet, where she do's much time bestow,
> Is both her Library and Chappel too,
> Where she enjoys society alone ... (PR, 13)

That she singles out the enjoyment of books and friends underscores what is clear throughout *Poetical Recreations*: she is a pleasure seeker and her calculus leads her to the conclusion that the joys she finds in her intellectual pursuits and in her male and female friends are not merely threatened by sex and marriage but would certainly and inevitably be destroyed by marrying. While she is clear-eyed about what this means given her gender, the advice she gives her male friends shows she believes the consequences of marriage, if different, are as dire for men as for women.

Barker's embrace of celibacy is not "for necessity" but a self-bestowed power that allows her to negotiate the world as if "she'd liv'd I'th' pristine days" (PR, 13):

> A Virgin bears the impress of all good,
> Under the Name, all Vertue's understood. (PR, 13)

A "Virgin bears" not love-killing, crying children but the *impress* in the sense of a stamp or seal of what is good in place of the biological stamp of bodies sexually impressed upon each other.[19] While the "Heavens" are thanked in the later version of the poem, Barker first thanked "ye Pow'rs" for bestowing on her "a kindness for Virginity," signaling an array of powers that must include the speaker's own choice that links celibacy to "all Vertue." This virtue she attains in relation to her books and her friends resonates directly with the "bright and fair" virtues she attributes in her opening poem to the pursuit of knowledge of her male friends.

Poetry to Anatomy

The only thing Barker rejects in the *Poetical Recreations* with more vehemence than marriage is, surprisingly, poetry itself. Before finishing the volume with a sonnet to her deceased brother, Barker includes a series of poems against poetry. In one she addresses poetry directly as her "kind Friend" but begs it to "cease t' infest" her heart, while another carries the disease metaphor a good way further by describing poetry as an itch turned by the scratch of praise into "a spreading *Leprosie*" (PR, 96–97). Barker's choice of medical imagery is not arbitrary.

On the contrary, her penultimate poem continues the tribute to her brother while figuring the speaker's personal embrace of the study of anatomy with a fervor that parallels her earlier enthusiasm for the virgin life. Barker's anatomy poem begins as a sort of Dear John letter to poetry in which the speaker announces she will be transferring her heart—an arresting metaphor in the context of organs being removed from bodies for dissection—to this "new Acquaintance" (PR, 99). Unlike the poetic uses of *anatomy* she might have known from John Donne or others or the philosophically discursive use of the term in Burton's *Anatomy of Melancholy* (1621), Barker is using the term literally to refer to medical anatomy. Set before her in the poem is Caspar Bartholin's *Anatomicae Institutiones Corporis Humani* (1611), a standard Renaissance anatomy text, updated through the century and translated to English in 1660, that signals a serious interest in anatomy.[20] But Barker begins with this text to signal a departure: she will move in the course of the poem from this standard text to the very vanguard of science in her own moment. Figured as a journey through—that is, inside—the human body, Barker's first-person narrator is accompanied by no less than William Harvey, who, of course, explains the circulation of the blood, the anatomical discovery of the century. Also along in the 1688 version of the poem is Joannes Walaeus, a professor at Leyden who championed Harvey's controversial discovery, and Walaeus is updated in the *Patch-Work* version to Thomas Willis, sometimes known as the Harvey of the nervous system.[21] The journey ends with a surprise encounter inside the body with the great anatomist Richard Lower, Willis's student and assistant, who went on to do his own groundbreaking work on the heart, including pioneering blood transfusions.

What in the world is going on here? Why is the virgin poet transferring her heart, which had been possessed by poetry, to the study of anatomy and by extension to the great anatomists of heart and brain who guide her study? The answer is, I think, far from clear from the poem itself, but context can bring this into focus. Her favorite contemporary male poet (at least beyond the circle of her friends) is the poet laureate Abraham Cowley, an edition of whose poems she presents to a friend in *Poetical Recreations,* saying Cowley's wit "O're-tops all that our *Isle* Brough forth."[22] Her Pindaric ode to her brother in that volume is also clearly influenced by Cowley.[23] While Cowley himself never announced an intention to give up poetry, he did invidiously compare poetry with the new science, deriding the insubstantial "desserts of poetry" before the "solid meats"—that is, contributions to knowledge—already emanating from a recently established Royal Society.[24] Like Barker, he turned directly to Harvey as the exemplar of what the new approaches to knowledge could deliver. In his "Ode upon Doctor Harvey," nature is a "beauteous Virgin" who has been "injoy'd by none, / Nor seen unveil'd by any one"—until, that is, Dr. Harvey, inflamed by "violent passion," pursues her.[25] While Apollo, Cowley's mythic model for Harvey, stopped himself, accepting that Daphne had changed into a tree, Harvey "stopt not so," pushing into bark and root and the "smallest fibres," forcing virgin nature to flee still farther, leaping "at last into the Winding streams of blood." But as the virgin sits reasoning that she is safe because the "Heart of Man ev'n from itself [does] conceal," she is suddenly accosted:

> Harvey was with her there,
> And held this slippery *Proteus* in a chain,
> Till all her mighty Mysteries she descry'd. . . .

This strikes the ear today as a scientific rape fantasy, but in the context of the period's gendered iconography of the new science,[26] the doctor's passion is for knowledge. Analogous to William Harvey's and other medical theories of generation at the time, the male Baconian scientist brings active principles to passive nature, and this meeting is productive of truth. Bensalem, the locus of Francis Bacon's fictionalized scientific utopia, is "the virgin of the world," and Joabin, a resident Jew called on as an objective witness, testifies that there is "not under the heavens so chaste a nation as this."[27] Furthermore, when Cowley drafted plans for the proposed Royal Society, he specified that "no Professor shall be a married man."[28] While we do not have any direct commentary by Barker on this sexist iconography, we see her reworking its elements: nature is neither virginal nor female but dynamically sexual; the chaste single lives of Cowley's professors and Bacon's utopia scientists obviate the significance of gender among students of nature. Barker's Harvey is not some priapic Apollo pursuing a terrified female virgin but a gregarious genius welcoming a virgin female associate into the humbling work of discovering the mysteries of human bodies not identified by gender. Barker's *Poetical Recreations* maps her journey from one who had turned poet in lieu of existing as an amatory/maternal object, affording her intellectual pleasures generally available only to men, until she recognized in her educational legacy from her brother a higher calling that demands (or at least calls for) the chaste, single-life passion of seekers after natural knowledge for whom gender is logically (if ideally) irrelevant to the pursuit.[29]

Barker's tour of the body begins traditionally enough with her Renaissance atlas of the body leading her to consider

> how from the Brains
> The Nerves descend; and how they do dispence,
> To ev'ry Member, Motive Pow'r and Sence;
> He [Bartholin] shews what Windows in this Structure's fix'd,
> How tribly Glaz'd, and Curtains drawn betwixt,
> Them and Earths objects: all which proves in vain,
> To keep out *Lust*, and *Innocence* retain . . . (PR, 100)

Oddly, Barker is reading the complex structure of the eye—thrice-glazed windows—not as anatomical mechanisms of sight but as protective screens between innocence and the allure of the sensual world. In the later version of this poem, this leads to a very conventional reflection on Eve's lust for the apple and its promise of divinity, as well as the consequence for women, that "Ignorance, e'er since, became our Share" (*GT*, 87). But in the *Poetical Recreations* her complaint is pointedly feminist: men, who share in the transgression as "Counter-parts," use the fact that Eve "*precipitated* first" (*PR*, 101) to debase and keep women ignorant.[30] In both poems, these Edenic reflections, prompted by her study of traditional anatomy, are suddenly interrupted by the appearance of the great contemporary anatomists—Walaeus and Harvey in the early text and Willis and Harvey in the revisions. "[F]ollow us," they "cry'd," leading Barker on a tour of the body organized as three "courts" corresponding to the organs of the abdomen, the heart, and the brain that—with Harvey as a guide—allows her to perceive "how all this *Magick* stood / By th' *Circles* of the *circulating* Blood" (*PR*, 103). The great modern anatomists call the virgin student away from her traditional anatomy lesson and away from reflections on her commonalities with Eve to share in the perception—"Then I perceiv'd" (*PR*, 103)—of new anatomically revealed truths.

The most interesting leg of the tour is in the brain:

> here methought I needs must stay,
> And listen next to what the *Artists* say:
> Here's *Cavities*, says one, and here, says he
> Is th' Seat of Fancy, Judgment, Memory:
> Here, says another, is the *fertile* Womb,
> From whence the Spirits *Animal* do come,
> Which are mysteriously *ingender'd* here,
> Of *Spirits* from *Arterious* Blood and Air.
> Here, said a third, *Life* made her first approach,
> Moving the Wheels of her Triumphant Coach:
> Hold there, said *Harvey* that must be deny'd,
> 'Twas in the deaf Ear on the *dexter* side. (*PR*, 103)

In dramatic contrast to the static anatomy she began with, Barker is now inside the brain itself, witnessing the current debates among scientists about not only how the brain functions but also the secret mechanisms of life itself. Do the arteries carry both blood and air to the brain? Are the animal spirits generated in the ventricles or other cavities? Is life traceable to these spirits or, as Harvey emphatically insists, to some vital spark added to the blood in the right auricle ("the deaf Ear on the dexter side")?[31]

Traditional men, in order to "Exalt themselves, insultingly will say, / Women know little, and they practise less" (*PR*, 101). But these men of science—Harvey,

Walaeus, Willis, and the other anatomical "artists"—have called Barker into what Willis had referred to as "the chapel of the deity" and, even more strikingly, into the center of their learned research and debates. What she finds there is not the patriarchal roles of Adam and Eve but the "fertile Womb" of the source of life itself, theories of "ingender'd" spirits and vital blood, and a complete absence, in these discussions of circulatory systems of life and sense, of the significance of biological gender distinctions. With the help of anatomists like Willis arguing that "women's bodies were governed by their brains and nerves and not their wombs,"[32] Barker is able to figure her own intellectual interests within a changing culture where "women-haters," as she refers to men who disparage women's medical knowledge (PR, 32), are being replaced by men of science like her deceased brother who encourage women's education and recognize their capacities and contributions.[33]

It is a check on, but not a contradiction of, Barker's optimistic representation of the new scientific culture that she expects her new anatomist acquaintances to "receive [her] for the sake of" her brother "from whom they did expect to see / New Lights to search *Nature's* obscurity" (PR, 99). Indeed, her brother's role in this poem and throughout the collection is instructive of both her enthusiasm and her melancholy sense of limits. Her anatomical tour ends with a surprise encounter with Willis's assistant, the great anatomist Richard Lower:

> But here we find great *Lower* by his Art,
> Surveying the whole *Structure* of the Heart;
> Welcome, said he, sweet Cousin, are you here,
> Sister to him, whose Worth we all *revere*?
> But ah, alas, so cruel was his Fate,
> As makes us since almost our Practice hate;
> Since we cou'd find out nought in all our Art,
> That cou'd prolong the motion of his Heart. (PR, 104)

The losses occasioned by her brother's death compound over the course of the volume: her private loss of one she loves, her loss of one who shared his precious education and perhaps his educated friends with her, the loss to science and medicine of his promised discoveries, and now, in Lower's melancholy assessment, the evidence that even the great discoveries of Harvey, Willis, himself, and others cannot preserve the single beating heart of one they revere. Through her representation of Lower, Barker drives home her sense of the humility of the new-science project that has opened a space for her—"Welcome, said he"—in the anatomical conversation. Sharing her sense of personal loss,

Lower draws a line to the mortal limits of "all our Art."[34] These men of science are in Barker's representation, unlike the men who fear and restrain women, modest practitioners whose debates and hearts are open because they recognize the human limits of their work.[35]

In the prayer to her brother that ends her "Farewell to Poetry," Barker returns to these limits:

> Should'st thou, my Dear, look down on us below,
>   To see how busie we
>   Are in *Anatomie,*
> Thoud'st laugh to see our Ignorance;
> Who some things miss, & some things hit by chance,
> For we, at best, do but in twilight go ... (*PR,* 105)

Notice that the "we" she employs includes her in the discoveries of Harvey, Willis, and Lower; she has written herself into the work of her heroes by making them all laborers "in twilight," laughable when looked at from the perspective of the divine. Here it is clear that she admires these men not only for the remarkable work they have done but also because the art they practice recognizes an authority beyond themselves. Of course, recourse to the divine had always supported, not inhibited, patriarchal authority. What is different here is that the experimental engagement with nature has opened a space in which debate is in relation to anatomical research rather than authority: Harvey's great discoveries put him in conversation and debate with his student Willis and Willis in conversation with his student Lower. Lower is represented in the poem mourning the loss of the expected light of Barker's brother. Perfect truth may reside in the afterlife, but Barker portrays herself as welcomed and invited into a space where even the greatest of men recognize their provisional roles.

This inclusive new-science conclusion to Barker's 1688 volume is perfectly in keeping with the volume's virgin epistemology. That is, the volume opens with the speaker inviting men to join her in nature, where the contemplation of a "few Objects" can lead to a "full inlargement of the Mind" (*PR,* 2). The kind of men she is inviting are those like her brother who share in her complaint that women's minds are starved of knowledge's "culture" (*PR,* 3) while men's minds are fed "Luxurious Banquets." If death is the "mighty *Levelller*" who in the end reduces us all to "One in equality" (*PR,* 55), then nature—whether in hills and valleys or in the bloodstreams and brain cavities of the body—is the ground on which limited humans, male and female, can explore and enlarge their equally capable and equally mortal minds. Eschewing heterosexual unions, both for herself and in her emphatic advice to her male friends, Barker hopes to rise above the cultural forms that have obscured common human mental abilities and limited the pleasures attendant to the pursuits of the intellect. Her anatomist friends and heroes and, by extension, the new science generally promise a community whose focus on truths available in nature must include the evident truth that brains and nervous systems are, apart from cultural conditioning, "relatively nonsexed."[36]

As noted above, Haraway made it her aim in *Modest_Witness@Second_Millennium* (1997) "to queer" the modest witness, a new-science ideal that had been presented, particularly with regard to Robert Boyle, by Steven Shapin and Simon Schaffer in their *Leviathan and the Air-Pump* (1985). Haraway goes about this by exploring how "Boyle—urbane, celibate, and civil—avoided the fate of being labeled a *haec vir,* a feminine man, in his insistence on the virtue of modesty" and also by adopting "the FemaleMan©" as her "surrogate, agent, and sister" to disrupt the "gendered categories" that have operated in "techno-science."[37] One need not read Barker as a proto-Haraway theorist[38] to acknowledge that Barker's personal rejection of heteronormative behavior extended beyond the traditional concept of female chastity and into a powerful argument for male and female celibacy of the sort associated with Boyle, the prime mover of the institution of the Royal Society, and Isaac Newton, the society's most iconic leader.[39] Barker calls her male friends away from the dire consequences of marriage not as a rejection of this world in favor of the rewards of the next but rather as an invitation to the joys of study and friendship within a lavishly rewarding natural world. Without illusions about the "women-haters" who seek power through exclusion, Barker thought she recognized allies in her brother, in her Cambridge coterie, and in Harvey, Walaeus, Willis, and Lower. The emergence of this modest male culture was made possible by its focus on what might be learned from nature rather than on traditional authorities. All this allows Barker to tell a story in *Poetical Recreations* in which she is a welcome member of a community of truth-seekers who are traveling, viscerally, through the nonsexed body itself, excitedly uncovering its secrets.

Notes

1. Astell also makes the argument from comparison to other species, noting that a "She Ape" is every bit as capable of imitation as a male ape. Astell, *Essay in Defence,* 12–13.
2. References to *Poetical Recreations* are indicated parenthetically in the text and abbreviated PR.
3. On the modest witness, see Shapin and Schaffer, *Leviathan and the Air-Pump,* 65–69; and Haraway, *Modest_Witness@Second_Millennium,* especially 21–39.

4. I am referring here to both Bosvil from *Love Intrigues* and Lysander from *Patch-Work Screen for the Ladies*. See *The Galesia Trilogy and Selected Manuscript Poems of Jane Barker*, edited by Carol Shiner Wilson, which includes *Love Intrigues* (1713), *A Patch-Work Screen for the Ladies* (1723), and *The Lining of the Patch-Work Screen; Design'd for the Farther Entertainment of the Ladies* (1726). References to *The Galesia Trilogy* are indicated parenthetically in the text and abbreviated GT.
5. King, "Jane Barker, Poetical Recreations," 561–62.
6. Bowers, "Jacobite Difference," 867; McArthur, "Jane Barker," 597.
7. McGrath, "Unaccountable Form," 354. *Tristram Shandy* is mentioned in the conclusion of this essay (368).
8. King, *Jane Barker, Exile*, 66.
9. King, 69–76.
10. Keiser, *Nervous Fictions*, 18–36; Mann, "Jane Barker, Manuscript Culture," 50–75.
11. The 1704 date for the completion of the Magdalen Manuscript comes from the chronology in King's *Jane Barker, Exile*, which is the most complete source of biographical information and publication history on Barker.
12. Among the complexities of reading PR on its own terms is Barker's much later claim that the volume was published without her consent. King offers substantial, though hardly conclusive, evidence that this could indeed be the case (*Jane Barker, Exile*, 30–38). King does not seem to fully consider, however, the possibility that Barker and her friends agreed that it was in her best interest to preserve deniability for publication given the fraught circumstances of women publishing at the time. King imagines a preface for the volume modeled on Aphra Behn's epistle to the reader for *The Dutch Lover* (1673), perhaps precisely the association Barker and her friends would aim to avoid.
13. The penultimate poem of the volume contains both the anatomical trip through the body and a reflection on anatomy from her deceased brother's perspective in heaven, while the final poem, "On the Death of my Brother. A Sonnet," is a summary work "To all these things, which here I name," including, presumably, anatomy.
14. Kahan, *Celibacies*, 3–4.
15. Haraway, *Modest_Witness@Second_Millennium*, 35.
16. PR has twenty-three unnumbered pages (including the table of contents and errata) before numbering begins with Barker's first poem. Orinda is invoked at the eighth of these pages, Shakespeare and Johnson at 10, and the Almighty at 13–14.
17. King calls it "a bitter little feminist parable about female exclusion from the sites of learning." King, *Jane Barker, Exile*, 52.
18 "What stupid enemy to Wit and Sense, Dares to dispute your Sexes Excellence?" (PR, 2.35). This couplet, by "a gentleman of St John's College, Cambridge," appears in one of a number of poems in the second half of *Poetical Recreations* addressed to Barker.

19. Dale, *Printed Reader*, discusses Barker's Dorinda from *The Lining of the Patch-Work Screen* (1726) as a representation of the "Miss reader," a female reader too strongly impressed by romance fiction. "The Virgin Life," on the other hand, represents a female reader imprinted by a celibate life in such a way that "her books" are central to the joy of "her whole Lives business" (*PR*, 13).
20. King notes the text was published in an edition for Oxford students in 1633, and the original was updated to include Harvey's discoveries in 1641 in an edition published by Caspar's son Thomas Bartholin (1616–80). King, *Jane Barker, Exile*, 85–86. See also Mann, "Jane Barker, Manuscript Culture," 63–65.
21. On Walaeus and Barker, see Mann, "Jane Barker, Manuscript Culture," 66–67. On Barker's replacement of Walaeus with Willis, see King, *Jane Barker, Exile*, 93–96. Zimmer, "Distant Mirror," 44.
22. "To Sir F.W. presenting him Cowley's first Works" (*PR*, 28).
23. "On the same. A Pindarique ODE" (*PR*, 51). Cowley similarly used the Pindaric to celebrate and mourn the deaths of scientific heroes: see Sawday, "Bodies by Art Fashioned," 323–28.
24. Boyle, *Swift*, 81–93.
25. Page numbering in Cowley's *Works* begins anew in each section: the "Ode upon Doctor Harvey" appears in "Verses written on several occasions" (12–14).
26. Carolyn Merchant's *The Death of Nature* (1980) was a foundational feminist reading of the new science. For considerations of how that discourse has developed, see Worthy, Allison, and Bauman, *After the Death of Nature*.
27. Bacon, *Major Works*, 476.
28. Cowley, *Works*, 48. "A Proposition for the Advancement of Experimental Philosophy" appears in the section that begins with "Verses written on several occasions" (43–51).
29. "Before the late seventeenth century," Lisa Forman Cody tells us in *Birthing the Nation*, "knowledge about sex," including pleasure, reproduction, and childcare, "belonged to the private world of women and midwives" and was not "considered a discipline, let alone a science" (16). Cody documents and explicates the dramatic changes the new science will bring to this area of knowledge. Barker, living in this transitional moment, imagines a new-science path for herself to join others for whom the pursuit of knowledge supersedes sex and its physical and social consequences.
30. The poem reads,

    Ev'n by our Counter-parts, who that they may
    Exalt themselves, insultingly will say,
    Women know little, and they practice less:
    But pride and Sloth they glory to profess. (*PR*, 101)

    My reading here may appear to be directly at odds with that of King, who reads the references to the Fall in this poem as a "misogynistic endorsement"

(97). The reason for the difference is that King is quoting the version published in 1723 (though in a chapter discussing the period up to 1688). In fact, the later revisions thoroughly eradicate the feminist complaint, removing any mention of men from this section of the poem.
31. Harvey never accepted the idea of animal spirits, believing the existence of the soul could be traced to the blood. See Martensen, *Brain Takes Shape*, 11–25.
32. Martensen, 153.
33. King speculates that Barker substituted Willis for Walaeus because the "Willisian brain-centered model" installed "the sexless brain at the centre of the hitherto womb-propelled female body." King, *Jane Barker, Exile*, 95.
34. Almost a decade and a half after Barker published this poem, Anne Finch, in "The Spleen," draws a portrait of the very same Lower, also at her poem's conclusion but with an aim nearly opposite to that of Barker. That is, rather than figuring Lower, with Harvey and Willis, as modest men of science not bound by the senseless rules of patriarchy, Finch, according to Desiree Hellegers in *Handmaid to Divinity*, represents Lower, and by extension "Harvey, Willis, and the other physician-virtuosi of the Royal College of Physicians," in a way that links "animal experimentation with the violence that the male physician/virtuoso perpetrates upon women's bodies in the name of medicine" (163–64). If Hellegers's reading of Finch is correct (she does not discuss Barker), the contrast suggests how completely the hopeful moment we see in Astell and Barker passes on to the persistent realities of patriarchy.
35. Keiser argues that Barker recognizes the limits of the scientific enterprise particularly with respect to the brain. *Nervous Fictions*, 22–27.
36. Martensen, *Brain Takes Shape*, 161.
37. Haraway, *Modest_Witness@Second_Millennium*, 29, 70.
38. Linking the "epistemological humility" she finds in Barker's patchwork novels to the author's single-life ethos and depiction of "non-normative sexuality" (19), McGrath discusses Barker's usefulness in "understanding the queerness of eighteenth-century archives." "Unaccountable Form," 364, 357, 368. I am showing how these ideas developed in Barker's earlier engagements with medicine and science.
39. For a discussion that complicates the connection between Boyle's celibacy and his scientific method, see Sargent, "Robert Boyle." For Newton's celibacy, see Christianson, *In the Presence*, 258–59.

Bibliography

Astell, Mary. *An Essay in Defence of the Female Sex in Which Are Inserted the Characters of a Pedant, a Squire, a Beau, a Vertuoso, a Poetaster, a City-Critick, &c.:*

In a Letter to a Lady / Written by a Lady. London: Printed for A. Roper and E. Wilkinson at the Black Boy, and R. Clavel at the Peacock, in Fleetstreet, 1696.

Bacon, Francis. *The Major Works.* Edited by Brian Vickers. Oxford: Oxford University Press, 2002.

Barker, Jane. *The Galesia Trilogy and Selected Manuscript Poems of Jane Barker.* Edited by Carol Shiner Wilson. New York: Oxford University Press, 1997.

———. *Poetical Recreations Consisting of Original Poems, Songs, Odes, &c. with Several New Translations: In Two Parts / Part I, Occasionally Written by Mrs. Jane Barker, Part II, by Several Gentlemen of the Universities, and Others.* Early English Books, 1641–1700 / 52:03. London: Printed for Benjamin Crayle . . . , 1688.

Bowers, Toni. "Jacobite Difference and the Poetry of Jane Barker." *ELH* 64, no. 4 (1997): 857–69.

Boyle, Frank. *Swift as Nemesis: Modernity and Its Satirist.* Stanford University Press, 2000.

Christianson, Gale E. *In the Presence of the Creator: Isaac Newton and His Times.* New York: Free Press, 1984.

Cody, Lisa Forman. *Birthing the Nation: Sex, Science, and the Conception of Eighteenth-Century Britons.* Oxford: Oxford University Press, 2005.

Cowley, Abraham. *The Works of Mr. Abraham Cowley Consisting of Those Which Were Formerly Printed and Those Which Be Design'd for the Press, Now Published Out of the Authors Original Copies.* London: Printed by J. M. for Henry Herringman . . . , 1680.

Dale, Amelia. *The Printed Reader: Gender, Quixotism, and Textual Bodies in Eighteenth-Century Britain.* Transits: Literature, Thought & Culture 1650–1850. Lewisburg, PA: Bucknell University Press, 2019.

Haraway, Donna J. *Modest_Witness@Second_Millennium. FemaleMan©_Meets_OncoMouse™: Feminism and Technoscience.* New York: Routledge, 1997.

Hellegers, Desiree. *Handmaid to Divinity: Natural Philosophy, Poetry, and Gender in Seventeenth-Century England.* Norman: University of Oklahoma Press, 2000.

Kahan, Benjamin. *Celibacies: American Modernism and Sexual Life.* Durham: Duke University Press, 2013.

Keiser, Jess. *Nervous Fictions: Literary Form and the Enlightenment Origins of Neuroscience.* Charlottesville: University of Virginia Press, 2020.

King, Kathryn R. *Jane Barker, Exile: A Literary Career 1675–1725.* Oxford: Clarendon Press, 2000.

———. "Jane Barker, Poetical Recreations, and the Sociable Text." *ELH* 61, no. 3 (1994): 551–70.

Mann, Rachel. "Jane Barker, Manuscript Culture, and the Epistemology of the Microscope." *Eighteenth-Century Life* 43, no. 1 (January 1, 2019): 50–75.

Martensen, Robert L. *The Brain Takes Shape: An Early History.* Oxford: Oxford University Press, 2004.

McArthur, Tonya Moutray. "Jane Barker and the Politics of Catholic Celibacy." *Studies in English Literature, 1500–1900* 47, no. 3 (2007): 595–618.

McGrath, Alice Tweedy. "Unaccountable Form: Queer Failure and Jane Barker's Patchwork Method." *Eighteenth Century: Theory and Interpretation* 60, no. 4 (Winter 2019): 353–73.

Merchant, Carolyn. *The Death of Nature: Women, Ecology, and the Scientific Revolution.* 1st ed. San Francisco: Harper & Row, 1980.

Sargent, Rose-Mary. "Robert Boyle and the Masculine Methods of Science." *Philosophy of Science* 71, no. 5 (December 2004): 857–67.

Sawday, J. H. "Bodies by Art Fashioned: Anatomy, Anatomists, and English Poetry 1570–1680." PhD diss., University of London, 1988.

Shapin, Steven, and Simon Schaffer. *Leviathan and the Air-Pump: Hobbes, Boyle, and the Experimental Life.* Princeton, NJ: Princeton University Press, 1985.

Worthy, Kenneth, Elizabeth Allison, and Whitney A. Bauman. *After the Death of Nature: Carolyn Merchant and the Future of Human-Nature Relations.* New York: Routledge, 2019.

Zimmer, Carl. "A Distant Mirror for the Brain." *Science* 303, no. 5654 (January 2, 2004): 43–44.

Margaret Cavendish, a Sensitive Witness

KRISTIN M. GIRTEN

> And now this *Iron Age*'s so *rusty* grown,
> That all the *Hearts* are turn'd to hard *flint-stone*.
> —MARGARET CAVENDISH, "Of the death and buriall of Truth"

When it comes to her feelings about the use of modern observational technologies like the microscope and the telescope, the seventeenth-century English philosopher Margaret Cavendish does not mince words. Her critique of them and the philosophers who rely on them could hardly be more scathing. As Todd Andrew Borlik explains, the challenge Cavendish poses to the use of such technologies associates her with other "detractors" from "Francis Bacon's quest to inaugurate a universal science," including the likes of John Aubrey and Michel de Montaigne.[1] I share Eve Keller's view that Cavendish took aim at "claims of methodological rigor, value-neutrality, and objectivity" of Bacon and his followers and that she perceived these qualities "not as monolithic conduits for achieving certainty, but as social constructions that are endorsed as much because they advance the needs of their adherents as because they are deemed to be scientifically effective or true."[2] However, while Keller focuses primarily on Cavendish's critique of Baconian empiricism and how this critique is informed by her "idiosyncratic philosophy of organic materialism," I focus instead on the empirical method that Cavendish presents as an alternative to Bacon's and how this method is particularly indebted to Lucretius's Epicurean poetic masterpiece *De Rerum Natura*.[3] I argue that as she challenges dependence on instruments that augment vision, Cavendish develops a new approach to obtaining natural knowledge that, informed by Epicurean materialism, privileges feeling and an affective openness to the objects of one's investigations over technologically enhanced vision and a detachment from such objects.

In recent years, scholars have become intrigued by the influence of materialism—specifically, the Epicurean materialism poeticized by Lucretius—on Cavendish. I would argue that, for Cavendish, Epicureanism provides an important foundation for both her own philosophical and literary authority and the philosophical methods she would develop and promote.[4] My analysis builds on that of Line Cottegnies and Emma Rees, both of whom demonstrate that, to quote Cottegnies, the "woman-friendly" dimension of Epicureanism provides Cavendish with a useful vehicle for asserting her own feminine philosophical as well as literary authority. Though some scholars argue that she ultimately abandoned materialism, I join Londa Schiebinger and Stephen Clucas in maintaining that materialist philosophies continued to influence her throughout her career.[5] Clucas offers a particularly in-depth, nuanced, and therefore persuasive analysis: "Margaret Cavendish's natural philosophical works are essentially atomistic, although we see in her theories a progressive attempt to refine and reconcile atomic ideas within a philosophical framework which opposes itself to the simple mechanism of 'classical atomism.'"[6] Though Clucas documents various ways in which, over the course of her career, Cavendish revises and refines the materialism that she espouses, he demonstrates that Epicureanism maintains a persistent appeal for her. Her publishing career begins with "a thorough-going Epicurean atomism," expressed in the first forty-five pages of *Poems and Fancies*.[7] However, even when she publishes "A Condemning Treatise of Atomes" in *The Philosophical and Physical Opinions* of 1655, what she presents "is not an *anti*-atomist statement, but rather a refinement of the crudities of an updated Democritean-Lucretian atomism" that "retains a residual attachment to the broad principle of atomic structure. Her objections, it seems, are to the idea of *mechanical* atomism," not to atomism generally.[8] As Clucas explains, in her later works, "Cavendish's atomism is a synthesis of materialism and vitalism."[9] Cavendish began her career standing on the shoulders of Lucretius; then, as she matured, she developed a more complex relationship to him, though she would nevertheless continue to take inspiration from the Epicureanism he espoused even as she interrogated some of its central principles.

In this essay, I aim not so much to tease out the nuances of Cavendish's atomism, as Clucas does so brilliantly, but rather to recognize how Epicurus's enduring influence on Cavendish contributes in significant ways to her development of an innovative philosophical method that she presents as an alternative to the more conventional variety promoted by the Royal Society. Though scholars have, in the last few decades, begun to take Cavendish more seriously not only as a literary author but also as a philosopher, it has nevertheless proved difficult to establish a coherent philosophical program across her career. By exploring her methods with a new appreciation for how Lucretian Epicureanism helped inspire them, this essay seeks to refine the story we tell about Cavendish's scientific innovations. Though her philosophical method may not have been as systematic as Bacon's, she nevertheless teaches how to do science by her own example and in a purposeful fashion.[10] I seek to encourage scholars to devote more attention to the epistemological genealogy of Cavendish's empiricism and specifically to its engagement with Epicurean materialism. In this essay, I analyze the empirical implications of passages from several of her works: her collection of verse *Poems and Fancies* (1653), her miscellany *The Worlds Olio* (1655), and her major scientific works *Observations upon Experimental Philosophy* (1666) and *Grounds of Natural Philosophy* (1668). I argue that, inspired by Epicureanism and over the course of her career, Cavendish develops a distinctive empirical practice of *sensitive witnessing* according to which knowledge is made by recognizing one's own material nature and by pursuing observations that are so profoundly embodied and engaged that they defy separation between scientist and specimen, subject and object.

Early Enlightenment philosophers relied heavily on optical technologies to establish their scientific authority, and Cavendish was determined to challenge this authority. Replicable experiments were a hallmark of the new science, and such technologies were perceived to extend the range and thus value of such experiments. They certainly were instrumental in enabling new and surprising discoveries.[11] Robert Hooke's *Micrographia* (1665) offers a particularly illustrative expression of how enamored new scientists were with optical technologies and their potential to reveal hidden truths. As Hooke explains, through microscopy, he seeks nothing less than "an inlargement of the dominion, of the Senses." So enthusiastic is Hooke about his "magnifying glasses" (including both the microscope and the telescope) that he devotes eight pages and a full-page engraving to enumerating their various parts, treating them as if they are specimens themselves.[12] Though Cavendish shared Hooke's profound and enduring curiosity about the inner workings of the material world, she nevertheless became a staunch critic of the employment of optical technologies to satisfy this curiosity. As she argues in a chapter from *Observations upon Experimental Philosophy* entitled "Of Micrography, and of Magnifying and Multiplying Glasses," optical technologies like the microscope and the telescope distort rather than reveal: "Magnifying, multiplying, and the like optic glasses, may, and do oftentimes present falsely the picture of an exterior object; I say, the picture, because it is not the real body of the object which the glass presents but the glass only figures or patterns out the picture presented in and by the glass, and there mistakes may easily be committed in taking copies from copies."[13] As this passage suggests, according to Cavendish, magnifying technologies manipulate specimens and therefore present a highly misleading and unreliable mediated depiction of them. She goes on to say, "Wherefore those that invented microscopes, and such like dioptrical glasses, at first, did, in my opinion, the world more injury than benefit; for this art has intoxicated so many men's brains, and wholly employed their thoughts and bodily actions about phenomena, or the exterior figures of objects, as all better arts and studies are laid aside."[14] For Cavendish, such magnifying instruments "intoxicate" empirical philosophers like Hooke. Furthermore, even when they give the appearance of accessing material depths, they nevertheless limit philosophers to the surface of things ("exterior figures of objects"), superficialities as opposed to deeper, internal, and thus valuable truths.

It is precisely such deeper truths that Cavendish seeks in her empirical practice, and she suggests that her methods of observation and knowledge-making are more effective at accessing these truths than the more conventional methods employed and encouraged by Royal Society members like Hooke. Cavendish's critique of philosophers' reliance on optical technologies is part of a broader challenge she poses to contemporary empirical methods that had initially been inspired by Bacon in his *Novum Organum* (1620). It is not the act of looking closely through magnifying glasses in and of itself that troubles Cavendish. Rather, it is what this act represents—namely, philosophers' delusional faith in their ability to see rightly at a distance and in what Keller portrays as their "perceptual passivity."[15] Baconian empiricism, which would inform the practices of early English scientists for years to come, was distinguished by a technique of purification that was thought to enable the scientist to transcend matter. According to Bacon, the new science promised to allow men to "recover" their "right over nature." However, to achieve scientific authority and thus fulfill this promise, Bacon insisted that men needed to perform an "expurgation of the intellect."[16] Men "would be sufficient of themselves," Bacon observes, "if the human intellect were even, and like a fair sheet of paper with no writing on it. But since the minds of men are strangely possessed and beset, so that there is no true and even surface left to reflect the genuine rays of things, it is necessary to seek a remedy for this also."[17] Men's minds are plagued by "idols, or phantoms" that threaten to prevent them from perceiving clearly and accurately. However, by teaching men how to cleanse their minds of these blots, Bacon "perform[s] the office of a true priest of sense."[18] The mind must, according to Bacon, be "guided at every step; and the business . . . done as if by machinery."[19] For Bacon, "the human understanding is like a false mirror, which, receiving rays irregularly, distorts and discolours the nature of things by mingling its own nature with it."[20] Consequently, if the new science is going to produce the reliable knowledge that Bacon seeks, then philosophers must, according to Bacon, develop a means by which to resist what he portrays as the inherent tendency of humans to "mingle" with matter. Their minds had to be "free and cleansed"—detached from as well as transcendent above the material world.[21]

This variety of philosophic mind has come to be known by scholars as Baconian "modesty"; a "modest witness" is a Baconian empiricist who has been successful at cleansing his mind of matter and thus at becoming detached from it.[22] I argue that it is skepticism about this form of witnessing that motivates Cavendish to challenge scientists' Baconian techniques. As Steven Shapin and Simon Schaffer explain, "A man whose narratives could be credited as mirrors of reality was a *modest* man."[23] The modest scientist was expected to keep himself out of his own experimental pursuits: he was to be "'a drudge of greater industry than reason'"; he was to write in a "plain, ascetic, unadorned" and "functional" style; and he was to distinguish, in his writing, between "experimental findings" and "their interpretation."[24] In these ways, he would rise to the level of "a disinterested observer."[25] Optical technologies were perceived to contribute to the new scientist's efforts at achieving disinterestedness. As Tita Chico demonstrates, by employing such instruments, scientists like Hooke made their own bodies into instruments, technologizing themselves and thus denying their own physicality. In so doing, they essentially appeared to objectify witnessing.[26] According to Cavendish, though, Baconian modesty is a false presumption that encumbers scientific progress because it distracts philosophers from worthier methods that promise to produce real discoveries. "As boys that play with watery bubbles or fling dust into each other's eyes, or make a hobbyhorse of snow, are worthy of reproof rather than praise, for wasting their time with useless sports," Cavendish explains, "so those that addict themselves to unprofitable arts, spend more time than they reap benefit thereby": "Nay, could they benefit men either in husbandry, architecture, or the like necessary and profitable employments; yet before the vulgar sort would learn to understand them, the world would want bread to eat, and houses to dwell in, as also clothes to keep them from the inconveniences of the inconstant weather."[27] For Cavendish, Baconian modest witnessing causes philosophers to waste their time; "addicted" to their own childish amusements, they lose touch with what really matters ("necessary and profitable employments"). This sentiment flies directly in the face of the enthusiastic assertions of public utility for which the early Royal Society was known. Take, for instance, Thomas Sprat's claim in his foundational *History of the Royal Society* that, as a result of its fellows' modest pursuits, England "has been able in a short time so well to recover it self: as not onely to attain to the perfection of its *former* Civility, and Learning, but also to set on foot, a *new* way of improvement of Arts, as *Great* and as *Beneficial* (to say no more) as any the wittiest or the happiest Age has ever invented."[28] Cavendish directly challenges the practical benefits of the new science, indicating that if true utility is to be had, then empirical philosophers must find methods for making knowledge other than those that Bacon sets out.

To delve further into Cavendish's *Observations*, with an eye to its expression of key Epicurean principles, is to discover why it is that Cavendish finds the practice of Baconian modesty so ineffectual. I would suggest that, for Cavendish, such modesty is based on a lie and is therefore infeasible. The modest "man conceits himself to be above nature," Cavendish observes, when in truth "all creatures of nature are produced but out of one matter, which is common to all, and ... there are continual and perpetual generations and productions in nature, as well as there are perpetual dissolutions."[29] As Liza Blake has shown, this passage and others like it from *Observations* evoke what might be seen as a "commonality of matter" or *common nature*, as Cavendish expresses a strong allegiance to vitalist materialism.[30] I would contend that she evokes Lucretius's *De Rerum Natura* specifically as she proceeds to elaborate her concept of common nature. Her assertion that "flowing atoms enter and issue" via "pores" in all material beings directly corresponds to Lucretius's Epicurean portrayal of all beings as "confluence[s] of seeds" or atoms.[31] For Cavendish as for Lucretius, all bodies are composites, or in the words of Cavendish, "Composed Figures."[32] Cavendish also shares Lucretius's Epicurean conviction that "motions doe / Dissolve all creatures, and their frames renew; / Whither first bodies, by impulsive force, / Or moving faculties, maintaine their course, / Wandring for ever in th'unbounded space."[33] For Cavendish, as for Lucretius, atoms are in a constant state of transcorporeal movement, and this movement continually "renews" the "frames" of "all creatures": "Creatures must be produced by Creatures. . . . Wherefore, all Natural Creatures are produced by the consent and agreement of many Self-moving Parts, or Corporeal Motions, which work to a particular Design, as to associate into particular kinds and sorts of Creatures."[34] Composed not only of atoms but also of "vacuitie" or void, all bodies are porous and, with the perpetual movement of atoms, are therefore perpetually penetrated by the atoms of other bodies. Thus, there can be no autonomous being in the Epicurean universe: "No fixt seate can be for bodies found."[35] All beings are inseparable from their environments, and as Aaron Hanlon observes, "humans and nature [are] of the same cloth."[36]

The philosopher is no exception to this tenet; consequently, for Cavendish, Baconian modesty is based on the false impression that the observer exists outside of the moving matter that Cavendish describes. According to Cavendish, not even the philosopher's intelligence or soul fundamentally distinguishes him from nonhuman nature: "There is [even] a soul and intelligence in the loadstone . . . as that there is a soul in man."[37] In other words, humans are of nature, not above nature, and only a "fool" believes otherwise.[38] Thus, Cavendish implies, it is foolish (and even deceptive) to cultivate the modesty that Bacon recommends as a means of ensuring the authority of one's empirical practice, for it is predicated on detaching oneself from, and thereby transcending, matter, which is an impossibility according to Cavendish. None of us—not even the modest philosopher—has "power at all over natural causes and effects": "Neither can natural causes nor effects be overpowered by man so, as if man was a degree above nature, but they must be as nature is pleased to order them; for man is but a small part, and his powers are but particular actions of nature, and therefore he cannot have a supreme an absolute power."[39] Empirical philosophers like Hooke and Henry Power, another microscopist and early member of the Royal Society Cavendish repeatedly targets, believe their modesty allows them to gain special access to nature: detached from and dominant over her, they are capable of, to quote Bacon, "entering and penetrating into [her] holes and corners" or subjecting her to "constraint" and "vexation," all in the service of fruitful experimentation.[40] In accordance with her persistent espousal of Epicurean principles, Cavendish disputes not only the practical benefits of this empirical approach but also its ontological foundation. For Cavendish, even the most virtuosic philosopher is fundamentally a material being and is therefore fundamentally incapable of "expurgating" his mind of matter as Bacon advises him to do.[41] As Keller explains, Cavendish insists "that man is inextricably *a part* of the nature he seeks to know: there simply exists no outside vantage point from which to view and thereby to control some object called nature."[42]

Though Keller reveals the challenge Cavendish poses to Baconian modesty by highlighting her insistence on man's entanglement with matter, she neglects to consider the Epicurean dimensions of Cavendish's philosophy and therefore stops short of recognizing that, in fact, Cavendish presents and promotes an empirical practice that uses one's material nature to one's advantage. Rather than distancing herself from matter as Bacon suggests, informed by Epicureanism, she not only acknowledges her material nature but cultivates an affective openness to her objects of investigation. Cavendish summarizes the method she proposes thus: "The best optic is a perfect natural eye, and a regular sensitive perception; and the best judge, is reason; and the best study is rational contemplation joined with the observations of regular sense."[43] Cavendish advocates a bipartite philosophy comprising both "sensitive" and "rational" perception. "A regular sensitive perception" provides the foundation, while reason is the arbiter. Though, as Lisa Sarasohn observes, Cavendish emphasizes "perception['s] inability to comprehend the 'interior, proper, and innate actions' of a creature," rather than eschewing sense perception and the physical engagement with matter that it entails, she insists on the value and importance of engaged affective observation, in spite of its limitations.[44] Whereas Bacon promotes an empiricism that is pure of material entanglement and therefore aloof and sober, Cavendish's empiricism is thoroughly enmeshed in matter and therefore receptive and vulnerable to stimuli. The Baconian witness is modest; Cavendish's witness is *sensitive*.

Take, for instance, her "Observation X. Of a Butterfly." Its opening is indicative of its approach: "Concerning the generation of butterflies, whether they be produced by the way of eggs, as some experimental philosophers do relate, or any other ways; or whether they be all produced after one and the same manner, shall not be my task now to determine; but I will only give my readers a short account of what I myself have observed."[45] This is far from a deductive or speculative method. Indeed, if anyone is speculative in their reasoning in this passage, it is the "experimental philosophers" whose theories seem to Cavendish unsubstantiated—based purely on what has been "related"—in comparison to hers, which she portrays as firmly rooted in personal ("I myself") experience. Though she does not demand replication for corroboration and validation as Bacon does, implying that her sensitive witnessing is in and of itself enough to verify truths, Cavendish's method is distinguished by induction much as Bacon's is.[46] This becomes all the more apparent as she proceeds to provide her account, which begins thus: "When I lived beyond the seas in banishment with my noble lord, one of my maids brought upon an old piece of wood, or stone (which it was I cannot perfectly remember) something to me which seemed to grow out of that same piece; it was about the length of half an inch or less, the tail was short and square, and seemed to grow out of that same piece."[47] She proceeds to amass multiple physical details about this and another similar specimen. She explains, "I laid by upon the window; and one morning I spied two butterflies playing about it; which, knowing the window had been close shut all the while, and finding the insect all empty, and only like a bare shell or skin, I supposed had been bred out of it."[48] As she refers to her political "banishment," Cavendish shows that she perceives herself as embedded in, rather than transcendent over, her social environment. Furthermore, with her acknowledgment "I cannot perfectly remember," she expresses a recognition that as material rather than transcendent beings, humans have incomplete and fallible memories. These passages also illustrate, though, that in spite of her acknowledgment of the limitations of human sense perception, Cavendish nevertheless relies at least as heavily on "observations of regular sense" as Baconian empiricists do.

In fact, her refusal to employ a microscope makes such unaccoutred observations all the more important. Whereas Hooke or Power have recourse to the microscope, Cavendish instead trains her "perfect natural eye" sensitively on the objects of her investigation, readily acknowledging the limit of her sight and thus the reach of her observation: "But yet this latter I will not certainly affirm, for I could not discern them with my eyes, except I had had some microscope, but a thousand to one I might have been also deceived by it: and had I opened this insect, or shell, at first; it might perhaps have given those butterflies an untimely death, or rather hindered their production."[49] Cavendish here sets clear limits for her empirical practice, eschewing both microscopic investigation, on the grounds of what she perceives as its empirical inaccuracy, and dissection, on moral grounds. She then proceeds to decline either to trust or to challenge empirical hearsay: "I have heard also that caterpillars are transformed into butterflies; whether it be true or not, I will not dispute."[50] Instead, she concludes by returning to her own sense perception, conveying an attentive observation of caterpillars' behaviors: "Only this I dare say, that I have seen caterpillars spin, as silkworms do, an oval ball about their seed, or rather about themselves."[51] We might be tempted to presume that the emphasis Cavendish places on her self and her own observations in such passages implies a firm if not fixed subjectivity that contradicts the openness and vulnerability that her practice of sensitive witnessing entails. However, analyzed through the framework of her sustained Epicureanism, her repeated use of the "I" takes on a different valence. It marks her as merely a creature among creatures whose physical and psychological boundaries are provisional at best. Hers is a humble rather than assertive subjectivity, which is evinced by how she prefaces her observations—"only this I dare say"—and this humility extends to the sensitive perception she practices and promotes. She is engaged with and responsive to rather than detached from the specimens she observes. Indeed, whereas such engagement may be perceived to undermine the modest witness's authority, within the context of her unorthodox experimental philosophy, it bolsters her authority, attesting to the precision and veracity of her observation.

As Cavendish's empirical observations convey, though she may at points emphasize the crucial role she believes reason to play in philosophy, "sensitive perception" nevertheless retains a prominent position within her philosophic practice. Throughout her writings, Cavendish demands that we resist the temptation to polarize matter and mind, sense and reason, insisting on their inseparability. While she may enjoy imagining a "Blazing World" that is "composed only of the rational" parts of matter, we betray the nuances of her philosophical method (not to mention the creative liberty asserted by her fiction) if we impose her fantasy of the fictional duchess's world-making on her *Observations upon Experimental Philosophy*.[52] These works were originally published in the same volume, but their differences are at least as instructive as their continuities are. In fact, in an essay entitled "Of the Senses and Brain" included in *The Worlds Olio*, she directly refutes the plausibility of the fiction she presents in *The Blazing World*: "Some say, That there is such a nature in Man, that he would conceive and understand without the *Senses*, though not so clearly, if he had but Life, which is Motion. Others say, There is nothing in the *Understanding*, that is not first in the *Senses*: which is more probable: for the *Senses* bring all the Materials into the *Brain*; and then the *Brain* cuts and divides them, and gives them quite other Forms than the *Senses* many times have presented."[53] Here again we see Cavendish presenting a bipartite method of observation. In reality, according to Cavendish, sense and reason are dialectically intertwined and thus fundamentally resistant to any attempts to polarize them. As she goes on to observe in *The Worlds Olio*, "We cannot call that a perfect knowledge which our *Reason* singly tells us; but what our perfect and healthful *Senses*, joined with our *Reason*, distinguish to us."[54] So intimate is their collaboration that, in Cavendish's philosophical practice, sense and reason not only operate in conjunction with one another but mutually inform and even help constitute one another.

We have thus far confined our attention to the sensitivity with which Cavendish engages in physical observation. However, I would contend that her rational explorations are sensitive as well, albeit differently so. In *The Worlds Olio*, she portrays these explorations, which she pursues via "discuss[ion]" or self-dialogue "that increases or begets Knowledge" as informed by "conjectures" or "probabilities" and as "demonstrative" in outcome.[55] In *Observations*, she celebrates the superiority of rational "discourse" to "deluding arts" (such as microscopy), attributing to the former the ability to "find or trace nature's corporeal figurative motions" while accusing the latter of being incapable of sufficiently "inform[ing] the senses."[56] With the word "trace," Cavendish imbues reason with physicality, even making it seem like a tactile and/or kinesthetic activity: to employ reason well is not simply to manufacture ideas out of thin air but, rather, to follow the motions of nature as if by foot or by hand.[57] In other words, reason must, like sense, be sensitive to nature for it to fulfill its promise of providing insight. Cavendish's impulse to physicalize reason and to promote a cultivation of its sensitivity extends to *The Worlds Olio* as well. There, she likens reason to the eyes: "For, the *Brains* are like Eyes, of which some are so quick, that they cannot fasten upon an Object to view the perfection of it. Others so dull, that they cannot see clearly; or so slow, they cannot untie themselves soon enough, but dwell too long upon it. So it is the discussing of the Object well, that encreases or begets Knowledg."[58] We must "discuss" and reason about "the Object well" if we are going to make knowledge effectively. To do so, we must aspire to a reason that is neither so quick that it is unable to "fasten upon an Object" nor so slow that it is unable to "untie" itself from the object. The proposal here is for a methodical and yet softened reason: a reason that, operating in a relaxed and open fashion, is highly receptive and responsive to the objects of its explorations. Here, as in *Observations*, Cavendish celebrates and advocates a sensitive reason that closely corresponds to and even echoes the sensitive perception she also recommends—a reason that is "acutely affected by external stimuli."[59]

While the Royal Society may have periodically professed a "come one, come all" attitude, thus appearing to encourage wide participation in its empirical pursuits, its ban on women attending its meetings is an indicator of the exclusionary nature of its methods.[60] In fact, Cavendish was the only woman to have been allowed to visit the exclusively male Royal Society until centuries after its inception. Distinguished by the practice of modest witnessing, early English science associated reliable knowledge with the notion of a detached and contained self, and this means of authorizing knowledge was one of the Royal Society's tactics for preserving its exclusiveness. As Donna Haraway explains, "The issue was whether women had the independent status to be modest witnesses, and they did not."[61] Feminine and masculine modesty have been historically perceived as binary opposites. The former is "of the body," enacted by women's preservation of their virginity; the latter is "of the mind," enacted by men's philosophical detachment from matter.[62] Consequently, "within the conventions of modest truth-telling, women might watch a [scientific] demonstration; they could not witness it."[63] From the outset, the Baconian modest witness was presumed to be a gentleman.

Margaret Cavendish and other early modern female philosophers were therefore obliged to find other ways of securing their authority. According to Bacon's portrayal of the modest witness, her close association with the body and thus with matter would put the female philosopher at a significant disadvantage, even potentially precluding her from engaging in respectable philosophical practice altogether. However, Cavendish instead chose to view such material association through the framework of the Epicureanism that Lucretius poeticized. As a result, her affinity with matter becomes something altogether different. Epicurean philosophy normalizes the irresistible pull of one's material environment and insists that all beings (whether female or male) are equally exposed to it. Women are simply more accustomed to acknowledging this pull because society demands that they do so. I agree with Keller's observation that Cavendish "did not believe that *as a woman* she had access to any privileged epistemological standpoint."[64] However, I would suggest that because of her recognition of how society constructs gender and associates women with matter, she did view herself to be better positioned and equipped as a woman to reap the great benefits of sensitive witnessing than male philosophers were. It is a privilege to deny one's material nature—a privilege to which only free men had access. Rather than defying gender expectations and attempting to claim this privilege for herself, Cavendish chose instead to reveal such privilege to be unwarranted and founded on a deceptive fiction about the status of man in relation to matter. Her Epicureanism substantiates her insistence that men are no different than any other creature—they too are porous composites, entangled with their environments through transcorporeal atomic exchange. She pursues and promotes sensitive rather than modest witnessing not as a compromise with patriarchal expectations and capitulation to the female gender role but rather because sensitivity is, for Cavendish, more compatible with, and thus equipped to access, the cosmos than modesty is. To return to this essay's epigraph, it is the empiricist who cultivates rather than eschews softness—openness, vulnerability, sensitivity—who has the potential to restore a "rusty" age and is therefore equipped to enact a truly "great instauration."[65]

Notes

Portions of this essay originally appeared in my book Sensitive Witnesses: Feminist Materialism in the British Enlightenment, *© 2023 by Kristen M. Girten. All rights reserved. Used by permission of the publisher, Stanford University Press, sup.org.*

1. Todd Andrew Borlik, "The Whale under the Microscope: Technology and Objectivity in Two Renaissance Utopias," in *Philosophies of Technology: Francis Bacon and His Contemporaries,* ed. Claus Zittel, Gisela Engel, Nanni Romano,

and Nicole C. Karafyllis (Leiden: Brill Academic Publishers, 2008), 239. Botlik cites John Aubrey, *Brief Lives* (1669–96) (London: Secker and Warburg, 1949), 130; and Michel de Montaigne, *Apology for Raymond Sebond*, ed. Roger Ariew and Marjorie Grene (Indianapolis: Hackett, 2003).
2. Eve Keller, "Producing Petty Gods: Margaret Cavendish's Critique of Experimental Science," *ELH* 64, no. 2 (1997): 451.
3. Keller, 449.
4. Line Cottegnies, "Generic *Bricolage* and Epicureanism in Margaret Cavendish's Imaginative Works," in *A Companion to the Cavendishes*, ed. Lisa Hopkins and Tom Rutter (York: Arc Humanities Press, 2020), 327; Emma L. E. Rees, "'Sweet honey of the Muses': Lucretian Resonance in *Poems, and Fancies*," *In-Between: Essays & Studies in Literary Criticism* 9, nos. 1 & 2 (2000): 3–16.
5. Londa Schiebinger, "Margaret Cavendish, Duchess of Newcastle," in *A History of Women Philosophers*, ed. Mary Ellen Waithe (Dordrecht: Kluwer Academic Publishers, 1991), 3:1–20; Stephen Clucas, "The Atomism of the Cavendish Circle: A Reappraisal," *Seventeenth Century* 9, no. 2 (1994): 259–60. See also Liza Blake, "After Life in Margaret Cavendish's Vitalist Posthumanism," *Criticism* 62, no. 3 (2020): 433–56; and Anne M. Thell, "'[A]s lightly as two thoughts': Motion, Materialism, and Cavendish's *Blazing World*," *Configurations* 23 (2015): 1–33. For an alternative view, see Lisa Sarasohn, "A Science Turned Upside Down: Feminism and the Natural Philosophy of Margaret Cavendish," *Huntington Library Quarterly* 47, no. 4 (1984): 299–307; and John Rogers, *The Matter of Revolution: Science, Poetry, and Politics in the Age of Milton* (Ithaca: Cornell University Press, 1996), 188.
6. Clucas, "Atomism," 259–60.
7. Clucas, 260.
8. Clucas, 261.
9. Clucas, 261.
10. Aaron R. Hanlon persuasively argues that Cavendish "was more systematic—more methodological—than critics have acknowledged." Aaron R. Hanlon, "Margaret Cavendish's Anthropocene Worlds," *New Literary History* 47 (2016): 51.
11. Marian Fournier, *The Fabric of Life: Microscopy in the Seventeenth Century* (Baltimore: Johns Hopkins University Press, 1996); Lisa Jardine, *Ingenious Pursuits: Building the Scientific Revolution* (New York: Anchor Books, 2000); Catherine Wilson, *The Invisible World: Early Modern Philosophy and the Invention of the Microscope* (Princeton: Princeton University Press, 1995).
12. Robert Hooke, *Micrographia, or, Some physiological descriptions of minute bodies made by magnifying glasses* (London, 1665), preface.
13. Margaret Cavendish, *Observations upon Experimental Philosophy*, ed. Eileen O'Neill (Cambridge: Cambridge University Press, 2001), 50–51.
14. Cavendish, *Observations*, 51.

15. Keller, "Producing Petty Gods," 455.
16. Francis Bacon, *Novum Organum,* in *The Works of Francis Bacon,* ed. James Spedding, Robert Leslie Ellis, and Douglas Denon Heath (New York: Hurd and Houghton, 1864), 8:45.
17. Bacon, 44–45.
18. Bacon, 44.
19. Bacon, 60–61.
20. Bacon, 77.
21. Bacon, 99.
22. See especially Steven Shapin and Simon Schaffer, *Leviathan and the Air-Pump: Hobbes, Boyle, and the Experimental Life* (Princeton: Princeton University Press, 1985), 65–69. Other scholars have more recently recognized the enduring legacy of such "modest" empiricism. For example, Tita Chico, *The Experimental Imagination: Literary Knowledge and Science in the British Enlightenment* (Stanford: Stanford University Press, 2018); Lorraine J. Daston and Peter Galison, *Objectivity* (Cambridge, MA: MIT Press, 2007); and Donna J. Haraway, *Modest_Witness@Second_Millennium. FemaleMan©_Meets_OncoMouse™: Feminism and Technoscience* (London: Routledge, 1997).
23. Shapin and Schaffer, 65.
24. Shapin and Schaffer, 65–67; Robert Boyle, "A Proëmical Essay . . . ," in *The Works of the Honourable Robert Boyle,* ed. Thomas Birch (J. & F. Rivington, 1772), 1:300.
25. Shapin and Schaffer, *Leviathan and the Air-Pump,* 69.
26. According to Chico, the chemist Robert Boyle "takes Hooke's figuration a step further by removing the human body altogether." Chico, *Experimental Imagination,* 39.
27. Cavendish, *Observations,* 52.
28. Thomas Sprat, *The History of the Royal Society* (London, 1667), 3.
29. Cavendish, *Observations,* 67.
30. Blake, "After Life," 444. See also Keller, "Producing Petty Gods."
31. Cavendish, *Observations,* 56; Lucy Hutchinson, *Lucretius de Rerum natura,* in *The Works of Lucy Hutchinson,* ed. Reid Barbour and David Norbrook (Oxford: Oxford University Press, 2012), 1:29, l. 179. Though there is no definitive evidence that Cavendish was familiar with Hutchinson's translation of Lucretius, which was begun in the early 1650s but not published until 1675, I have nevertheless chosen to use this translation because (1) it appears to have been the first English translation of *De Rerum Natura* and (2) I believe there are connections to be made between Hutchinson's and Cavendish's interests in Lucretian Epicureanism and the gender implications of these interests. (David Norbrook even suggests that Hutchinson's translation may have been an attempt to compete with Cavendish.) Cottegnies provides a useful overview of

the various Epicurean works that appear to have directly influenced Cavendish. David Norbrook, introduction to *The Translation of Lucretius*, in *The Works of Lucy Hutchinson*, ed. Reid Barbour and David Norbrook (Oxford: Oxford University Press, 2012), xxxiii–xlii; Cottegnies, Generic Bricolage, 327–28.

32. Margaret Cavendish, *The Grounds of Natural Philosophy* (London, 1668), 2.1.17. See also Cavendish, *Observations*, 35.
33. Hutchinson, *Lucretius de Rerum natura*, 1:87, ll. 60–64.
34. Cavendish, *Grounds*, 3.4.30–31. The term *transcorporeal* is borrowed from Stacy Alaimo, who uses it as a framework for theorizing the way in which "the human is always intermeshed with the more-than-human world" such that "the substance of the human is ultimately inseparable from 'the environment.'" Stacy Alaimo, *Bodily Natures: Science, Environment, and the Material Self* (Bloomington: Indiana University Press, 2010), 2.
35. Hutchinson, *Lucretius de Rerum natura*, 1:89, l. 93.
36. Liza Blake shows how Cavendish's belief in "the commonality of matter across creatures" allows her to, in the tradition of Lucretius, encourage her readers to "be happy about death" rather than to fear it. Blake, "After Life," 444; see also Aaron R. Hanlon, "Margaret Cavendish's Anthropocene Worlds," *New Literary History* 47 (2016): 64.
37. Cavendish, *Observations*, 56.
38. Cavendish, 49.
39. Cavendish, 95, 49.
40. Bacon, *Novum Organum*, 8:48, 71.
41. Cavendish, *Observations*, 49.
42. Keller, "Producing Petty Gods," 457.
43. Cavendish, *Observations*, 53.
44. Lisa Sarasohn, *The Natural Philosophy of Margaret Cavendish: Reason and Fancy during the Scientific Revolution* (Baltimore: Johns Hopkins University Press, 2010), 160; Cavendish, *Observations*, 141.
45. Cavendish, 61.
46. Though it is true, as Ann Battigelli observes, that Cavendish's recourse to reason distinguishes Cavendish's program from Baconian empiricism, I would challenge her polarization of their methods. Anna Battigelli, *Margaret Cavendish and the Exiles of the Mind* (Lexington: University Press of Kentucky, 1998), 88–89.
47. Cavendish, *Observations*, 61.
48. Cavendish, 61.
49. Cavendish, 62.
50. Cavendish, 62.
51. Cavendish, 62.
52. Margaret Cavendish, *The Blazing World & Other Writings* (London: Penguin, 1994), 188–89. John Rogers presents a persuasive argument that Cavendish

saw in reason the possibility of an escape from "the tyranny of sexual subjection." However, I would suggest that he neglects to honor the challenge Cavendish poses to binaristic thinking and thus extends his argument too far when he contends that she identifies women with rational matter and men with "the sensitive particles of matter." John Rogers, "Margaret Cavendish and the Gendering of the Vitalist Utopia," in *The Matter of Revolution: Science, Poetry, and Politics in the Age of Milton* (Ithaca: Cornell University Press, 1996), 201–2. Keller too asserts that *The Blazing World* takes aim at such binaristic thinking. Keller, "Producing Petty Gods," 461.
53. Margaret Cavendish, *The Worlds Olio* (London, 1671), bk. 1, p. 42.
54. Cavendish, bk. 1, p. 45.
55. Cavendish, bk. 1, pp. 44–45.
56. Cavendish, *Observations*, 49.
57. *Oxford English Dictionary Online*, s.v. "Trace," https://www.oed.com.
58. Cavendish, *Worlds Olio*, 44.
59. *Oxford English Dictionary Online*, s.v. "Sensitive," https://www.oed.com.
60. In his *History of the Royal Society*, Thomas Sprat comments that the Royal Society "embrac[es] all assistance"—"not only by the Hands of learned and profess'd Philosophers: but from the Shops of *Mechanics;* from the Voyages of *Merchants;* from the Ploughs of *Husbandmen;* from the Sports, the Fishponds, the Parks, the Gardens of *Gentlemen.*" Thomas Sprat, *History of the Royal Society* [1667], ed. Jackson I. Cope and Harold Whitmore (St. Louis: Washington University Press, 1958), 67, 72.
61. Haraway, *Modest_Witness@Second_Millennium*, 27.
62. Haraway, 30.
63. Haraway, 31.
64. Keller, "Producing Petty Gods," 450.
65. "Instauratio Magna," or "Great Instauration," is the title of Bacon's greatest philosophic work and includes *Novum Organum*, cited earlier.

Maria Edgeworth's Avian Entwinements
Experimental Science and the Cultivation of the Female Mind in Practical Education *and* Belinda

ADELA RAMOS

I n *Practical Education* (1798), Maria Edgeworth redefines the "art of education" as an "experimental science."[1] She centers the experiment in the pedagogical repertoire of the parent through vivid narrative examples that call to mind Joseph Wright of Derby's famous candlelight painting *An Experiment on a Bird in the Air-Pump* (1768). In Wright's iconic painting, as in Edgeworth's treatise, experimental science summons a multigenerational collective of women and men to witness knowledge production. Interpretations of the painting are as varied as the attitudes of Wright's subjects, which move between the attentive wonder of the male participants and the anguished expressions of the young women, who either refuse to witness the dying Tanimbar cockatoo or do so in dismay. While some depart from the assumption that Wright implicitly celebrates the dominant narrative that eroticized empirical discovery, recent readings find that such a definitive interpretation oversimplifies the dynamic gender and interspecies relationships at the center of the painting. In particular, Diana Donald argues that the juxtaposition of the reluctant female witnesses and the cockatoo at once portrays the period's assumptions about women's proclivity to sympathize with animal-kind and existing efforts to integrate women into scientific pursuits through air pump demonstrations.[2] That is, while Wright's female witnesses confirm misogynistic assumptions about women's animality, their presence at the table also attests to existing beliefs about experimental science's edifying potential.

For Edgeworth, experimental science certainly held this power. It was a liberatory practice of national import, a means to cultivate reason and industry. Consequently, the marginalization of women from scientific pursuit—a possibility that the painting does not foreclose—was deeply concerning. In her treatise and metropolitan novel *Belinda* (1801), Edgeworth addresses this concern by formulating a domestic science whereby she corrects the association of women with animals while providing a means for them to become scientific agents at home. The juxtaposition of the reluctant female witnesses and the flailing cockatoo of Wright's painting evinces how, in the eighteenth-century imaginary, women's minds and bodies were entwined with those of birds through potent metonymical associations. Edgeworth's awareness of how the bird-woman connection served to sexualize and infantilize women and mark them as irrational is evident in the assemblage of birds that flutter through the pages of her treatise and novel. Birds have remained tangential in recent critical discussions of *Belinda* that center a bowl of goldfish as the emblem of her experimental science.[3] In what follows, I argue that birds aid Edgeworth as she outlines the empirical and feminist aims of her educational project by examining what I refer to as "avian entwinements." Avian entwinements are pedagogical and affiliative practices whereby Edgeworth grants scientific agency to children and women at home. With this concept, I trace how Edgeworth domesticates two empirical methodologies, the experiment and observation, through the stories of three birds—a cuckoo, a macaw, and a bullfinch—in the service of familial happiness and female rationality.

In Edgeworth's avian entwinements, as in Wright's painting, exotic birds are at the service of European knowledge production. In this sense, Edgeworth's entwinements might be considered in relation to Thomas van Dooren's concept of "avian entanglements." For van Dooren, "entanglement" connotes the "coevolution and ecological dependency" of human and avian "ways of life." These ways of life comprehend how each individual shares, nurtures, and produces their species' present, past, and future. Neither solely biological nor social, "multispecies entanglements" are deep relationships wherein "learning and development take place" and "social practices and cultures are formed." van Dooren situates these entanglements "*inside*" a time of extinctions that begins with Edgeworth's time of imperial expansion, when she describes the "hundred and eighty beautiful birds," including an "abundance" of red macaws, "collected after great labor and expence" in a voyage from Cayenne to Brest.[4] Edgeworth was likely unaware of how the imperial project her writing supported placed the avians she imagines at risk of extinction. Therefore, although *entangled* has of late served as an important theoretical means of interrogating human-animal relations, I use *entwined*. *Entangled* carries a long view of coevolution that was not accessible to Edgeworth in our own terms. However, the entwinements I examine foreground how for her domestic scientists, learning and development were supported by a multispecies way of life.

The birds of *Practical Education* and *Belinda* can be mechanical or living beings from which experimental knowledge is derived. I conceive of the cuckoo, macaw, and bullfinch as "practical birds." Their sounds—the cuckoo's two-note call, the macaw's screeching, and the bullfinch's song—summon readers and characters to observe, collect facts, and pay attention, which for Edgeworth is "absolutely essential to successful instruction."[5] By focusing my own attention on these birds, I chart how Edgeworth's entwinements render experiment and observation safe modes of knowledge production for children and women. Their application at home demonstrates how eighteenth-century women writers drew on experimental science not only to imagine new educational models but also to reformulate ideas of femininity. In the treatise, the cuckoo emblematizes the domestication of science and experiment's power to both cultivate the mind and strengthen affiliative bonds. While Edgeworth aspires to render the experiment gender-neutral, her awareness that women "cannot rectify the material mistakes of their conduct" leads her to recommend that their education "be more the result of reasoning than experiment."[6] As an alternative, she prescribes a kind of experimental prudence founded on reasoning, which she exemplifies in scenes where women derive knowledge from the practical observation of birds.

With experimental prudence, Edgeworth's formulation of the experiment as a pedagogical tool for children develops into a broader feminist project in the novel, where it corrects the sympathetic attachments between women and animals that undercut female rationality. *Belinda* models practical observation, a distinctive form of female scientific practice, in scenes of "bird love," of women's overidentification with fashionable pet birds. A symptom of an uncultivated mind, "bird love" subsumes eighteenth-century questions about women's mimicry and distractibility, which the novel solves in favor of their rationality. As Edgeworth corrects this sympathetic attachment, she joins her feminist contemporaries in drawing tactical boundaries between humans and animals. Jane Spencer has shown how, for thinkers from Margaret Cavendish to Mary Wollstonecraft, these efforts were supported by a nuanced understanding of species difference and multispecies kinship that made room for the "animality in human life" and promoted ethical consideration for other species. Though not a radical, Edgeworth was influenced by Wollstonecraft's feminism; like her, she anchors the cultivation of the female mind in perfectibility, the capacity of the human mind to improve.[7] Although anthropocentrist, this approach is committed to instilling humanity, a benevolence that begins with learning kindness to animals. In her writing, the prudent female scientist relates to animals from this understanding of humanity and offers a reformed version of bird love.

I open my examination of avian entwinements with the figure of the tattling girl in *Practical Education,* an example of the dangers of unregulated sympathy.

This section explains how Edgeworth's experimental science responds to the masculinist associations of children and women with animals and situates her project in the context of eighteenth-century feminist arguments about reason and the period's culture of sensibility. Then I turn to the first of three practical birds: the mechanical cuckoo. By examining the dissection of this wooden toy, I show, first, how Edgeworth reconceives the experiment as a gender-neutral pedagogical tool and, second, how she formulates experimental prudence and the kind of practical observation that it cultivates as alternative practices for young women. Next, I examine how the female characters of *Belinda* model prudence through the observation of two practical birds, a blue macaw and a bullfinch. Through this rational practice, Edgeworth corrects expressions of culpable sympathetic attachment to pet birds or bird love. As I conclude, I explain how, notwithstanding its radical potential, Edgeworth's domestic science follows the characteristic reformism of her oeuvre, which educates children and servants alike for the good of the nation. As in her children's tales and Irish novels, it coexists with imperial paternalism: in the end, reformed bird love is absorbed by the enlightened family, where rational women apply it to ensure order in the multispecies family.[8]

The Tattling Girl

As *Practical Education* describes boys and girls turning into "little philosophers" and "young mechanics," it breaks away from strictly gendered toys and imagines girls participating in male knowledge production forms—as her stepmother, Honora Edgeworth, and her siblings did—with the goal of exercising "their reasoning and inventive faculties upon every object which surrounds them."[9] The limits of Edgeworth's gender-neutral project are signaled in her prefatory announcement that volume 2's chapters on "Temper," "Female Accomplishments," and "Prudence and Economy" address female education, although stipulations for women also appear in volume 1's "Toys" and "Sympathy and Sensibility." Anne Chandler's illuminating analysis of the treatise details how as Edgeworth curtails women's experimentation, she undermines her belief in the social benefit of placing self-interest over propriety, which she draws from Jean-Jacques Rousseau and family friend Erasmus Darwin. In preparing older girls for womanhood, she offers a "sobering caveat" that delimits their "claim to intellectual and expressive autonomy" by recommending reasoning rather than experimentation. These caveats, as Joanna Wharton has demonstrated, were common in women's science of the mind. Although their theories overstepped the gender boundary by integrating science, they unequivocally preserved the separation of the public and private spheres.[10]

Edgeworth's caveat, I argue, is informed by beliefs about the proximity of both children and women to animals. Along with working people and the enslaved, as Spencer explains, they appeared like animals in their "comparative irrationality and closeness to mere sensual life."[11] The bird-woman association can be traced to the early uses of *byrd* as a term for the young of other animals, including humans, and "a maiden, a girl."[12] In the eighteenth century, bird-woman closeness proliferated in iconographic representations, like Wright's painting. Donald explains how these representations were oftentimes informed by patriarchal ideas of sexual mastery and provided a means to ridicule women's sympathetic relationships with animals. Even where women's sympathy for animals was endorsed, it called for regulation because it was deemed infantile. To curb this sympathetic attachment, educators and scientists, including Darwin, recommended that girls attend experimental philosophy lectures or air pump demonstrations, which often concluded with the bird's death.[13]

Edgeworth did not recommend that girls attend public demonstrations, but her concerns about women's sympathy are implicit in the warning that "peculiar caution is necessary to manage female sensibility." She recommends pairing the cultivation of "reasoning powers" with the repression of "enthusiasm for fine feeling" and specifically counters the Rousseauvian prescription of dolls in terms that conjure up women's association with birds. Dolls promote a "love of finery and fashion" and an aspiration for "tattling and visiting" instead of inquisitiveness.[14] *Tattling*—a common term for idle talk, chatter, or gossip, so frequently attributed to women—is rooted in the Germanic trope of the gabbling goose.[15] Bird noises—tattling, gabbling, and parroting—were associated with the susceptibility to mimicry of women, children, colonial others, and the enslaved. Resorting to an avian metaphor to overturn the association between women and birds, Edgeworth follows Wollstonecraft's use of parroting in *A Vindication of the Rights of Woman*'s chapter "On National Education," when she laments how woman's deficient education teaches them "to recite what they do not understand," finding blame in the mother, who listens "with astonishment to the parrot-like prattle."[16] The tattling girl of *Practical Education* intimates Edgeworth's belief that a deficient education left women to wallow in animality. She resorts to the experiment as the kind of "proper object" that children need to engross their sympathy and sensibility through their reasoning powers. Crucially, it counters their proclivity to imitation and affectation, an "involuntary effect" of unregulated sympathy. In girls specifically, unregulated sympathy breeds mimicry and distractibility and a tendency to dissipation and romance, qualities that, along with ornamental beauty, prompted their association with birds.[17]

Edgeworth's anthropocentric project does not merely tolerate sympathy for animals. She teaches humanity in *Practical Education,* and by establishing practical observation as a means for women to relate to animals, she makes room to imagine avians as family in *Belinda.* Edgeworth argues that early kindness toward animals does not organically translate into humanity in adulthood. Children learn capriciousness from adults, who show benevolence toward their favorites while tolerating cruelty toward "ugly and shocking" animals. Therefore, she recommends that animals be placed under young people's care only once they "have *fixed* habits of benevolence," can find beauty even in a frog, and avoid cruelty toward all species.[18] To be sure, this does not mean she follows the contemporary sensibility for what Tobias Menely has termed "creaturely voice," the manifestation of interspecies "vocal and bodily expressivity" that came to represent a political obligation toward animals.[19] The line she drew between human and animal expressivity can be gleaned from the third-person narration of her tales for children, which avoids any conflation between human and animal expression. Nevertheless, Edgeworth exemplifies how, as Heather Keenleyside has explained, animals were "a powerful figure for conceiving social and political community" and "a potential member of it."[20] In Edgeworth's writing, this manifests in how animals can be kin and are a means to make kin in the enlightened family that represents the nation's character and supports its imperial interests. Human-animal kinship in Edgeworth brings to mind Donna Haraway's recent call for feminists to stretch and recompose kin, to "unravel the ties of both genealogy and kin, and kin and species." Edgeworth does the opposite: in *Belinda,* bird love is corrected in support of patriarchal relations. Yet in cultivating sympathy for "oddkin," like frogs, and sanctioning reformed bird love, Edgeworth admits a messy multispecies codependence.[21]

The Mechanical Cuckoo

The dissection of a mechanical cuckoo emblematizes the multidimensionality of "experiment" in *Practical Education.* As Edgeworth recounts a time when her father engaged his children in the "dissection of a wooden cuckoo" to reveal the origin of the plaything's two-note call, she aims to illustrate the unexplored potential of toys that imitate "the notes or noises of birds and beasts." While "ingenious in their construction," they have a passing effect because objects such as the cuckoo can't be unassembled. Transience is compounded by parents who opt to preserve the toy rather than cultivate children's understanding when they inquire "how the bird cried cuckoo." The dissection provides at least two lessons: first, in its immediate application, it teaches the child a lesson in mechanics; second, the parent or preceptor learns to allow children's "experiments upon objects within their reach" so that their ideas run their course and they can acquire "knowledge by their own experience." Read in the greater scope of Edgeworth's work, where virtuous citizenship is inexorably tied to industry, experimentation contributes to the cultivation of national character.[22]

In its conception as an experimental science and how it invites children to emulate it, *Practical Education* exemplifies Tita Chico's conceptualization of the experiment as "an act of imaginative work" and of science itself as a trope. Chico explains how the promotion of experimental knowledge was founded on the scientist's ability to turn their findings into a coherent narrative revealing the extent to which Enlightenment science relied on representation and literariness.[23] As a cross-disciplinary writer, Edgeworth understood well the literariness of science. Wharton has demonstrated how her children's stories show her awareness of the "value of the literary imagination to bring science to life." This awareness manifests in an object's "conversability," which is captured in how characters converse about objects and how this instills discernment.[24] The cuckoo's dissection convenes this conversability: Edgeworth explains how domestic experiments lead "the mind to reason upon them," inducing children to "judge of the different conclusions which are drawn from them by different people." This is captured in the "many guesses that were made by all the spectators" about its "internal structure." Particularly revealing is the significance that Edgeworth lends to this method when she states that "the astonishment of the company was universal, when the bellows were cut open, and the simple contrivance was revealed to view."[25] The move from universal astonishment to a simple contrivance could appear ironic, especially in how the lesson builds up to the opening of the bellows through the discussion that precedes. But the simplicity of the contrivance should be read in relation to Edgeworth's critique of parents who give more import to a plaything's integrity than to their children's understanding. In fact, the dissection itself is led not by the curious child but by the father, who models inquisitiveness for his children. The experiment not only provides boys and girls with knowledge of mechanical toys but also resolves the unproductive tension between parents and children. In other words, experiments become affiliative opportunities for the well-ordered family.

The narrative strategies Edgeworth employs in the dissection intimate her awareness of the gendered subjectivity of the experimental scientist. As Chico explains, the gendered authority that undergirds the modest witness unites the qualities of masculinity, gentility and privilege, morality and knowledgeability, all of which distinguished him from women and laborers. Through literary figuration, the modest witness erases himself from the processes of observation and experimentation in the name of "objectivity and universal transparency."[26] Although Edgeworth does not erase her subjectivity, she appears to tentatively draw on the modest witness in how she narrates the dissection from a communal perspective: she uses the "we" that would refer readers to the coauthors as witnesses, and she filters the details of the experiment from the perspective of "a large family of children." Although these are the Edgeworth children, as young Honora's redacted name confirms, the connection between the narrators of the experiment and the family is hidden, lending the lesson a sense of objectivity.[27]

Through this authorial persona, a seemingly disembodied expression of a familial collective, Edgeworth extracts herself from the bind of female modesty without denying the delicacy of her position as a female writer. To be sure, Edgeworth's "we" evinces her dependence on patriarchal authority to publish her educational method. But she addresses the political implications of pronoun use in her chapter on "Female Accomplishments." She unapologetically recommends that girls and women learn the general principles of science *well* rather than coursing superficially through natural history and chemistry. Noting that in previous chapters, examples of science education employ "the masculine pronoun *he*" to refer to the pupil, she clarifies that this has been "for grammatical convenience, not at all because we agree . . . 'that the masculine is the more worthy gender.'" Again here she must use "we," yet the "Preface" acknowledges explicitly that she authored most chapters.[28]

While the cuckoo's dissection is sanctioned for girls, Edgeworth draws clear limits around experimentation for women. Although women should partake of scientific knowledge production in the rearing of children, their "prudence must be more the result of reasoning than experiment." Experimentation requires "promptitude of choice" and decisiveness, which in women requires regulation. Instead, Edgeworth infuses prudence with the practicality of experimentation by emphasizing reasoning and self-examination in its application. In particular, she addresses the need for women to "question their own minds" and "reason about their feelings" in their choice of friends and all matters of taste. Crucially, Edgeworth connects the value of this self-examination to love. If women better understood "the reasons of their 'preferences and aversions,'" she argues, "there would not probably be so many love matches, and so few love marriages." This brand of prudence informs *Belinda*'s eponymous heroine and plot, which, as in many of her Irish novels, provides the protagonist with opportunities to reason about her feelings for potential suitors.[29]

For courtship to end in a "love marriage," it requires the application of empirical methodologies: practical observation, self-examination, the collection of facts.[30] The language Edgeworth uses to explain prudence in the treatise is echoed in descriptions of how Belinda conducts "self-examination" and reflects about how to "command" her "affections."[31] Belinda and her flawed mentor, Lady Delacour, discuss at length the importance of love marriages. Belinda pronounces herself in favor of rational love, a kind of love that is ruled not by passion but by habit and that entails an in-depth examination of character as she chooses between the West Indian Mr. Vincent, who loves her ardently, and Mr. Hervey, whose taste for science and literature proves more compatible with hers. For Lady Delacour, learning to love by habit carries the risk of tolerance, which is a danger to national integrity when it accommodates reprehensible foreign customs. In the end, the novel sanctions love that derives from desire, not habit, while raising questions about the infallibility of empirical observation. My focus here, however, is not on Belinda's love interests but on how the novel's bird love teaches the importance of pairing reason and sympathy in the cultivation of the female mind. The practical birds of *Belinda* center prudence to counteract unregulated sympathy and its corollaries, mimicry and distractibility.

Bird Love

Unregulated sympathy is exemplified by two characters in *Belinda* who are afflicted with bird love. One is Marriott, Lady Delacour's maid, whose "prodigious affection" for her screeching macaw leads her to choose the azure avian over her mistress. The other is Rachel Hartley (Virginia St. Pierre), an inexperienced young woman who appears to prefer her bullfinch, Bobby, over her protector, Clarence Hervey. Edgeworth's choice of species for the novel's beloved birds derives from at least two of their shared qualities: beautiful plumage and imitation of human expression. The macaw was a coveted pet, as suggested in natural histories like Albin Eleazar's *A Natural History of Birds* (1734), which describes them in relation to their "good Price" of "ten Guineas." While the blue macaw is native to the Caribbean and South America, the bullfinch is identified in Thomas Bewick's *History of British Birds* as "common in every part of this island." They were also valued for their plumage and their remarkable ability to whistle and sing. Music and science dovetailed in the study of bird song during the Enlightenment, when keepers used special flutes to teach birds favorite tunes, eagerly deciphering their imitative abilities. In addition, both macaws and bullfinches can learn to speak. The talking bird fueled period debates about species identity, famously so in John Locke's *Essay concerning Human Understanding,* where the talking parrot illustrates the difference between *speaking like* a human and *being* human.[32] In their beauty and imitative abilities, the macaw and the bullfinch inflect Edgeworth's concerns about the relationship between sympathy and reason in women's education. In the novel, the bird-woman parallel calls for practical observation and self-examination to regulate sensibility and to discern between untested innocence and mimicry.

Far from talking, Marriott's macaw has only his beautiful plumage to recommend him. His screeching is so loud that it can "penetrate eleven doors" and drives Lady Delacour to command Marriott to give him up, which she vehemently refuses to do, defending him "with as much eagerness as if it had been her child."[33] Her maternal bond with the bird alludes to the common eighteenth-century refrain that women's excessive love for pets posed a threat to the social order, specifically to the place of husbands and children in the domestic hierarchy.[34] In *Belinda,* the danger resides in how Marriott's misdirected sympathy challenges the master-servant hierarchy and how her bird love arises from Lady Delacour's own culpable imitation of fashionable ladies. Belinda resolves the conflict by coaxing Marriott, who is initially "incapable of listening to reason," into self-examination. She helps Marriott lay out the facts of her relationship to her mistress by comparing, for instance, her short time with the macaw to the length of time she has served the Delacours. Thus, Marriott's sympathy is redirected to its proper object—her mistress.[35]

Marriott's reformed "passion for birds" turns her into a kind of naturalist for domestic purposes. As she employs her knowledge of bird song in support of Lady Delacour's matchmaking, scientific practice is at the service of domestic happiness. By following Bobby's unique song, Marriott finds Rachel, whom Clarence Hervey keeps secluded in his attempt to educate her according to Rousseau's system of education. Marriott engages in practical observation as she attempts to discern whether Rachel is truly innocent or if she is merely performing. By comparing what she sees before her to what she knows about how innocence is feigned, she examines Rachel's fondness for Bobby as she kisses the bird and rests him on her bosom. This act aligns Rachel with eighteenth-century pairings of female innocence and sexualized vulnerability. Although some accept that she "is innocence itself," Marriott struggles to believe what she sees. Accordingly, the narrator expresses her critique of Hervey's misguided idealization of innocence in recounting how Rachel's "absolute ignorance of the world frequently gave an air of originality to her most trivial observations." Her observations merely appear to have substance and, like a singing bird, render Rachel "at once interesting and entertaining." Marriott remains unsure about her innocence but discerns that she is "all nature,"[36] a remark that, in the context of *Practical Education*, underscores her proximity to children and her animality. In children, "artless expressions of sympathy" are celebrated because they provide a counterpoint to "the world." These expressions, Edgeworth argues, convey "the genuine language of nature."[37] Far from genuine, though, Rachel's sole exposure to romance novels proves that despite being sheltered, she can express "haughty indignation," a performance of virtue.[38]

Rachel exemplifies how unregulated sympathy also leads to distractibility as she manifests markers of "bird-wittedness," a term Edgeworth draws from Francis Bacon's *Advancement of Learning* (1605) in the treatise's explanation of attention and abstraction. For Bacon, bird-witted children have "not the faculty of attention," a problem that can be remedied through mathematics.[39] For Edgeworth, this "disease of the mind" requires practice so children learn to withdraw their attention from all external objects to focus on particular ideas. Although unable to name her condition, Rachel describes the unruliness of her mind as having "only confused ideas, floating in [her] imagination." These are the ideas of an absorbed reader of romance novels deprived of the cultivation of taste through experimentation. While absorption might imply abstraction unaccompanied by industry, it does not allow Rachel to practice concentrating on a particular set of ideas. Her ideas, like her affections, flutter like birds from one subject to the next. Rachel proves that without experimentation, innocence is ignorance, and she further elevates prudent Belinda as a woman of "cultivated tastes" and "active understanding."[40]

The Caged Practical Bird

Edgeworth's practical birds convey lessons on the moral implications of cultivating sympathy in moderation through reason by way of observation and experimentation. In the broader scope of the novel, Marriott's reformation supports Lady Delacour's collection of facts about Clarence Hervey to aid Belinda's eventual love match. Hervey's failed experiment elevates women from the position of ornamental objects of entertainment to rational humans. While Bobby's fate is unexplored, Marriott's blue macaw finds a new keeper, Helena Delacour, who is espied leaning over his cage, declaring, "Yes, though you scream so frightfully, my pretty macaw, I love you as well as Marriott ever did."[41] Helena's bird love must be conceptualized in relation to the novel's imperial backdrop and the inextricable connection between human trafficking and the exotic pet trade. Along with the goldfish, the macaw might be conceived in relation to what Suvendrini Perera argues are the seized colonial resources that adorn the novel.[42] Yet understood in relation to Edgeworth's domestic science, whose outcomes are at the service of the well-ordered British family, they do more than adorn the novel. When the blue macaw is banished from the Delacour household, in Belinda's prudent hands, he transitions from screeching nuisance to "bird of good omen" as she and Helena set in motion the reconciliation of the Delacour and Percival households.[43] Helena shares with Belinda the kind of judgment that repairs fractured bonds, exemplifying what Marilyn Butler describes as Edgeworth's communities of "active rational beings bent on ordering and leading their own small sphere."[44] The avian entwinements of Edgeworth's work provide women with the means to be scientific agents. This scientific agency holds radical potential for the reformulation of the roles of mothers, wives, daughters, and relatives. At the same time, the novel's practical birds are experimental objects and companions, colonial and domestic animals that strengthen the dominant culture through affiliative bonds in the service of Edgeworth's paternalism. Put differently, Edgeworth's practical birds produce a way of life where women can make kin with birds from the dominant subject position.

Notes

This essay has benefited from the invaluable feedback I received at the Indiana University, Bloomington, Center for Eighteenth-Century Studies Workshop (2014), especially from Scott Juengel and Richard Nash, and later have received from invaluable colleagues Wendy Call, Christopher Loar, and Jodi Wyett. The incisive and supportive feedback of this volume's editors, David Alff and Danielle Spratt, and the anonymous reviewers improved this essay immensely. I want to thank the Wang Center for Global Education (Pacific Lutheran University) for the funding that allowed me to research Edgeworth's work and eighteenth-century natural histories at the Bodleian Library, Oxford, for a few weeks in the fall of 2018.

1. Edgeworth and Edgeworth, "Preface," *Practical Education*, 1:v.
2. For a thorough overview of the predominant readings of Wright's painting and for her argument, see Donald, prologue to *Picturing Animals*, 23, 14–18.
3. Edgeworth, *Belinda*, 98–99, 161. For discussions of lunar science in *Belinda*, see Douthwaite, *Wild Girl*, esp. 184–86; Chandler, "Edgeworth and the Lunar Enlightenment," 87–104; and Wright, "Opening the Phosphoric 'Envelope,'" 509–36.
4. Van Dooren, *Flight Ways*, 4, 22, 27, 40; Edgeworth, *Belinda*, 161–63.
5. Edgeworth and Edgeworth, "On Attention," in *Practical Education*, 1:112, 115.

6. Edgeworth and Edgeworth, "Prudence and Economy," in *Practical Education*, 2:699.
7. Spencer, *Writing about Animals*, 109–43. For additional noteworthy discussions about Wollstonecraft's influence on Edgeworth, see Myers, "My Art," 104–45; and Weiss, "Intellectual Rules: The Extraordinary Ordinary *Belinda*," in *Female Philosopher*, 169–219.
8. Butler, *Jane Austen*, 142–53. For valuable articulations of Edgeworth's paternalism, see Dunne, *Maria Edgeworth*, 8–21; Kaufman and Fauske, introduction to *Uncomfortable Authority*, 15; and Murphy, "'Fate of empires,'" *Young Irelands*, ed. Mary Shine Thompson (Dublin: Four Courts, 2011), 22–30.
9. Edgeworth and Edgeworth, "Toys," in *Practical Education*, 1:31.
10. Wharton, *Material Enlightenment*, 23, 217; Chandler, "Maria Edgeworth," 95, 114, 112.
11. Spencer, *Writing about Animals*, 77–78.
12. *Oxford English Dictionary Online*, s.v. "Bird, n.," accessed June 9, 2021, http://www.oed.com/view/Entry/19327.
13. Donald, prologue, 18–25.
14. Edgeworth and Edgeworth, "Sympathy and Sensibility" and "Toys," in *Practical Education*, 1:296–97, 3–4.
15. *Oxford English Dictionary Online*, s.v. "Tattle, v." accessed April 27, 2021, http://www.oed.com/view/Entry/198116.
16. Wollstonecraft, *Vindication*, 257. For the connection between parrots and enslaved Africans, see Nussbaum, *Limits of the Human*, 136–38.
17. Edgeworth and Edgeworth, "Sympathy and Sensibility," in *Practical Education*, 1:267, 270, 298.
18. Edgeworth and Edgeworth, 1:283–84. See Spencer's analysis of Edgeworth's third-person narrator in *Writing about Animals*, 74–108.
19. Menely, *Animal Claim*, 50, 5–6, 9.
20. Keenleyside, *Animals and Other People*, 21, 17.
21. Haraway, *Staying with the Trouble*, 2, 103, 102.
22. Edgeworth and Edgeworth, "Preface" and "Toys," in *Practical Education*, 1:2, 5, 16, 8–9, v–vi, 14; Wharton provides a thorough analysis of how Edgeworth's method was mythologized in relation to Honora Edgeworth: *Material Enlightenment*, 73–111.
23. Chico, *Experimental Imagination*, 21, 3.
24. Wharton, *Material Enlightenment*, 220.
25. Edgeworth and Edgeworth, "Toys," in *Practical Education*, 1:28, 16.
26. Chico, *Experimental Imagination*, 37.
27. Edgeworth and Edgeworth, "Toys," in *Practical Education*, 1:16.
28. Edgeworth and Edgeworth, "Female Accomplishments" and "Preface," in *Practical Education*, 2:552, 1:x.

29. Edgeworth and Edgeworth, "Prudence and Economy," in *Practical Education,* 2:699–701. On Edgeworth's Irish novels, see Robert Tracy, "The Cracked Looking Glass of a Servant: Inventing the Colonial Novel," in *Unappeasable Host,* esp. 25.
30. Edgeworth and Edgeworth, "Prudence and Economy," 2:700–701.
31. Edgeworth, *Belinda,* 139.
32. Eleazar, *Natural History of Birds,* 2:2, 10; Bewick, *Natural History,* 176; Rothenberg, *Why Birds Sing,* 19; Locke, *Essay,* 171.
33. Edgeworth, *Belinda,* 141, 158–59.
34. Brown, *Homeless Dogs,* 71; Wyett, "Lap of Luxury," 275–301.
35. Edgeworth, *Belinda,* 160–61.
36. Edgeworth, *Belinda,* 329–30.
37. Edgeworth and Edgeworth, "Sympathy and Sensibility," in *Practical Education,* 1:265.
38. Edgeworth, *Belinda,* 366, 372, 384.
39. Edgeworth and Edgeworth, "Attention," in *Practical Education,* 1:100–101; Bacon, *Major Works,* 242.
40. Edgeworth, *Belinda,* 381, 379.
41. Edgeworth, *Belinda,* 403.
42. Perera, *Reaches of Empire,* 24. For a brief discussion of the exotic pet trade and enslavement, see Tague, *Animal Companions,* 52–53.
43. Edgeworth, *Belinda,* 163.
44. Butler, *Jane Austen,* 127.

Bibliography

Bacon, Francis. *Francis Bacon: The Major Works.* Edited by Brian Vickers. New York: Oxford University Press, 2008.

Bewick, Thomas. *A Natural History of British Birds: Thirty-Five Engravings on Wood.* Alnwick: Printed and sold by W. Davison, 1814.

Brown, Laura. *Homeless Dogs & Melancholy Apes: Humans and Other Animals in the Modern Literary Imagination.* Ithaca: Cornell University Press, 2010.

Butler, Marilyn. *Jane Austen and the War of Ideas.* Oxford: Oxford University Press, 1990.

Chandler, Anne. "Maria Edgeworth on Citizenship: Rousseau, Darwin, and Feminist Pessimism in *Practical Education.*" *Tulsa Studies in Women's Literature* 35, no. 1 (2016): 93–122.

Chandler, James. "Edgeworth and the Lunar Enlightenment." *Eighteenth-Century Studies* 45, no. 1 (Fall 2011): 87–104.

Chico, Tita. *The Experimental Imagination: Literary Knowledge and Science in the British Enlightenment.* Stanford: Stanford University Press, 2018.

Donald, Diana. *Picturing Animals in Britain, 1750–1850.* New Haven, CT: Yale University Press for the Paul Mellon Centre for Studies in British Art, 2007.
Douthwaite, Julia V. *The Wild Girl, Natural Man, and the Monster: Dangerous Experiments in the Age of Enlightenment.* Chicago: University of Chicago Press, 2002.
Dunne, Tom. *Maria Edgeworth and the Colonial Mind.* O'Donnell Lecture. Cork, Ireland: University College, Cork, 1984.
Edgeworth, Maria. *Belinda.* Edited by Kathryn J. Kirkpatrick. New York: Oxford University Press, 1994.
Edgeworth, Maria, and Richard Lovell Edgeworth. *Practical Education.* 2 vols. New York: Cambridge University Press, 2012.
Eleazar, Albin. *A Natural History of Birds. Illustrated with a Hundred and Four Copper Plates, Engraved from the Life. Published by the Author Eleazar Ablin, and Carefully Colour'd by His Daughter and Himself, from the Originals, Drawn from the Live Birds.* Vol. 2. London, 1734.
Haraway, Donna. *Staying with the Trouble: Making Kin in the Chthulucene.* Durham: Duke University Press, 2016.
Kaufman, Heidi, and Chris Fauske, eds. *An Uncomfortable Authority: Maria Edgeworth and Her Contexts.* Newark: University of Delaware Press, 2004.
Keenleyside, Heather. *Animals and Other People: Literary Forms and Living Beings in the Long Eighteenth Century.* Philadelphia: University of Pennsylvania Press, 2016.
Locke, John. *An Essay concerning Human Understanding.* Edited by Peter Niditch. London: Oxford University Press, 1975.
Menely, Tobias. *The Animal Claim: Sensibility and the Creaturely Voice.* Chicago: University of Chicago Press, 2015.
Murphy, Sharon, "'The fate of empires depends on the education of youth': Maria Edgeworth's Writing for Children." In *Young Irelands,* edited by Mary Shine Thompson. Portland, OR: Four Courts, 2011.
Myers, Mitzi. "My Art Belongs to Daddy? Thomas Day, Maria Edgeworth, and the Pre-texts of *Belinda*: Women Writers and Patriarchal Authority." In *Revising Women: Eighteenth-Century "Women's Fiction" and Social Engagement,* edited by Paula R. Backscheider, 104–45. Baltimore: Johns Hopkins University Press, 2000.
Nussbaum, Felicity. *The Limits of the Human: Fictions of Anomaly, Race, and Gender in the Long Eighteenth Century.* Cambridge: Cambridge University Press, 2003.
Perera, Suvendrini. *Reaches of Empire: The English Novel from Edgeworth to Dickens.* New York: Columbia University Press, 1991.
Rothenberg, David. *Why Birds Sing: A Journey into the Mystery of Bird Song.* Boulder: Basic Books, 2006.
Spencer, Jane. *Writing about Animals in the Age of Revolution.* Oxford: Oxford University Press, 2020.

Tague, Ingrid H. *Animal Companions: Pets and Social Change in Eighteenth-Century Britain*. University Park: Pennsylvania State University Press, 2015.

Tracy, Robert. *The Unappeasable Host: Studies in Irish Identities*. Dublin: University College Dublin Press, 1998.

van Dooren, Thom. *Flight Ways: Life and Loss at the Edge of Extinction*. New York: Columbia University Press, 2014.

Weiss, Deborah. *The Female Philosopher and Her Afterlives: Mary Wollstonecraft, the British Novel, and the Transformations of Feminism, 1796–1811*. Palgrave Studies in the Enlightenment, Romanticism and Cultures of Print. Cham: Springer International Publishing AG, 2017.

Wharton, Joanna. *Material Enlightenment: Women Writers and the Science of Mind, 1770–1830*. Woodbridge Rochester, NY: Boydell Press, 2018.

Wollstonecraft, Mary. *A Vindication of the Rights of Woman*. Edited by Sylvana Tomaselli. New York: Cambridge University Press, 1995.

Wright, Nicole M. "Opening the Phosphoric 'Envelope': Scientific Appraisal, Domestic Spectacle, and (Un)'Reasonable Creatures' in Edgeworth's Belinda." *Eighteenth-Century Fiction* 24, no. 3 (Spring 2012): 509–36.

Wyett, Jodi L. "The Lap of Luxury: Lapdogs, Literature, and Social Meaning in the 'Long' Eighteenth Century." *LIT: Literature Interpretation Theory* 10, no. 4 (2000): 275–301.

PART IV

Environment

The Hoe and the Plow

Plantation Labor under the Somatic Energy Regime

RAMESH MALLIPEDDI

It is greatly to be regretted that machinery is not more generally employed [in the Caribbean colonies]. The plough is almost unknown; and the hoe and tray and bill-hook are nearly all the utensils which are used in agriculture.
—SYLVESTER HOVEY, *Letters from the West Indies* (1838)

During a visit to the West Indies in 1818, Matthew Lewis attempted to introduce the animal-powered plow in the cultivation of sugar on his Jamaica estate. In the eighteenth century, canefields in the Caribbean were prepared for planting mostly with handheld hoes rather than plows even though hoeing was hard labor, more onerous than any other task in the crop cycle. Although hoe culture was regarded as wasteful and inefficient, the plow was never fully substituted for the hoe because of the planters' lack of interest in technical improvements. But in the final decades of the eighteenth century, several prominent planters, including Samuel Martin and Edward Long, successfully experimented with animal-powered instruments. Lewis viewed the measure, the substitution of "the labour of animals for that of slaves," as humanitarian, necessary for relieving the burden of farmwork. However, due to the absence of experienced English plowmen to train and superintend unskilled field-workers and the lack of expert carpenters and ironsmiths to maintain and repair broken implements, this well-meaning effort turned out to be impracticable. Perhaps more significantly, Lewis ascribes the failure to Africans' putative racial incapacity, to "the awkwardness, and still more the obstinacy," of his workers who "broke plough after plough, and ruined beast after beast, till the attempt was abandoned in despair."[1]

These efforts to replace the hoe with the plow and hand tools with animal-hauled machinery were undertaken rather late in the colonial period, nearly a century and a half after sugar cultivation commenced in the Caribbean. While farmers in Europe and North America universally used cattle plows to open land, Caribbean planters continued to rely on handheld hoes. The discrepancy in methods of cultivation between the metropolis and the colony is all the more surprising in view of the seventeenth-century Scientific Revolution that transformed English husbandry. For instance, Walter Blith's *The English Improver Improved* (1652), one of the most influential texts on agrarian innovation, lists a variety of plows designed for particular soils, including the plain trenching plow, the trenching wheel plow, the single wheel plow, the Hertfordshire wheel plow, and the double plow in addition to a range of other farming implements such as spades and harrows (see figs. 1 and 2). In contrast, Black laborers in the Caribbean had four rudimentary tools: the hoe, the tray, the billhook, and the ax (see fig. 3). Although the plantation complex is variously characterized as the "primordial site of Atlantic modernity" and the "advanced front of modern capitalism," agrarian technology in the British Caribbean

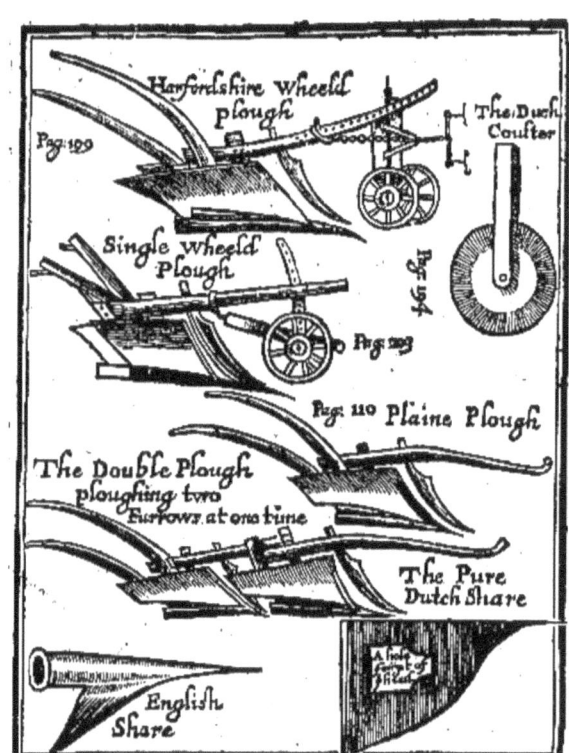

FIGURE 1. Wheeled plows. From Walter Blith, *English Improver Improved* (1652). (Author's personal collection)

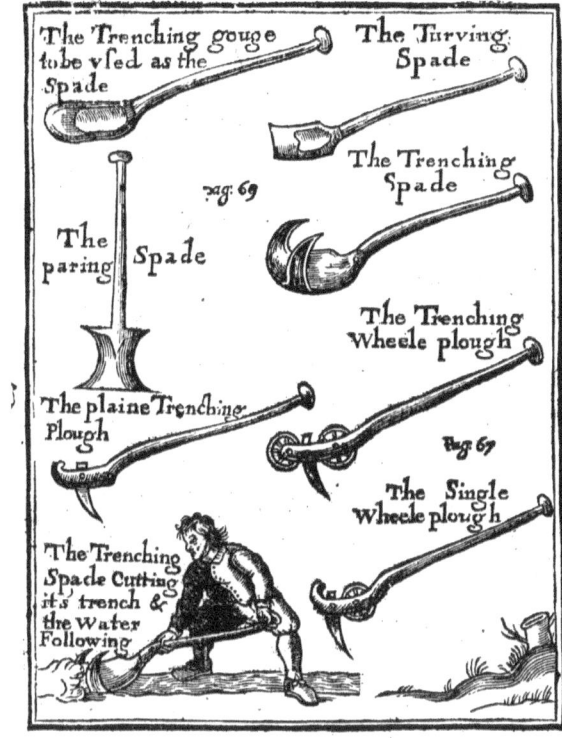

FIGURE 2. Plow designs. From Walter Blith, *English Improver Improved* (1652). (Author's personal collection)

remained archaic and rudimentary from the establishment of sugar colonies in the 1650s until emancipation in 1838.[2]

Commentators on New World slavery from Karl Marx to Eric Williams viewed slavery as inimical to technological progress, given the ready availability of cheap, unskilled labor and the lack of incentives for making the workforce more productive or productive technologies more efficient. Alternatively, economic historians have argued that the planter class showed a keen interest in mechanization and technological developments, especially after the campaign to end the slave trade threatened to cut off labor supplies.[3] Caitlin Rosenthal has recently shown in her business history of plantation societies that chattel slavery was not an impediment to innovation but in fact the source of several advanced methods of scientific management, including quantitative techniques of labor organization and valuation practices that are hallmarks of the factory system of production under industrial capitalism.[4] But existing work on slavery and technology has not adequately considered the recalcitrance of the physical environment. For instance, prominent planters in the Caribbean were enthusiastic about new, improved methods of cultivation, but

FIGURE 3. "Field Negro, Sugar Cane in the Background." From Richard Bridgens, *West India Scenery . . . from sketches taken during a voyage to, and residence of seven years in . . . Trinidad* (London, 1836), plate 14. (Yale Center for British Art, Paul Mellon Collection)

the plow was not widely used because of the stiffness of tropical soils and the scarcity of strong draft animals, which could not be maintained on the small (and sometimes nonexistent) pastures of many Caribbean islands. Moreover, plowing was thought to overexpose the soil, leading to desiccation, a real concern where sugar monocropping quickly depleted fertility. In the wake of soil denudation and the diminution of grassland, slaves were made to do the work of livestock. Indeed, any effort to introduce new technologies involved human beings, animals, and things in a complex network of relationships.

The association between slaves and tools has a long genealogy since antiquity, beginning with Aristotle's categorization in the *Politics* of the slave as the master's instrument and Marcus Terentius Varro's threefold division in his agricultural treatise *Res Rustica* of the slave as the speaking implement (*instrumentum vocale*), the animal as the semimute implement (*instrumentum semivocale*), and the plow as the mute implement (*instrumentum mutum*). In these canonical accounts, slaves and animals are tools, integral to the productive process, yet occupy a subordinate place because of the physicality of their labor. In characterizing enslaved people and livestock as the "nerves of a plantation," the success of which consists, "as in a well constructed machine, upon the energy and right disposition of the main springs, or primary parts," the Antiguan planter Samuel Martin underscores their economic function.[5] But proponents of new science, affiliated with Samuel Hartlib, the influential philosopher and educationist who maintained extensive correspondence with leading thinkers in England and Europe, also understood human beings and animals to be part of what the environmental historian J. R. McNeill calls a "somatic energy regime"—that is, as sources of mechanical power.[6] In the preindustrial period, before fossil fuels became the principal source of energy, living bodies supplied more than 70 percent of kinetic power. Since draft animals were expensive in the Caribbean colonies and large tracts of grassland were required for their maintenance, slaveholders quickly realized that it was economically more advantageous to substitute human labor for animal power and to convert freed-up pastures into canefields. The planters' choices concerning labor and methods of cultivation—that is, whether to use hand hoes or animal-drawn plows—were thus guided by technical, social, and economic imperatives. Indeed, they persisted in the use of the hand hoe, notwithstanding its known detrimental effects, because of the ready availability of Black laborers. Current work on chattel slavery has documented the role of socioeconomic institutions—racial ideologies, legal codes, and coercive labor practices—in the subjugation of Africans. But attention to agrarian implements such as the hoe and the plow—and technology and energy systems more generally—not only reveals what Penny Harvey terms the "lithic vitality" of nonorganic matter in human affairs but also illuminates how the violence and privations of chattel slavery were produced at the nexus of material and social forces.[7]

Agrarian Improvement in an Uneven World

Although mechanized farming has largely eliminated the use of hand tools and animal labor in much of Europe and North America over the last century, the hoe and the plow were the quintessential implements of soil preparation for millennia—and continue to be so in most non-Western communities around the world. The cultural and economic significance of these implements is reflected in Fernand Braudel's division of preindustrial populations into "the people of the hoe" and the "people of the plough."[8] Lands cultivated with the hoe, according to Braudel, form a belt, encompassing Oceania, pre-Columbian America, Central Africa, and parts of South and Southeast Asia, where tillage by the plow and the hoe often coexisted, but for the most part, hoe culture is characteristic of areas with vast stretches of fallow land, low population densities, absence of domesticated livestock, and a uniformity of plants, animals, and tools. In contrast, the presence of domesticated plants and animals was a sine qua non of plow culture, whose beginnings are traced to Mesopotamia in the fourth millennium. The early scratch plow (or the ard) was a rudimentary tool fitted with a fire-hardened share, but in the Iron Age (after the tenth century BC), plowshares came to be encased in metal. In his survey of agricultural tools, E. M. Jope views "the protection of the cutting point with an iron shoe" as one of the most important technological developments, inasmuch as the iron plow brought the heavier but more fertile clay soils under cultivation for growing cereal grasses initially in the Near East and Asia and subsequently in Europe.[9] Historians consider the plow instrumental in humankind's transition from seminomadism to sedentarism. Indeed, the establishment of large, permanent settlements, intensive exploitation of land, and dramatic population growth, not to mention the rise of political institutions such as the state, would have been impossible without fixed-field agriculture brought into existence by the plow.

The principal feature of the digging stick or hoe is that it is a percussive tool whose pointed end or blade is removed from the earth in a series of repeated but discontinuous strokes to move the soil so as to create holes or build mounds for the reception of seeds and plant cuttings. Plots of land tilled with the hoe tend to be small, semicircular in shape, and located near dwellings or, as in conuco and swidden agriculture, on hillslopes—and even between tree stumps in forests. They contain a diversity of root and vegetable crops, including maize, beans, yams, and squash. In contrast, the plow makes a continuous furrow, cutting and turning over the soil. The fields prepared by it are larger and elongated, laid out as such to avoid the frequent turning of draft animals dragging the plow. Whereas hoe culture is typified by a mixture of plants, plowed fields are planted with a single crop. Moreover, the plow pulverizes the soil, creating a smooth, well-worked surface in contrast to the uneven mounds or hills characteristic of conucos and swidden culture. Indeed, it is one of the most important devices in mankind's transformation of natural environments. In book 1 of his *Georgics,* Virgil celebrates the plow as one of the "farmer's weapons / Without which the crops could not be sown or sprouted" and goes on to provide a structural account of its various parts, including the stock, share, beam, and yoke:

> For your plow, a living elm bent by force in the woods
> is tamed into its stock and accepts the curved share.
> To its stem a shaft extending eight feet is fitted,
> along with two moldboards and a double-backed beam.
> Beforehand, a linden is felled for the yoke, and a tall beech

for the handle, which turns the wheeled chariot's bottom from the rear.[10]

Manufactured thus, the plow is a sturdy implement, a formidable tool in mankind's struggle against nature. Earlier in the poem, Virgil celebrates the deep, long furrow created by the ox-drawn plow: "I'd have my ox groan as he pulls the plow deep / and my plowshare glisten, polished by the furrow."[11] The plow has been celebrated as a symbol of culture and a civilizing instrument since antiquity. In its ability to impose order and symmetry, the plow's significance is not only economic but also cultural.

Farmers in Europe have used the traction plow, fitted with iron shares, since 200 BCE. Although there were modifications, such as adding moldboards, coulters, and wheels, its basic design remained unchanged for centuries. The historian G. E. Fussell notes that "for sixteen and a half centuries after the birth of Christ, and no one knows how many centuries before that date, men had ploughed the land as they thought fit and as experience directed. None of them, so far as is known, had bothered about the theory of the job. They were strictly practical men, and, if they thought some slight change in pattern an improvement, they made it and tried it."[12] But after the 1640s, several reformers, inspired by the Baconian empirical method, emphasized the need for a scientific study of the plow. Hartlib stressed agrarian improvement. The drainage of the fens, enclosure of farmland, floating meadows, crop rotation, and animal breeding all date to this period.[13] In 1651, Hartlib insisted that scientific study of the instrument was necessary for improving efficiency: "It would be an extraordinary benefit to this Country, if that 1 or 2 horses could plough and draw as much as 4 or 6."[14] Bemoaning the wide variety of plows in use all over the nation and the ignorance of the husbandman in mechanical principles, Hartlib wondered why "so many excellent Mechanicks who have beaten their brains about the perpetual motion and other curiosities . . . should never so much as honour the Plough (which is the most necessary Instrument in the world) by their labour and studies." Blith, in *English Improver Improved*, attempted to address Hartlib's charge: "I shall endeavour the facilitating the great weight and burthen of the Plough, and give you the description of some forms most suitable unto ease and speede, and hope thereby to take off a considerable strength and charge from the Husbandman's daily toyle."[15] The crowning achievement of eighteenth-century scientific husbandry was of course Jethro Tull's mechanical inventions and the publication of his book *Horse-hoe Husbandry* (1732), but the seventeenth-century Scientific Revolution laid the foundation for agrarian innovation in the following century.

Notwithstanding the resurgence of interest in agrarian improvement and invention of labor-saving machines at this time, cultivation in the Caribbean was undertaken with the hoe. Indeed, as Chris Evans has argued, the hoe was the instrument that "pushed the frontier of Britain's Atlantic empire forward."[16] It was the preferred tool in the early days because of the physical terrain. When colonists landed in Barbados in 1627, the island was covered with mature tropical forest. Using Indigenous tree-felling techniques like ringbarking and burning, the newcomers cut down large trees, but the removal of trunks and extraction of stumps proved a formidable challenge. Indeed, during the first decade, the ground was covered with decaying stumps, logs, and branches. As Richard Ligon notes, "When they [the fallen trees] were laid along, the branches were so thick and boisterous, as required more help, and those strong and active men, to lop and remove them off the ground. At the time we came first there, we found both Potatoes, Maize, and Bonavists, planted between the boughs, and the Trees lying along upon the ground; so far short was the ground then cleared."[17] The arrivants even cultivated commercial staples such as indigo and cotton in that fashion: "We found Indigo planted, and so well ordered, as it sold in London at very good rates, and their Cotton wool, and Fustic wood, proved very good and staple commodities." During the first one and a half decades of settlement, or what historians termed the tobacco era (1627–40), smallholders cultivated tobacco and cotton with the help of indentured servants on small patches of cleared forestland near the coast. Although a few prominent planters had titles to large tracts, the majority of holdings during the tobacco period averaged five acres. However, during the following two decades (1640–60), sugar replaced tobacco as the commercial staple with help from Dutch traders and financiers, principal suppliers of sugar to European markets from Brazil.

The capital-intensive nature of sugar cultivation led to the absorption of small farms by large estates and a monopolization of land ownership. The transition to sugar generated spectacular profits, but concerns about soil deterioration began to emerge by 1665. The exposure of tropical soil following deforestation led to nutrient loss. The following advice to prospective planters by Thomas Tryon, the London-based merchant who lived in Barbados between 1663 and 1669, captures the economic and ecological crisis caused by soil erosion:

> Your Island being but of a small extent, and you being often necessitated to Plant the same Vegetations in the same Ground, must have worn it out extremely in respect to Virtue and Strength for them, which is farther

and more clearly demonstrable by the depth thereof at the time of the first Settlement, which was very considerable, whereas now the whole Island is become a kind of a Rock, the proper and natural Earth being no more than 2 or 3 Foot deep, before you come to a whitish Marle, somewhat like our chalky Ground in England, tho' much hotter, so that the Land in general is mightily wasted, not only in respect to its quantity, but also quality.[18]

The language Tryon uses ("wasted" and "worn out") to describe the rapid exhaustion of the island's soil base recurs in many contemporary accounts by colonial officials and planters. Dislodged soil not only reduced crop yields but also clogged up the colony's ports and harbors.[19] Acknowledging that "many of the Channels, Wharfs and Landing-places at the Port of Bridge-Town, as well as the Mole-head, having for some years past been in a great measure rendered useless" by the "Sand and Mud gathered in great quantities," the Barbados colonial assembly in 1748 made provisions to "clear the Harbour of the Mud, Dirt, and Sand; to deepen the Channel and render it navigable for Sloops, Shallops, and other Craft."[20] Soil erosion thus threatened not only plantation lands but also the colony's transportation infrastructure.

Contemporaries recommended diversity and crop rotation to combat soil loss caused by monoculture. For instance, Tryon urged owners, "Plant for the future but one half of the sugar canes you have been hitherto wont to do, and Manure the other half of your Land for provisions," as "changing and planting [the ground] with other Vegetations cannot but be extremely Beneficial."[21] But given the limited availability of arable land and the owners' determination to plant all land in cane, the advice proved impractical. Colonial planters adopted several conservationist measures to arrest, if not reverse, the loss of soil fertility, including manuring and erecting protective walls. Replenishing the soil with dung became common practice, although purchasing it from cattle farms (founded expressly to meet growing demand for manure) added to the cost of production. Thus, Edward Littleton complained, "We make high and strong walls or wears to stop the mould that washes from our grounds: which we carry back in carts or upon Negroes heads. Our Negroes work at it like Ants or Bees."[22] In his 1690 letter to the Lords of Trade and Plantations, the governor of Barbados, Christopher Codrington, urged the English government to assume full control of St. Christopher (a joint possession of England and France at the time) because sugar production in his own colony had become unprofitable: "Barbados lives chiefly by trade, for the soil is so miserably poor that it scarce anywhere produces without dung, and I dare aver that the same quantity of goods could be made in this island with less than half of the labour and expense."[23]

The most far-reaching measure that the planters devised to contain soil loss was holing, or turning up the ground with hand hoes, a method of planting in which enslaved people dug rectilinear trenches of six inches in depth and about five square feet in size for the reception of cut canes (fig. 4). The land to be planted was marked by a row of wooden pegs. With bookkeepers and overseers "carrying the line," gangs of sixty to eighty slaves worked in unison. Upon completion of the first row of holes, the gang stepped back to the next row to excavate the second. On average, an individual was expected to dig seventy holes a day. This method of labor organization was highly regimented, allowing for the optimal extraction of labor inasmuch as the whole gang had to perform "the same amount of work, and in the same time, in order to 'keep line.'"[24] As James Stephens observed, the drivers were required not only to "urge forward the whole gang with sufficient speed, but sedulously to watch that all in the line, whether male or female, old or young, strong or feeble, work as nearly as possible in equal time, and with equal effect ... all must work, or pause together."[25] Introduced first in Barbados in the early eighteenth century, holing remained the preferred method for breaking the soil all over

FIGURE 4. "Holeing a Cane-Piece." From William Clark, *Digging Holes for Planting Sugar Cane*, Antigua, West Indies (1823). (Yale Center for British Art, Paul Mellon Collection)

the Caribbean until the abolition of slavery in 1838, as attested by Sylvester Hovey's remark quoted in the epigraph to this chapter.

Holing was physically demanding and toilsome, perhaps the most arduous task during the production process, a source of morbidity and mortality among Black laborers. Excavating the soil under the tropical sun in dried up, hardened lands—lands that "become a kind of a Rock," in Tryon's words—exacted a heavy toll on enslaved peoples' lives. The historian Justin Roberts has recently argued that while the whip as a "dehumanizing tool and instrument of torture" is extensively documented, the "incremental physical violence" of the hoe remains understudied, even as the "hoe was the tool that did the most destructive damage to enslaved bodies."[26] According to Henry De La Beche, labor on coffee plantations and cattle pens was deemed "lighter" than on sugar estates, because on the former establishments "they have no holes to dig."[27] "The period of planting," as one contemporary account put it, "is the season most abhorrent to the negroes, and it is the work which they all uniformly detest, as being the most severe which they experience."[28] Hence in his medical treatise *An Essay on the More Common West-India Diseases* (1765), James Grainger advised that new slaves must be protected from the rigors of hoeing during the seasoning period: "To put a hoe in the hands of a new Negroe, and to oblige him to work with a seasoned gang, is to murder that Negroe."[29] Grainger's advice is meant to mitigate the harshness of fieldwork on the new arrivals, but African migrants were expert hoe cultivators at home, as Olaudah Equiano recalls in the opening chapter of his autobiography: "Our tillage is exercised in a large plain or common, some hours walk from our dwellings, and all the neighbours resort thither in a body. They use no beasts of husbandry; and their only instruments are hoes, axes, shovels, and beaks, or pointed iron to dig with."[30] The plantation complex grafted the methods and tools characteristic of subsistence farming onto large-scale agro-industrial enterprises for the production of commercial staples.

Implements of Labor, Instruments of Torture

It is on account of the deadly nature of plantation labor that the hoe and the plow became the centerpiece of antislavery activism. The first two recommendations for ameliorating the condition of enslaved people in the *Abstract of the Evidence . . . before the House of Commons* (1791), the pivotal publication in the Abolitionist Society's campaign against the slave trade, concerned agricultural implements. By recommending that the British Parliament "let the plough be introduced on every estate which will admit the use of it" and also "let the East Indian shovel be introduced, in the place of the hoe," antislavery campaigners underscored the importance of new tools to their program of amelioration.[31] But the extensive evidence compiled by the society also offers a glimpse into the effects of planter neglect on slaves' daily lives. Asked by the House of Commons Select Committee whether "the implements of husbandry and the mode of cultivation were such as to diminish as much as possible the labour of the Slaves," Robert Forster, who lived in the Caribbean between 1772 and 1778, replied,

> No, they were not. The plough might be introduced to advantage in performing several operations in their agriculture, and though perhaps not entirely to supersede the use of the hoe, yet might ease the Negroes of many difficult and laborious parts of their manual labour; also in grinding their own corn, which they were obliged to do in the night by hand, and which during crop time was a great hardship; they might be much relieved by some trifling mechanism applied to the sugar mills, and in many other instances. In general, they seem to have no idea of introducing any improvements to alleviate the labour of their Slaves.[32]

Sugar manufacture, involving crushing, boiling, and refining, was a complex industrial operation. Processing mills were elaborate technical apparatuses. The plantation, as Sidney Mintz claimed, was a "synthesis of field and factory."[33] Yet the owners' failure to provide a simple grain mill to ease the work of grinding corn reveals how the ruling oligarchy's indifference and cynical disregard extended to the most quotidian dimensions of slave life. Indeed, the privations caused by the absence of mechanical implements were not confined to plantation work but typified domestic life as well.

The plow as a labor-saving device was enthusiastically championed not only by antislavery sympathizers such as Forster but also by the apologists of slavery, including Edward Long, the prominent Jamaican planter, who argued that its use was both economical and humane. Based on personal observation, Long discovered how on several estates "one plough turned up as much ground in one day, and in a much better manner, than 100 Negroes could perform with their hoes in the same time."[34] Long did not call for the replacement of hand tools with animal-powered machinery but suggested that in lands previously tilled with the plow, slaves found it "uncommonly easy" to dig holes, as they had little more to do than clear the furrow, allowing for the work to be completed in half the time. The plow, used in conjunction with the hoe, was economical and cost-effective inasmuch as it enhanced the productivity of field hands. But Long also acknowledged, like his antislavery counterparts, the physical toll of cultivation by the hoe: "No other work on a plantation is so severe and so detrimental to them as that of holing."[35]

Notwithstanding the broad consensus on the use of the plow across the political spectrum, among opponents and defenders of slavery alike, the instrument never gained widespread acceptance. The obstacles, in part, were physical or topographical, since estates located on steep, hilly, or stony terrains were unsuitable for plowing. The presumed racial incapacity of Africans, their ostensible inability to handle machines, was considered another obstacle. In addition, managers cited the absence of skilled blacksmiths for routine maintenance as a hindrance. Thus, William Fitzmaurice, an overseer and bookkeeper who lived in Jamaica between 1771 and 1786, noted that he never used the plow, as "there was not a blacksmith's shop within fifteen miles to repair [it] when out of order."[36] Finally, the reasons for rejection were also social, as planters wished to keep slaves employed year-round to maintain discipline because labor rendered redundant had few other uses. This view is corroborated by contemporary observers. Asked by the Select Committee what he considered to be the chief obstacle to the introduction of plows, a witness replied, "I have understood, and I believe that it was the general opinion, that if Negroes were not constantly kept at hard labour, they would become insolent and unruly."[37] The hoe is an instrument for enforcing labor and social discipline, one that allowed planters to keep enslaved people "in line" both on and off the field.

Perhaps the biggest impediment to the widespread use of the plow was the scarcity of draft animals. From the early days, horses, oxen, and mules were not raised in the colonies but imported from the Cape Verde Islands, Madeira, England, and North America, mainly for milling and haulage, and estates set aside pastures to provide fodder. For instance, Colonel Modyford's 500-acre Barbados plantation in 1647 had 200 acres in sugar fields, 120 in woodlots, 80 in pasture, and 70 in provisions. But as sugar became a valuable export staple, planters began converting grasslands, forests, and provision grounds into canefields. In the absence of pasture, most estates had few animals. As Long admitted, horses and oxen were "dearest in purchase" and "chargeable in their maintenance." Domestic animals were expensive—or "something of a luxury"—not only in the Caribbean but in the preindustrial world more generally, in part because of the limited availability of arable land and the high mortality rates of the animals. The primary sources of mechanical power under the somatic energy regime were thus biological—human beings and domesticated animals. Ted McCormick has recently argued that minimizing or reducing the use of animal energy in agricultural production was one of the primary objectives of thinkers associated with the Scientific Revolution.[38] Hartlib's call for a plow with which "1 or 2 horses could plough and draw as much as 4 or 6" was meant to enhance productive efficiency—that is, to minimize energy input while holding the output steady. In fact, one of the most important, if fanciful, mechanical inventions undertaken by Cressey Dymock, a Hartlibian, was a perpetual motion machine for a range of agricultural and industrial operations, from grinding corn to drawing water from mines, as a substitute for somatic energy so that pasture lands thus freed up from the maintenance of animals could be converted to farmland. A pamphlet attributed to Dymock offered a different rationale for introducing the same machine in plantation colonies:

> If my engine bee made use of in the Barbados for the grinding of sugar there will nessesarily follow (besids all private benifitts) this publique advantage that whereas they are now forced to lett many acres ly for fother [i.e., fodder] for those draught cattle winter & somer the proffitt thence arrising beeing farre short of what the same land would yeild if planted with sugar canes, cotton, Indico, or the lyke, by this meanes all that land may bee converted to those more beneficiall uses, to the great increase & trade of those more staple comodityes.[39]

Of course, the perpetual motion engine as a panacea to labor shortages did not materialize in the seventeenth century, but the economics of land use envisioned here had remarkable staying power well into the nineteenth. Planters understood, like Dymock, that human labor was cheaper than animal power and that the cultivation of export staples was more profitable than raising food crops or keeping land under pasture. In the absence of engines on the one hand and livestock on the other, enslaved Africans were made to do the work of animals under the somatic labor regime characteristic of slavery—work that was tantamount to, as Samuel Martin put it, "degrading human nature to the toil of brutes."[40]

The eighteenth-century Scientific Revolution was thus a key, if underacknowledged, contributor to emergent discourses and practices of race, labor, technology, and energy utilization. McNeill has suggested that for the ruling classes before 1800, "the stock of human and domestic animal populations served as an energy store, a flywheel in the society's energy system."[41] Slavery was the linchpin in the somatic energy regime because it allowed the enterprising and powerful to regulate energy flows, to amass and exploit the power concentrated in human bodies, especially in situations where draft animals were scarce. Yet this scarcity in the Caribbean was the handiwork of planters, whose decision to plant all available land in cane led to the diminution of grazing land and hence of livestock. As Martin acknowledged, pulling the plow in Caribbean lands "required more strength of cattle and horses than [the] small pastures can sustain."[42] Planters were indifferent to enslaved people's well-being, as reflected in their failure to introduce new implements or substitute the hand hoe with the animal-powered plow, because they considered Black life, unlike the soils they cultivated, expendable, an inexhaustible energy store that could be depleted and replenished with fresh importations. The disastrous consequences of plantation agriculture are a by-product of the complex interplay between human and other-than-human forces, encompassing human bodies, animals, tools, and soil. Yet the stories we tell about plantation agriculture seldom grant agency to the dynamic vitality of matter. We cannot fully understand the catastrophic privations of chattel slavery without moving past the dichotomies between the material and the social, the inanimate and the living, that continue to structure many of our existing narratives.

Notes

1. Matthew Lewis, *Journal of a West-India Proprietor* (London, 1834), 325.
2. Hilary McD. Beckles, "Capitalism, Slavery and Caribbean Modernity," *Callaloo* 20, no. 4 (1997): 777; Bill Schwarz, "Breaking Bread with History: C.L.R. James and *the Black Jacobins*; Stuart Hall," *History Workshop Journal* 46, no. 1 (1998): 23.
3. R. Keith Aufhauser, "Slavery and Technological Change," *Journal of Economic History* 34, no. 1 (1974): 36–50.
4. Caitlin Rosenthal, *Accounting for Slavery: Masters and Management* (Cambridge, MA: Harvard University Press, 2018).
5. Samuel Martin, *An Essay on Plantership* (London, 1751), 1.
6. J. R. McNeill, *Something New under the Sun: An Environmental History of the Twentieth-Century World* (New York: Norton, 2001), 11; Vaclav Smil, *Energy and Civilization: A History* (Cambridge, MA: MIT Press, 2017).
7. Penny Harvey, "Lithic Vitality: Human Entanglements with Nonorganic Matter," in *Anthropos and the Material*, ed. Penny Harvey, Christian Krohn-Hansen, and Knut G. Nustad (Durham: Duke University Press, 2019), 143–60.
8. Fernand Braudel, *Civilization and Capitalism, 15th–18th Centuries: The Structures of Everyday Life*, 3 vol. (Berkeley: University of California Press, 1992), 1:174.
9. E. M. Jope, "Agricultural Implements," in *A History of Technology*, ed. Charles Joseph Singer (New York: Oxford University Press, 1955), 81–100, 83.
10. Virgil, *Georgics*, trans, Janet Lambke (New Haven: Yale University Press, 2005), 4.
11. Virgil, 4.

12. G. E. Fussell, "Ploughs and Ploughing before 1800," *Agricultural History* 40, no. 3 (1966): 177–86, 183.
13. For a helpful account of farming practices and land use in the seventeenth century, see Frances Dolan, *Digging the Past: How and Why to Imagine Seventeenth-Century Agriculture* (Philadelphia: University of Pennsylvania Press, 2000).
14. Samuel Hartlib, *Samuel Hartlib His Legacy of Husbandry* (London, 1651), 8.
15. Walter Blith, *English Improver Improved* (London, 1652), b2.
16. Chris Evans, "The Plantation Hoe: The Rise and Fall of an Atlantic Commodity, 1650–1850," *William & Mary Quarterly* 69, no. 1 (2012): 71–100, 71.
17. Richard Ligon, *A True and Exact History of Barbados* (New York: Hackett, 2010), 69.
18. Thomas Tryon, *Tryon's Letter, Domestik and Foreign, to Several Persons of Quality* (London, 1700), 190.
19. For two excellent accounts of the impact of agriculture on soil loss, see Daniel Hillel, *Civilization and the Life of the Soil* (Berkeley: University of California Press, 1992); and David R. Montgomery, *Dirt: The Erosion of Civilizations* (Berkeley: University of California Press, 2012).
20. Richard Hall, *Acts, Passed in the Island of Barbados: From 1643 to 1762* (London, 1764), 341, 342.
21. Tryon, *Tryon's Letter*, 189.
22. Edward Littleton, *The Groans of the Plantations* (London, 1689), 18.
23. "America and West Indies: August 1690," in *Calendar of State Papers Colonial, America and West Indies*, ed. J. W. Fortescue, vol. 13, *1689–1692* (London: Her Majesty's Stationery Office, 1901), 301–17.
24. *Marly; or, A Planter's Life in Jamaica* (Glasgow, 1828), 165.
25. James Stephen, *The Crisis of the Sugar Colonies* (London, 1802), 11.
26. Justin Roberts, "The Whip and the Hoe: Violence, Work and Productivity on Anglo-American Plantations," *Journal of Global Slavery* 6, no. 1 (2021): 108–30, 109.
27. Henry De La Beche, *Notes on the Present Condition of the Negroes in Jamaica* (London, 1825), 22.
28. *Marly*, 163.
29. James Grainger, *An Essay on the More Common West-India Diseases, by a Physician in the West-Indies* (London, 1764), 11.
30. Olaudah Equiano, *The Interesting Narrative and Other Writings*, ed. Vincent Carretta (New York: Penguin, 1995), 38.
31. *An Abstract of the Evidence Delivered before a Select Committee of the House of Commons in the Years 1790 and 1791* (London: J. Phillips, 1792), 135–36.
32. *Minutes of the Evidence Taken before a Committee of the House of Commons* (London, 1790), 130.

33. Sidney W. Mintz, *Sweetness and Power: The Place of Sugar in Modern History* (New York: Penguin, 1985), 47.
34. Edward Long, *The History of Jamaica* (London, 1774), 2:449.
35. Long, 2:448.
36. *Minutes of the Evidence,* 225.
37. *Minutes of the Evidence,* 130.
38. Ted McCormick, "Food, Population, and Empire in the Hartlib Circle, 1639–1660," *Osiris* 35 (2020): 60–83.
39. Cressey Dymock, "Memorandum about Engines" (undated), 62/8a, in *The Hartlib Papers,* ed. M. Greengrass, M. Leslie, and M. Hannon (Sheffield: HRI Online Publications), https://www.dhi.ac.uk/hartlib/view?docset=main&docname=62A_08, quoted in Ted McCormick, "Perpetual Motion: Technology, Slavery, and History," *Memorious* (blog), accessed October 2, 2020, https://memoriousblog.com/2018/06/24/technology-slavery-and-history/.
40. Martin, *Essay on Plantership,* 7.
41. McNeill, *Something New,* 12.
42. Martin, *Essay on Plantership,* 38.

Infrastructural Inversion at Clarens

St. Preux in the Garden

ERIC GIDAL

My aim in this essay is to focus on what Geoffrey Bowker and Susan Leigh Star have called "infrastructural inversion." For Bowker and Star, this term simply means a strategic foregrounding of the technologies and protocols of knowledge production. They adapt it as a foundational method in their 1999 study of contemporary systems of classification and standardization, systems that, they argue, "form a juncture of social organization, moral order, and layers of technical integration." Infrastructural inversion becomes, in their project, a procedure for "recognizing the depths of interdependence of technical networks and standards, on the one hand, and the real work of politics and knowledge production on the other."[1] They apply this methodology to the study of global systems of medical classification, specifically the International Classification of Diseases (ICD), racial classifications in South African apartheid, and the Nursing Intervention Classification system (NIC) developed at the University of Iowa. I want to trace this critical strategy as an operative principle in Jean-Jacques Rousseau's 1761 novel *Julie, ou La Nouvelle Héloïse*. Like Bowker and Star's project, Rousseau's novel combines a sociology of knowledge with technologies of design and may be seen to apply a model of infrastructural inversion as both a rhetorical technique and a principle of environmental reflection.

It's an anachronistic application—*infrastructure* emerges as a French term only in the 1870s in relation to railroad engineering and is brought over to English only in the 1920s—but my reasons for applying it to a reading of Rousseau's novel are at least twofold. First, I am interested in infrastructure as an important topic for environmental media studies and ecocritical literary history. Media theorists such as John Durham Peters, Nicole Starosielski, and

Shannon Mattern have argued that an attention to infrastructure, like an attention to media, offers a means "to make environments visible" and that such built environments have a history that is less one of linear progress and more one of "residual forms," a term that Mattern borrows from Raymond Williams and applies to questions of path dependency in both the technological and ecological realms.[2] From this perspective, turning back to the eighteenth century, a time that David Alff has usefully framed as "before infrastructure," not only helps uncover a prehistory of the term but also helps measure the historical interconnections between "infrastructure, environment, and modernity."[3] Second, an attention to "infrastructural inversion" as a rhetorical technique can provide a prehistory of "infrastructuralism" itself, broadly formulated by Peters as "the intentional violation of a social norm to bring the background out into the open."[4] Infrastructural inversion as a critical methodology has been taken up across a range of disciplines from multispecies ethnography and health care analysis to information systems (IS) and computer-supported cooperative work (CSCW).[5] By applying it to a reading of Rousseau's novel, I hope to test what kinds of contributions literary history may make to what Sara Pritchard, in her book on the engineering of the Rhône, articulates as "the study of historical interactions between ecological and technological systems, both materially and discursively."[6] What precedents to this critical and historical project can we find in literary and philosophical works of the eighteenth century? And how can its methods be applied retrospectively to better appreciate the infrastructures of aesthetic and cultural critique? In what follows I offer no definitive answers to such questions, just a brief sketch of the novel as an exercise in concealment and inversion and a few thoughts as to the implications of such a reading to ongoing inquiries into infrastructural engineering and environmental change.

Those familiar with the novel will remember its bifurcated structure: three-hundred-some pages of tortured adolescent lust followed by another three-hundred-some pages of moral regret, punctuated throughout with mountain wanderings, clandestine liaisons, sentimental expiration, and, of course, philosophical dialogues on everything from dairy production and European colonialism to Italian opera and postcoital embrace. The bifurcated structure is key, as Rousseau aims both to manifest and to cure the false desires born of a corrupt society through moral and material engineering. The novel repeatedly valorizes authentic sentiments, honest communication, and virtuous self-regard even as it emphasizes the impossibility of realizing such ideals and the necessity of cultivation to draw us closer to moral duty. What Nicholas Paige has identified as an "aesthetics of renunciation" in Rousseau's novel applies not only to the characters but to his readers, who are enjoined to participate and reflect on the mechanics of sentimental transport and moral reform.[7] Without such cultivated renunciation, Rousseau suggests, the love between Julie d'Etange and her tutor St. Preux can end only in destructive rebellion or mental slavery.

This seems to be their fate as the novel's first half comes to a dramatic conclusion. Exposed by the discovery of their letters and forced into marriage to her father's old associate, Monsieur de Wolmar, Julie comes to recognize the insufficiency of sentiments as moral guides when measured against the dictates of faith and duty. "Make all my acts conform to my constant will which is thine," she prays in supplication, "and no longer allow the error of a moment to prevail over the choice of my entire life."[8] With no less zeal, St. Preux seeks to justify self-murder to his English mentor Milord Bomston and argues that "to seek what is good and flee what is ill for oneself insofar as it offends no one else is the right of nature" (311). Yet both are reproached: Julie by the tears that blot her letter and reveal her true feelings and St. Preux by his English correspondent, who admonishes him not to "burn [his] house down to avoid the bother of putting it in order" (319). Instead of the easy escapes of submission or suicide, Rousseau offers us the utopia of Clarens, established under the guidance of Wolmar, and a program of moral reeducation that occupies the second half of the novel. Here "innocence and peace" (342) are established and maintained through "the confidence of beautiful souls" (346), a moral order that "follows and corrects" (375) nature, and a purified subordination recast as the realization of individual will.

Rousseau details the workings of this utopia through a sequence of infrastructural inversions. In his letters to Bomston, St. Preux details the "lovely and moving spectacle" of "a simple and well-regulated house where order, peace, and innocence reign." This spectacle has not grown organically from the soil of the Vaud but has been meticulously engineered, and much of the fourth and fifth parts of the novel concern St. Preux's education in its varied apparatuses. The main house on the estate has been transformed from ornament to utility, a model of architectural and landscape design that eschews the "old, gloomy, [and] uncomfortable" château of Julie's father for an estate where "everything . . . is agreeable and cheerful": as St. Preux puts it, "There is not a single room where one is not recognizably in the country, and where one fails to find all the conveniences of the city" (363). How have Wolmar and Julie created such a fusion of pastoral virtue and urban convenience? Infrastructure. The estate is almost entirely self-sufficient owing to the expansion of productive acreage and some decidedly restrictive labor policies.

Coach houses, billiard rooms, menageries, and flower gardens have given way to a wine press, a dairy, and an enlarged vegetable garden. Yews have given way to fruit trees, horse chestnuts to mulberries, lindens to walnut trees. "Everywhere they have replaced attractive things with useful things," writes St. Preux, "and attractiveness has almost always come out the better" (364). This utilitarian aesthetic depends on a system of labor regulations that prohibits immigration, depresses wages, and diminishes job security. It also depends on a system of ideological coercion that rewards informants and punishes insolence. "The master's whole art," St. Preux reveals, "consists in hiding this coercion under the veil of pleasure or interest, so that they think they desire all they are obliged to do" (373). "Confidence and attachment," "respect and authority," produce a mode of subordination that naturalizes deference to authority under the gendered signs of truth and beauty. Workers seek out Julie's benevolent grace as a counter to Wolmar's repeated proclamation: "Je vous chasse" (You're fired!). Conversely, St. Preux observes that while Wolmar may yield the power to dismiss, the possibility of Julie's disappointment holds even greater sway: "The former, giving voice to justice and truth, humiliates and confounds the guilty parties; the latter makes them feel a mortal regret for it, by expressing her own at having to deprive them of her good will" (383).

If the domestic economy of Clarens depends on such ideological interpolation, Rousseau's novel provides a kind of structural understanding, exploring the economic and agricultural mechanisms by which this bondage may be legitimized and sustained. In some sense, the economy at Clarens appears to be a rejection of infrastructure insofar as it aims at an ideal of autarky that minimizes dependence on external resources. It achieves by design the material and moral independence that St. Preux had admired in earlier treks through the upland hamlets of the Valais, where he witnessed "a surprising mixture of wild and cultivated nature" (63). As in the mountain hamlets, there is little money used at the estate, and a commitment to in-house production and local trade through bartered goods and services diverts the desire for luxuries and creates a model of sustainable living: "Here the fruit of past labor sustains the present plenty, and the fruit of present labor heralds the plenty to come; one enjoys at the same time what one spends and what one brings in, and the various time frames come together to consolidate the security of the present" (451). Such material sustainability in turn supports a system of moral sustainability: "With a sound soul, can one tire of discharging the dearest and most charming duties of mankind, and making each other's life happy? Every evening Julie, satisfied with her day, desires nothing different for the morrow, and every morning she asks heaven for a day like the one before. . . . Is not being content in the continuation of one's state a sure sign that one lives happily therein?" (453). This conjunction of material and moral sustainability in turn depends on ethical, epistemological, and economic reforms that unite pleasure with restraint. "And yet, Milord," writes St. Preux, "none of all this is apparent at first glance. Everywhere an air of profusion obscures the order that creates it; it takes time to perceive the sumptuary laws that lead to affluence and pleasure, and at first one has a hard time understanding how one can enjoy what one economizes" (451).

St. Preux thus describes in great detail the mechanics of this nearly autonomous system even as such details reveal its dependence on external textile factories, timber management, and public bonds in a Physiocratic model of agricultural growth:

> They follow the maxim of extracting from the land all it can yield, [writes St. Preux] not to obtain a larger gain from it, but to feed more men. Monsieur de Wolmar contends that land produces in proportion to the number of hands that till it; better tilled it yields more; this excess production furnishes the means of tilling it better still; the more men and beasts you put on it, the more surplus it supplies over and above their subsistence. It is not known, he says, where this continual and reciprocal increase in product and laborers might end. (364)

Echoing Rousseau's plans for the constitution of Corsica ("The taste for agriculture is advantageous to the population not only by multiplying men's means of existence, but also by giving the body of the nation a temperament and morals that cause them to be born in greater number"), Wolmar celebrates population growth and agricultural productivity as the foundation of his virtuous society, a counter to the slave economies in the Caribbean detailed by Ramesh Mallipeddi in the previous essay ("The Hoe and the Plow: Plantation Labor under the Somatic Energy Regime").[9] Thomas Malthus would famously repudiate this claim forty years after the novel's publication, training his "dismal science" on the Pays de Vaud, where he calculated that low mortality rates combined with the limitations of the soil placed a natural check on birth rates whose occasional breach had produced only increased poverty and emigration.[10] The ecological carrying capacity of the estate at Clarens may not be as expansive as Wolmar or St. Preux imagine. But by connecting moral virtues to an early form of biophysical economics, Rousseau underscores the dependencies of social reform on factors of energy inputs and labor efficiencies, resources and demographics.[11]

Which brings us to the question of irrigation. Julia Simon has richly explored the topic of water diversion throughout Rousseau's writings and his vision of hydraulic engineering as a necessary and ennobling modification of an inherently inhospitable world.[12] In his "Essay on the Origin of Languages," Rousseau asks, "How many arid lands are habitable only by means of the ditches and canals that men have drawn off from rivers! Almost the whole of Persia continues to exist only through this artifice. China swarms with People with the help of its numerous canals: without them the Low Countries would be inundated by rivers, as they would be by the sea without their dikes. Egypt, the most fertile land on earth, is habitable only by means of human labor."[13] Simon productively contrasts such valorizations with critiques of the monarchical absolutism and military engineering enabling and informing the gardens at Versailles and the Canal du Midi. Building on Chandra Mukerji's histories of these two engineering feats, Simon underscores how Rousseau persistently distinguishes between infrastructures meant to serve the people and those created for the benefit and assertion of a decadent aristocracy. A project such as the Canal du Midi offers "a model of impersonal rule," in Mukerji's words, "engineering the land as a form of government," and it emerges from "a political strategy that put territory at the center of politics and intelligence about the natural world in the heart of political administration."[14] The Canton of Vaud is as removed from such royal works as it is from the Nile Delta, and the water diversion most germane to the fictional estate at Clarens may be the mountain meadows of the Valais recorded in the novel's first section and the communal traditions of meadow irrigation practiced in the region at least since the eleventh century, developed more for soil improvement and pest control than royal pride or transnational trade.[15] No mention is made of irrigation techniques on the agricultural lands either in the mountain villages or at the estate at Clarens, but water diversion is central to the least productive plot of land on the estate, the garden or Elysium that Julie and Wolmar at last reveal to St. Preux.

A classic eighteenth-century set piece, the garden at Clarens manifests nature's generosity but with a "signature of human agency" that Robert Pogue Harrison has placed at the center of the literary garden in the Western tradition.[16] It is precisely this "signature" that is in question, as Rousseau emphasizes the opacity of the garden's genesis. Its very existence is concealed: "This place," St. Preux notes, "although quite close to the house, is so well hidden by the shaded avenue separating them that it cannot be seen from anywhere. The heavy foliage surrounding it does not allow the eye to penetrate, and it is always carefully locked" (387). Once inside, he is overwhelmed by its apparent wildness: "This place is enchanting, it is true, but rustic and wild; I see no human labor here. You closed the gate; water came along I know not how; nature alone did the rest and you yourself could never have managed to do as well." "It is true," Julie replies, "that nature did it all, but under my direction, and there is nothing here that I have not designed. . . . Take a few steps and you will understand." And understand he does as he begins to recognize how the plants (all local to the region) have been "arranged and combined," how the branches of the trees have "been made to bend back to the ground," and how garlands have been "cast from tree to tree" and even grafted to their roots. He perceives how the "clear, crystalline" streams and rivulets that run through the garden have been diverted from the "public spout" and circulated through the garden by "dividing and reuniting" its courses and attenuating the slope "to extend the circuit and allow for the babble of a few small waterfalls" (389–90):

> The bed of streams was made up of a layer of clay, covered with an inch of pebbles from the lake and scattered with shells. These same streams running intermittently under a few large tiles covered over with earth and grass at ground level formed where they emerged an equal number of artificial springs. A few trickles rose from them through syphons over rugged patches and bubbled as they fell back. And so the soil thus constantly refreshed and moistened yielded forth new flowers and kept the grass always verdant and lovely. (390)

Simon usefully reads this hydraulic system as the triumph of "a Horatian aesthetic of utility and pleasure" over aristocratic performances of wealth and power, a space of true community in a managed environment. Yet we may also observe that what St. Preux discovers most of all in this "artificial wilderness" is less the mechanics of hydraulic technology or bioengineering than "the pleasure of seeking and selecting." "I was more eager to see objects than to examine their impressions," St. Preux writes, "and I was happy to abandon myself to that enchanting sight without taking the trouble to think." As in the domestic economy of the larger estate, where "an air of profusion obscures the order that creates it," the garden has been designed to obscure the genesis of its own appearance. "Everything is verdant, fresh, vigorous, and the gardener's hand is not to be seen," wonders St. Preux. "I see no human footprints. Ah! [replies] Monsieur de Wolmar, that is because we have taken great care to erase them" (393). Wolmar terms such erasure "trickery" (393) or "feigned irregularity" (393), but we can understand it as a variant of what Lisa Parks calls "infrastructural concealment." "Infrastructures are often designed purposefully to be invisible or transparent," Parks observes, "integrated with the built environment, whether submerged underground, covered by ceilings and walls, or camouflaged as 'nature.'" The paradox of Julie's garden, of course, is that it is precisely such camouflage that creates the appearance of "nature's charms," charms that in turn legitimize the act of camouflage itself. As Wolmar explains, "Those who love [nature] and cannot go so far to find her are reduced to doing her violence, forcing her in a way to come and live with them, and all this cannot be done without a modicum of illusion" (394).

Jean Starobinski has called *La Nouvelle Héloïse* "an extended reverie on the theme of transparency and dissimulation," and we can perceive how this sequence of concealments and inversions constructs a moral society on the basis of an integrated infrastructure of economic and ecological systems, a process central to romantic programs of aesthetic education.[17] As Friedrich Schiller would argue in his Kantian reading of Rousseau's novel, this aesthetic education brings us from slavery to freedom, from subordinate to lawgiver, "superior to every terror of Nature so long as [we] know how to give form to it, and to turn it into [our] object."[18] Such aesthetic projects, as Clifford Siskin has recently argued, are not identical with infrastructure; rather, they "shape knowledge in a manner that supports the building of infrastructures" and, as Bowker and Star would insist, such infrastructures in turn support the shaping of knowledge.[19] For Rousseau, it is this dialectic between knowledge and infrastructure that holds the greatest interest. For while *Julie* has been understood as laying the groundwork for a range of social engineering projects from behavioral management to totalitarianism, we may consider its emphasis on inversion as a rhetorical principle that unites self and society through the cultivation of critical subjectivity. In this respect, what St. Preux witnesses at Clarens conjoins socialization with economic integration and technological engineering realized through a burgeoning romanticism of resource extraction and aesthetic compensation.

St. Preux's letters certainly explore what Nikhil Anand, Akhil Gupta, and Hannah Appel have identified in their collection of essays *The Promise of Infrastructure* as the "sociomaterial terrain": "an articulation of materialities with institutional actors, legal regimes, policies, and knowledge practices that is constantly in formation across space and time."[20] At the same time, Rousseau offers his readers a program of rhetorical inversion that anticipates what Annalisa Pelizza calls the "vectorial glance," a practice of "looking at the technicalities of infrastructures as key sites where shifts in institutional authority and accountability can become visible."[21] The shift in institutional authority and accountability made visible in Rousseau's novel concerns the movement from imposition to interpolation as both characters and readers come to participate in the creation of a second nature, both subjective and material. If, as Anne Deneys-Tunney and Yves Charles Zarka observe, Rousseau's philosophical project strove "to restructure nature, within the political, moral, and anthropological context of social man," infrastructural inversion would seem to provide a reflexive consciousness of that engineered environment.[22]

Observing such inversions helps us consider Rousseau's program of concealment and revelation in relation to more current reflections on infrastructure and environment, particularly the discourse of landscape urbanism, which Elizabeth Mossop usefully frames as strategies "to make ecological processes operational in design, harnessing natural phenomena such as erosion, succession, or water cycles in the generation of landscapes."[23] Rousseau was no urbanist, and the estate at Clarens, as I have observed, aims to minimize its economic and material dependencies while promoting a paternalistic social organization that is an unlikely fit for the more democratic and egalitarian ambitions of current work in landscape design. But the simultaneously ecological and pedagogical ambitions of Wolmar's projects, fit emblems for the novel that unfolds them, offer important precedents to landscape urbanism's emphasis on infrastructure as aesthetic performativity as much as functional sustainability. As James Corner observes regarding the reemergence of landscape as a category of design in the twenty-first century, "Materiality, representation, and imagination are not separate worlds; political change through practices of place construction owes as much to the representational and symbolic realms as to material activities."[24] Weaving engineering and aesthetics, design and poetics, Rousseau's deployment of infrastructural inversion forms a core element of his processual pedagogy and a significant literary predecessor to sustainable design not only as an ambition of engineering but as a template for social consensus in an increasingly fragmented world.

Notes

1. Geoffrey C. Bowker and Susan Leigh Star, *Sorting Things Out: Classification and Its Consequences* (Boston: MIT Press, 1999), 33–34.
2. John Durham Peters, *The Marvelous Clouds: Toward a Philosophy of Elemental Media* (Chicago: University of Chicago Press, 2015), 38; Nicole Starosielski, *The Undersea Network* (Durham: Duke University Press, 2015); Shannon Mattern, *Code and Clay, Data and Dirt: Five Thousand Years of Urban Media* (Minneapolis: University of Minnesota Press, 2017), xxviii.
3. David Alff, "Before Infrastructure: The Poetics of Paving in John Gay's Trivia," *PMLA* 132, no. 5 (2017): 1134–48; Nikhil Anand, Akhil Gupta, and Hannah Appel, introduction to *The Promise of Infrastructure* (Durham: Duke University Press, 2018), 7.
4. Peters, *Marvelous Clouds*, 35.

5. See, for example, Atsuro Morita, "Multispecies Infrastructure: Infrastructural Inversion and Involutionary Entanglements in the Chao Phraya Delta, Thailand," *Ethnos* 82, no. 4 (2017): 738–57; Casper Bruun Jensen, "Power, Technology and Social Studies of Health Care: An Infrastructural Inversion," *Health Care Analysis* 16, no. 4 (2008): 355–74; Annalisa Pelizza, "Developing the Vectorial Glance: Infrastructural Inversion for the New Agenda on Government Information Systems," *Science, Technology, & Human Values* 41, no. 2 (2016): 298–321; and Jesper Simonsen, Helena Karasti, and Morten Hertzum, "Infrastructuring and Participatory Design: Exploring Infrastructural Inversion as Analytic, Empirical and Generative," *Computer Supported Cooperative Work* 29, nos. 1–2 (2020): 115–51.
6. Sara B. Pritchard, *Confluence: The Nature of Technology and the Remaking of the Rhône* (Cambridge, MA: Harvard University Press, 2011), 13.
7. Nicholas Paige, "Rousseau's Readers Revisited: The Aesthetics of *La Nouvelle Héloïse*," *Eighteenth-Century Studies* 42, no. 1 (2008): 131–54.
8. Jean-Jacques Rousseau, *Julie, or The New Heloise*, trans. Philip Stewart and Jean Vaché (Hanover, NH: Dartmouth College Press, 2014), 294. All quotations from the novel are from this edition, with subsequent page numbers offered in parentheses.
9. Jean-Jacques Rousseau, "Plan for a Constitution for Corsica," in *The Plan for Perpetual Peace, On the Government of Poland, and Other Writings on History and Politics*, trans. Christopher Kelly and Judith Bush (Hanover, NH: Dartmouth College Press, 2013), 123–55, 126.
10. Thomas Malthus, *An Essay on Population* (London, 1803), 267–84.
11. Ironically, Rousseau's novel was published toward the beginning of a century-long population explosion in Europe from the 1750s to the 1850s. C. Pfister has documented how the Canton of Bern, to the north of Rousseau's imagined estate, did achieve a relative measure of self-sufficiency during this period, "although items such as cloth, wine, coffee, and sugar were imported in exchange for cattle, timber, and cheese." Christian Pfister, "The Early Loss of Ecological Stability in an Agrarian Region," in Peter Brimblecombe and Christian Pfister, *The Silent Countdown: Essays in European Environmental History (London: Springer, 1990)*, 37–55, 43.
12. Julia Simon, "Diverting Water in Rousseau: Technology, the Sublime, and the Quotidian," *Eighteenth Century: Theory and Interpretation* 53, no. 1 (2012): 73–97.
13. Jean-Jacques Rousseau, "Essay on the Origin of Languages," in *Essay on the Origin of Languages and Writings Related to Music*, trans. John T. Scott (Hanover, NH: Dartmouth College Press, 1998), 313–14.
14. Chandra Mukerji, *Impossible Engineering: Technology and Territoriality on the Canal du Midi* (Princeton: Princeton University Press, 2009), 5, 22. See also

Mukerji, *Territorial Ambitions and the Gardens of Versailles* (Cambridge: Cambridge University Press, 1997).

15. As Chris Leibundgut explains, "Meadow irrigation led to an increase in the growth of grass and hay allowing intensification of dairy and meat production," while "increased biomass produced from irrigated meadows increased the production of manure." Leibundgut, "Historical Meadow Irrigation in Europe—a Basis for Agricultural Development," in *The Basis of Civilization—Water Science?*, ed. J. C. Rodda and Lucio Ubertini (Wallingford, Oxfordshire, UK: IAHS—International Association of Hydrological Science, 2004), 77–87, 79.
16. Robert Pogue Harrison, *Gardens: An Essay on the Human Condition* (Chicago: University of Chicago Press, 2008), 7.
17. Jean Starobinski, *Jean-Jacques Rousseau: Transparency and Obstruction*, trans. Arthur Goldmanner (Chicago: University of Chicago Press, 1988), 81.
18. Friedrich Schiller, *On the Aesthetic Education of Man*, trans. Reginald Snell (London: Dover, 2004), 121.
19. Clifford Siskin, *System: The Shaping of Modern Knowledge* (Cambridge, MA: MIT Press, 2017), 7.
20. Anand, Gupta, and Appel, introduction to *Promise of Infrastructure*, 12.
21. Pelizza, "Developing the Vectorial Glance," 305.
22. Anne Deneys-Tunney and Yves Charles Zarka, eds., *Rousseau between Nature and Culture* (Berlin: De Gruyter, 2016), 3.
23. Elizabeth Mossop, "Landscapes of Infrastructure," in *The Landscape Urbanism Reader*, ed. Charles Waldheim, (Princeton: Princeton Architectural Press, 2006), 163–77, 165.
24. James Corner, "Terra Fluxus," in Waldheim, *Landscape Urbanism Reader*, 21–33, 32.

Taxonomic Subversion and Vegetal Expansion in Charlotte Smith's "Beachy Head"

ANNA K. SAGAL

Stories of plants in the Western literary tradition are ubiquitous—rooted in myth or folklore, wound through children's didactic fables, and woven into cautionary tales for adult audiences. Despite their formal differences, these stories share, perhaps surprisingly, an element of violence. Greek myth routinely transmogrifies men and women into plants, often in response to sexual threat—Daphne's change into a laurel to escape Apollo's attentions or Smyrna's shift into a weeping myrrh tree upon the discovery she'd been enchanted into committing incest—although there are milder examples of bodily transgression, including Narcissus being changed into his namesake flower as a punishment for vanity. Children in fairy tales and fables are tempted by flowers, vegetables, or fungi into various disastrous scenarios that result in brutal physical deformity and torture. One of the most enduring examples of this is Christina Rossetti's *Goblin Market* (1862), in which the consumption of fairy fruit sexually and morally corrupts a young woman. Similarly, nineteenth-century French and German fairy tales and their oral antecedents included poisonous apples ("Snow White") and vengeful trees ("The Juniper Tree").

While classic European fairy tales and myths associated vegetal life with violence, eighteenth-century English writing entwined plants with human sexuality. Such writing typically incorporated plants in relation to female sexuality: sometimes pornographically, as in *The Natural History of the Frutex Vulvaria* (1731), but more often in a disciplinary framework that discouraged sexual expression or desire, such as the use of "blooming" to reference female beauty or virginity and satirical verse like Richard Polwhele's *The Pursuits of Literature* (1794).[1] In short, when humans tell stories about plants, we use vegetal life as a didactic force to control undesirable behavior. This manipulation of plant life for our own cultural purposes is facilitated by a mindset in which we understand plants to be passive, mostly inert forces we can control in literary and organic ways. However, this mindset is flawed; furthermore, this idea that plants are ours to control arises precisely from the underlying suspicion that we cannot truly ever control them.

Humans have long relied on plants for necessities like food, shelter, and medicine but also (and more importantly for this argument) for cultural symbolism including family devices, political allegiances, and religious observation. Yet some deep, fundamental aspect of plant life remains beyond our ability to fully comprehend. It might be argued, in fact, that the rise of botanical taxonomic anxieties in the seventeenth and eighteenth centuries is an indirect result of the stubborn illegibility of plants as living beings with their own sets of needs, desires, and motivations. To read plants as a literary scholar is, unavoidably, to read human attempts at encountering vegetal life through literature. As part of my own efforts to reread plant life in literary history, I focus this essay on an author whose work explores the profound ambivalence in human-plant relationships: Charlotte Smith.

Smith's naturalist poetry is notable not only for her active interest in depicting plant life in painstaking detail but also for her inclusion of binomial nomenclature, the two-part taxonomic identifier system popularized by Linnaean acolytes in the mid-eighteenth century. The most famous example of this strategy is "Beachy Head" (1807), Smith's unfinished ode to seaside nature and sentimentalized British history.[2] Most critical interpretations of her binomial usage attribute Smith's careful attention to the technical minutiae of botanical study to an avid interest in concrete knowledge about plant life. However, the placement of the binomials in endnotes, already a secondary layer in the act of reading a poem, suggests that Smith deprioritizes scientific knowledge in favor of poetic ways of knowing and describing the natural world. I also contend that Smith intimately understood the profound alterity of vegetality, even as she concurrently offers an attempt at rendering plant life legible to her readers—a tension within "Beachy Head" that is epitomized in a single endnote. In one of the embedded sestains in the last section of the poem, Smith glosses "cuckoo-flowers" (590) with an unexpectedly pedantic correction of Shakespeare's botanical knowledge in *Love's Labours Lost*: "Shakespeare describes the Cuckoo buds as being yellow. He probably meant the numerous Ranunculi, or March Marigolds (Caltha palustris), which so gild the meadows in Spring; but *poets have never been botanists*" (590n; emphasis mine). The last clause incites multiple contradictory reactions in any regular reader of

Smith's poetry: amusement, disbelief, confusion, or irritation. Does she possibly imply that all poets who came before her failed to live up to exacting botanical standards? Or does she suggest that poets are incapable of becoming botanically precise?

In an endnote to line 512, Smith references John Aikin's influential *An Essay on the Application of Natural History to Poetry* (1777), a popular contemporary work of literary criticism that commended the use of botanical imagery in conjunction with a judicious and limited amount of technical information. This reference demonstrates that she was familiar with current debates about the role of technical scientific data in artistic productions and that her deviation from them is therefore intentional. While such deviation is perhaps an obvious move to make in a poem with literary aspirations, it is also true that the Linnaean binomials were more technical than most other scientific allusions in contemporary poetry. I suggest further that Smith destabilizes the authority of contemporary scientific knowledge by relegating binomials to a subsidiary section of the poem, by inconsistently deploying taxonomic terms, and by allowing plants the linguistic and conceptual space to do what they do best—proliferate in spite of human intervention.

I argue that her purposeful departure from those prevailing standards reveals an ambitious, radical agenda—to undermine the influence of technical scientific data on poetic labor and problematize the cultural primacy of humanity in relation to vegetality. I first discuss a few examples of notable plants within "Beachy Head" to highlight moments where Smith implies that taxonomic data is insufficient or inappropriate as a model for presenting localized populations of plants or singular organisms. Rather than adhering to the scientific hierarchy and ceding authority on nature to technical ways of knowing that are themselves partial, Smith counters this prevailing discourse by offering her readers a poetic model of vegetal study that more authentically depicts plant life. In the second half of this essay, I focus more specifically on the points at which plants seem to seize narrative control within the poem, lines or stanzas where individual plants act as agents rather than passive recipients of human interpretation (Smith's or the readers'). By writing the plants in "Beachy Head" as inimitable, active, and assertive, she provides a perspective on vegetal life that diverges from scientific consensus. I propose that Smith's critique of binomial nomenclature makes apparent the deeply rhetorical—and therefore artificial—nature of human-plant relationships. Ultimately, this essay highlights the complexities of human-plant interactions in eighteenth-century women's writing in order to nuance our historical understanding of the relationship among poets, scientists, and plants. Instead of characterizing the relationship as a hierarchical one, privileging human kinds of knowing, we must recognize their fundamental entanglement if we are to productively advance eighteenth-century science studies.

*   *   *

Charlotte Smith demonstrated an active interest in botanical study throughout her literary career, often in the context of pedagogical projects for young people. The linked practices of studying Linnaean terminology and drawing individual plants while strolling through nature are presented in *Rural Walks* (1795), *Rambles Farther* (1796), and *Minor Morals* (1800) as generative practices for cultivating emotional and intellectual skills in young women. Engagement with botanical study appears in some of her fiction as well, such as Althea Dacres's botanical sketches in *Marchmont* (1796). Regardless of the generic and critical shift away from pedagogical aims, "Beachy Head" has typically been labeled as her most scientifically invested botanical work due to the extensive use of explanatory endnotes that offer the Linnaean binomial for the plants and animals that populate the world of the poem.

The deployment of taxonomic data in creative works was unusual but not unheard of in the eighteenth century. Women authors included taxonomy in volumes that declared pedagogical ambition alongside the assertion of scientific knowledge, like Maria Jacson's *Botanical Dialogues* (1797) or Frances Arabella Rowden's *A Poetical Introduction to the Study of Botany* (1801). Likewise, the annotation of verse with scientific detail was a poetic strategy used by James Grainger in *The Sugar-Cane* (1764) and Erasmus Darwin in *The Botanic Garden* (1791) and theorized by John Aikin in his aforementioned *Essay on the Application of Natural History to Poetry*. Crucially, Smith diverges from both Grainger and Darwin in how she uses technical terminology in poetry. While Darwin utilizes the Linnaean classification system to laud new botanical discoveries via erotic, anthropomorphized plant ladies and Grainger uses scientific data to praise colonial bioprospecting in lieu of examining the brutality of West Indian chattel slavery, Smith questions the viability of taxonomic notation in the context of indigenous plant life in England. Unlike her contemporaries, who relied on the authority of taxonomic convention to bolster their poetic reputations, Smith challenges scientific authority by highlighting its insufficiency in both poetic and botanical realms.

In contemporary scholarship, Smith has drawn significant attention for her annotative poetic practices. For example, Judith Pascoe proposes that Smith's naturalist poetry evokes "a claim to scientific authority utilized in the service of a poetics of the botanically exact."[3] Kandi Tayebi elaborates by framing the naturalist endnotes as offering "the idea that there is an underlying order and classification" to the organic world.[4] Rosalind Powell also suggests that the concurrent uses of colloquial and binomial names for plants in Smith's verse "[forge] links between natural philosophy and fancy to didactic and aesthetic effect."[5] Most recently, Melissa Bailes reads Smith's endnotes as part of a larger pattern of textual integration that "equates poetic and scientific authority."[6] While one reading of "Beachy Head" might suggest that the incorporation of scientific nomenclature into poetic form is the author's way of equally valuing both kinds of knowledge, sustained attention to a few select plants within the poem reveals Smith's desire to question the primacy of the scientific model as a way of encountering nature.

An alternative mode of human-vegetal relations begins to emerge via Smith's purposeful omission of binomial names for some of the most compelling plants in "Beachy Head." Early in the poem, as Smith describes a lone farm struggling in the inhospitable chalk headland of Beachy Head, we are treated to a vignette of avian and vegetal abundance. Birds of various sorts—plover, linnet, finch—have ample opportunity to dine upon charlock (wild mustard) and thistle (220, 223). While both thistle and charlock are suitable for human consumption, they have chemical and morphological defenses against herbivorous animals. Charlock is poisonous to domesticated creatures like horses; depending on the species of thistle, the plant could be covered in minimal or extensive spininess to injure grazing sheep.[7] In addition to complicating the grazing habits of livestock, these two plants are also notorious in England as actively detrimental to agriculture (especially cereal crops), often overgrowing fields and redirecting vital nutrients in the soil toward themselves.[8] The frequently reprinted agricultural text *A Practical Treatise of Husbandry* warns of "*Thistles* and many other weeds, which greatly exhaust the earth," and charlock, called one of "the weeds which are feared the most"—a sentiment echoed in *A New Medicinal, Economical, and Domestic Herbal* (1808), which refers to charlock as "a very noxious weed."[9]

To refer to thistle and charlock as "weeds," as the agricultural guides of the period usually did, is to frame these plants in the context of their relationship to and value for a human community: as a corollary, technical botanical works used binomial nomenclature for the very same purpose. Modern ecological perspectives are less judgmental of weeds, defining them as "wild species that live near people and benefit from a close relationship with people and their artificially created habitats."[10] Within the farmland of "Beachy Head"—an artificial terrain, like all agricultural land, created by an array of disruptive processes, including uprooting indigenous plants, removing stones, leveling soil distribution, introducing foreign seedlings, and manipulating nutrient levels—the surviving undomesticated plants coexist with the cultivated plants. Many plants denigrated as weeds are called "invasive species" if their indigenous distribution is elsewhere geographically. Given that charlock and a few thistle species are indigenous to the British Isles, the irony here is that food crops like wheat and barley are the imported species (albeit they had been acclimatized to Britain centuries ago). Thus, native wild plants are competing with the real invasive/invading crops for resources like water, nutrients, or sunlight. Of course, the weeds simultaneously benefit from the attentiveness of farmers tending their source of food and income. In the discourse of treatises like *A Practical Treatise of Husbandry*, weeds compete with crops—bad plants overcome good plants—and humans suffer from it.

But what if, as Marijke van der Veen proposes, we reconfigure our worldview to understand weeds as "plants that contest with man for the possession of the soil"?[11] Thistle and charlock, then, are antagonistic not solely toward crops but also to humans. Their lack of taxonomic annotation in a poem that is otherwise attentively noting wild indigenous species that propagate without human cultivation, like woodbine, vetch, and bryony (250–52), is arguably purposeful—the absent binomials here are a profound lacuna, opening the possibility of plants that cannot or should not be understood only (or at all) within taxonomic frameworks. Furthermore, in the context of the agrarian focus of this section of the poem, surely Smith is aware of the painful associations some of her readers would have with these destructive plants. It seems especially meaningful, then, that both plants are thriving, dominant members of the Beachy Head ecosystem: thistle is "profusely spread" across the terrain, and the charlock is "unprofitably gay," suggesting a threatening level of vegetal expansion (220–23).

The poem also characterizes selected species that do bear taxonomic notation as resistant to scientific analysis by imbuing specific plants with a surplus of meaning—poetic, emotional, and botanical. For example, in the stanza beginning at line 320, the speaker shifts poetic focus from the contemporary biota on Beachy Head to the remembered flowers of their youth. A celebration of quotidian vegetal life is at the heart of these childhood memories, as they regale us with the attractions of a humble cottage garden featuring common indigenous plants like pansies, pinks (dianthus), rosemary, and roses (331–33). Early lines remind the reader that while domestic gardens are "for use design'd," they are "not of beauty destitute" and thus are still worthy to be cast as actors in a sweeping poetic vista in spite of their indebtedness to humans

(327–28). With this in mind, Smith's decision to single out the roses as "almost uncultured" bears critical significance (334). For one, her characterization of a plant that ordinarily requires quite a lot of horticultural attention as "almost uncultured" contributes to the poem's deferential stance toward plant life, eliding human intervention and labor to the point of favoring vegetal life over human life on Beachy Head. Smith's work to return roses to the wild is particularly notable in light of the fact that roses have also often served as polarizing cultural symbols in British history, suggesting again that she is interested in shifting the balance of power in the human-vegetal dynamic.[12] Cleverly, these half-wild roses also provide a transition between present/domestic and past/wild in this stanza, as the reader sinks into the speaker's half-remembered scenes of "fond regret" (340).

This idyllic landscape of past memory is a vibrant and evocative backdrop for the speaker's youthful memories—including the introduction of the ambiguous "pensive lover of uncultur'd flowers" (359). The figure of the pensive lover is both disembodied and intensely corporeal: they are indolently reclining on a piece of anonymous terrain while actively committing violence upon that terrain as they "[pluck] the wood sorrel" (361). While, of course, on one level a banal depiction of human interaction with nature (especially if we read dual meaning into "pensive lover" and imagine a lovelorn youth picking a flower in performative melancholy), this tiny moment of human presence in dozens of lines featuring plants in abundance is a harsh intrusion into vegetal stability. This wood sorrel plant is markedly fragile on the surface, with "light thin leaves" that Smith notes are trifoliate: "heart-shaped, and triply folded" (361–62). The pensive lover's pensive act is a violent one, tearing the stem, and perhaps also leaves and petals, and wrenching part of this vegetal entity from its coexistence with moss and anemones. Vitally, however, even the most assiduous plucker of wood sorrel will only rend a part of the plant; wood sorrel is a rhizomatous plant, the significance of which I expand on later.

Unlike charlock or thistle, wood sorrel is assigned its binomial tag in notes—*Oxalis acetosella*—albeit sans any kind of elaboration or further commentary. Donelle Ruwe insightfully marks the introduction of the wood sorrel as proof of Smith's respectful stance toward the "irreducible alterity [of the plant]," as the pairing of taxonomic term and lush description in the verse generates, in Ruwe's formulation, Smith's "botanic poetic."[13] Smith's evocative description of the delicate wood sorrel indeed paints a vivid picture for her readers, but the precise choice of plant in conjunction with the minimum scientific annotation is a meaningful critical move that bears further examination. *Oxalis acetosella* had been featured in the second volume of William

Curtis's *Flora Londinensis* (1777; see fig. 1) as a common indigenous plant with a caveat: "There are but few woods about us in which the *Wood-Sorrel* does not occur. It will not grow in a garden unless it has shade."[14] Wood sorrel was so ubiquitous in the eighteenth century that it was frequently included in salads, in antipyretic or anti-inflammatory medical recipes, and in a stain removal mixture with similar efficacy to lemon juice.[15] Yet as Curtis notes, it is difficult to intentionally cultivate, as wood sorrel prefers the untamed places, the "o'ershadowing woods of beech" and "whispering shade" of "Beachy Head" (356–58). The wood sorrel is both irreducible to its taxonomic annotation and unimaginably vast in terms of sheer size and scope. While reference texts like William Woodville's *Medical Botany* (1790) or scientific volumes like Curtis's *Flora Londinensis* aim to confine the wood sorrel solely within the bounds of human utility, Smith's "Beachy Head" makes it clear that this plant is elusive and quick to spread beyond (and away from) cultivated spaces with its "creeping [root]" (363).

These three plants—charlock, thistle, and wood sorrel—are notable for the meaning they infuse in the poem beyond (or instead of) technical botanical data. The poetic resonances of these plants bring some of the strongest emotional forces to the entire poem, including the implied threat to a

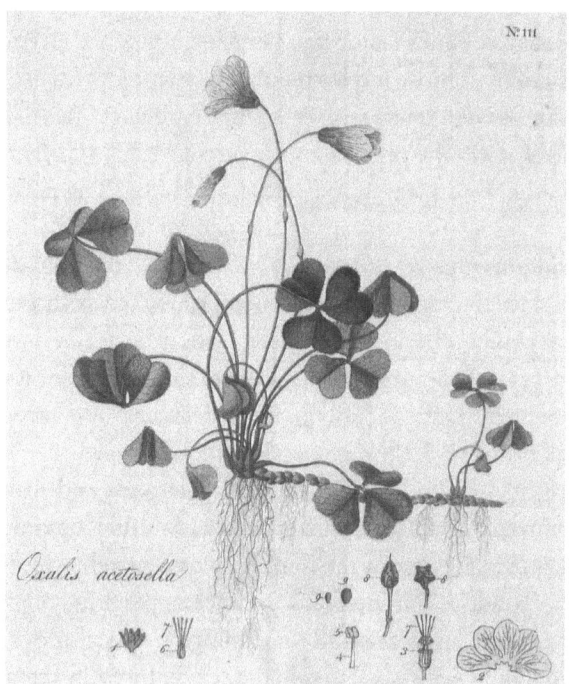

FIGURE 1. "Oxalis acetosella," *Flora Londinensis*, 1777. (Smithsonian Libraries and Archives)

farmer's livelihood and the sentimental potential of such a situation; the exhilarating sense of freedom and awe evoked by a landscape blanketed with these tiny blossoms; the romanticized, hazy vignette of a woodland bower; and the formlessly disturbing moment of living beings wrenched from the soil. By comparison, the multilayered reading experience of heavily annotated sections of "Beachy Head," such as the dry recitation of archaeological trivia in the notes to lines 390–94, seems less immersive (by nature of the endnote format) and less evocative (by design).

The act of the "pensive lover of uncultur'd flowers" plucking the wood sorrel is also a juncture in "Beachy Head" at which Smith's second critical aim becomes clear: to depict plants that not only belie the sufficiency of taxonomic description but seem themselves to actively reject human interference through linguistic and physical action. While charlock and thistle have no binomial assigned in "Beachy Head," many of the plants within the vegetal population of Beachy Head are weighted with an additional name (or two or three). Ostensibly, the multiple names for each plant are intended to give readers the tools to recognize the biota of "Beachy Head" in scientific, communal, and poetic discourses. Prior to the eighteenth century, many attempts to systematically account for vegetal diversity either relied on inaccurate and incomplete herbal records or utilized a multipartite naming convention. The Linnaean system, while not completely unique in either its adoption of the binomial or its choice of sexual dimorphism as an organizing feature, found success in England and was enthusiastically endorsed as easier to learn for amateur naturalists.[16] In theory, the Linnaean binomial was assigned to a species only after sustained study of countless specimens to prevent potential misidentification (which, of course, still occurred frequently). The binomial is thus more of an aggregate name, a name that represents a type of plant but not that exact plant. Much like botanical illustrations of the eighteenth century intentionally depicted the imagined ideal of a plant rather than an individual specimen, taxonomic endeavors were invested not in individuating plants but in anonymizing them.[17]

Smith's destabilization of such oppressive taxonomic authority is further advanced by the assertive growth of vegetal beings that occurs both in verse and in the notes. Attractive, fanciful plants like vetch, bittersweet, bryony, and the aforementioned wood sorrel are affixed with labels in what Michael Marder refers to as a "highly differentiated" organizational system that subsumes the individual plant within a "dead" network of taxonomic categories (350, 352, 361).[18] Yet the plants of "Beachy Head" don't stay "dead." Rather, the expansion of names for each type of plant results in the expansion of individual organisms: readers envision a riot of bindweed, harebell, and pagil in spite of the definite articles and singular grammatical framing that imply only one single plant (353–55). In one example, the poem swerves the reader from a particular evocation of a single plant (the vetch), to a generalized evocation of that type of plant (*Vicia sylvatica*), and back to a singular vegetal organism ("the tangling vetch"; 351 and 351n). Conjuring the branching terms and proliferating subcategories that compose a visual representation of taxonomic relationships, Smith's incorporation of Linnaean binomials linguistically expands the space occupied by each plant.

Recognizing the expansion of plant life in "Beachy Head" is crucial to understanding Smith's project as a radical one. The profusion and gaiety (suggesting, importantly, both joy and excess) of these plants also threaten to exceed an implied boundary, evoking what Marder calls the "bad infinity" of plant life.[19] An observant reader of the same stanza discussed earlier will note that these plants seem to defy the most basic assumption about vegetal life—contrary to expectations, they are not sessile. Careful attention to the trajectory of plants and their movement reveals what Isabel Kranz identifies as "an esoteric undercurrent . . . in which flowers convey a surplus of meaning that cannot be contained."[20] The woodbine is "clasping," the vetch "tangling" and "[interweaving]" with bittersweet and bryony (350–52). We might read this at first as a fairly conventional way to describe plant growth, but then we note also beech trees "o'ershadowing," wood sorrel "creeping," and the anemones "[gathering]" (356–64). As the plants accrete taxonomic data, they expand—not only linguistically but physically as well. Here, they undeniably manifest what Lara Farina calls the tendency of plants to "mass" in Western literature.[21] None of these plants are the solitary, isolated type of specimens used to scaffold the Linnaean taxonomic system. Instead, they are multiple, multiplying. While these are not the monster plants of nineteenth- and twentieth-century science fiction, the volitional movements of these plants, revealed through the subtly sinister word choice of Smith's verse, seem apparent. And where volition is present, agency may be possible.

As a way of further examining this elusive but intriguing concept of plant agency, I return again to lines 361–63. Recall that the "creeping" root of the wood sorrel is looming in a direct response to human violence—the "pensive lover" and their idle plucking. While the rhizomatous roots of the wood sorrel grow in the same laterally proliferating ways under most circumstances, in "Beachy Head" they creep, expand, and advance *specifically* because of an attack and an immediate threat to their well-being. The rhizome of this particular oxalis is brought to the reader's attention for an important critical purpose: to remind us of the unseen dimensions of vegetal life. The antagonist of these lines, the "lover of uncultur'd flowers," is ironically unable to appreciate the multiplicity of a rhizomatous plant like the wood sorrel—a plant that exists as a collective and cannot be divided into individual specimens. Instead, their act of casual violence lays bare both the "esoteric undercurrent" of cultural meaning that Smith values (and that taxonomic notation cannot represent) and the organic underpinning of the rhizome permeating the soil. By allowing a plant like the wood sorrel to act on its/their own behalf, Smith offers a glimpse of vegetal life existing on its own terms. Rather than lying passively in the shaded copse, looking decorative (a human aesthetic) and awaiting its assimilation into poetic meter (a human invention), the wood sorrel is ever-expanding, growing beneath the surface and waiting out their human antagonists.

I pause a moment here to admit to my anthropomorphizing of the wood sorrel in spite of my best efforts to avoid it. The "creeping" (growth) of the wood sorrel is in fact a biological imperative, and the direction of the "creeping" is determined by optimal soil conditions as perceived by the rhizome. The construction of a relationship between this pensive lover and this sprig of wood sorrel is a human invention yet again, devised by poets and literary critics. What wood sorrel knows of humanity is, ultimately, incomprehensible to us. Yet, as I have written elsewhere, there is real value in articulating a specific kind of nonhuman agency within critical plant studies to help us grapple with these moments of anthropomorphic impulse.[22] If agency, in its broadest sense, requires both the ability and the opportunity to determine one's own actions, then we must understand any concept of nonhuman agency as encapsulating the idea of free will—agency requires willpower and therefore intelligence. The question of plant intelligence is popular in contemporary botanical sciences, and countless studies have been conducted on plants' ability to react to exterior stimuli, take preventative action in the face of looming danger, and to communicate with other plants not of the same species.[23] It is also a meaningful topic in philosophy and cultural studies; Matthew Hall, for example, has elucidated a long history of vegetal agency from a cultural perspective, and Monica Gagliano has compellingly written about the connections between plant subjectivity and ethics.[24] The issue was important in the eighteenth century too. For example, naturalist and poet Erasmus Darwin was actively interested in questions of plant intelligence and explored the subject to varying degrees in *The Loves of the Plants* (1789), *Zoonomia* (1794), and *Phytologia* (1800). Through his careful examination of vegetal movement, irritability, and adaptive behaviors in *Phytologia*, he concluded that "[plants] must possess a brain" and that "there is indubitable proof of plants possessing some degree of voluntarity."[25] Given Smith's familiarity with Darwin's work, it is plausible to see his ideas about plant agency at the root, if you will, of her own critical project in this poem. Poetry, Smith and I agree, has the potential to enrich our ongoing conversations about vegetal agency.

While the wood sorrel reveals unseen dimensions of vegetal life to Smith's readers and thereby destabilizes the most basic botanical knowledge, the anemone (see fig. 2) evidences most clearly Smith's efforts to overtly contradict scientific authority. The beneficiary of one of the more metaphorical descriptors in the entire verse, the anemone is "the copse's pride. . . . With rays like golden studs on ivory laid / Most delicate" (364–66). Unlike the more botanically apt depiction of the wood sorrel, the image of the anemone evokes

FIGURE 2. "Anemone nemorosa," *Flora Londinensis*, 1777. (Smithsonian Libraries and Archives)

jewelry or fine decorative goods with an unexpectedly exotic origin.[26] Furthermore, the anemone also bears one of Smith's most blatant departures from taxonomic authority. Rather than omitting the binomial (as she did with charlock and thistle), Smith includes an intentionally incorrect version, explaining her deviation from scientific convention as a poetic decision: "It appears to be settled on late and excellent authorities, that this word should not be accented on the second syllable, but on the penultima. I have however ventured the more known accentuation, as . . . suiting better the nature of my verse" (364n). Linnaean botanical names were not frequently typeset with accents outside of pedagogical texts, with a few exceptions, so Smith's decision to include "anémone" in the verse for poetic reasons and acknowledge the intentional deviation with "anemóne" in the endnote seems unnecessarily specific. Yet the transparency of this writerly maneuver reinforces her critical aims; as Smith herself reminds readers, "poets have never been botanists." Indeed, the poetic approach to describing this particular anemone is more evocative of the actual individual plant in its real context. Readers don't envision a type specimen described in a textbook or an uprooted sprig depicted in technical illustration. We picture instead a brilliant, shining swathe of little white flowers, alluringly glowing with the light of a sun that rarely reaches them. Unsurprisingly, the anemone is also one of the overtly mobile plants in the poem—a rhizomatous plant, it "gathers" alongside the "creeping" wood sorrel, moving, growing, and swelling across the landscape of Beachy Head and beyond the lines of "Beachy Head."

The binomials—absent and present—in "Beachy Head" underscore Smith's critique of technical botanical knowledge: look at all these labels, she argues, and look how little they tell you about the plants to which they are affixed. The scientific vocabulary developed in the eighteenth century to more accurately represent plant life has in fact generated an extensive rhetorical tradition that most successfully obscures the plant itself and enables a more complete absorption into human culture. On the contrary, Smith's poetry offers the reader a new metric for assessing vegetality, a way of looking beyond technical terms and didactic aims. The wood sorrel and charlock of Beachy Head are immediate, urgent, and vitally, assertively present. We may not understand them, but they unmistakably tell their own stories—stories, like the plants themselves, that expand further than restrictive taxonomic classifications. "Poets have never been botanists," Smith argues, because technical botanical knowledge inevitably prohibits a more full, rich understanding of nature that poetic knowledge can more closely approximate. The inclusion of taxonomic data in "Beachy Head" instead contributes to the botanical and linguistic abundance of the poem, a flourishing vegetality that resists, even now, human intervention.

Notes

1. For a brilliant reading of the symbolism and critical significance of fruit in seventeenth- and eighteenth-century texts, see Bellamy, *Language of Fruit*.
2. Smith, "Beachy Head," 203–10. Hereafter in parentheses by line number.
3. Pascoe, "Female Botanists," 201.
4. Tayebi, "Undermining," 133.
5. Powell, "Linnaeus," 114.
6. Bailes, *Questioning Nature*, 100.
7. John Hill's *The Vegetable System* [. . .] (1759–63) reviews several different varieties of thistle, describing different species as bearing "dry and scaly" leaves, "pointed and edged, with perfect Thorns," which are "weak" in the dwarf distaff thistle while more notable in the elegant distaff thistle. Hill, *Vegetable System*, 8–9.
8. See Marshall, *Rural economy*, 192.
9. Duhamel du Monceau, *Practical Treatise of Husbandry*, 105; *New Medicinal, Economical, and Domestic Herbal*, 36.
10. van der Veen, "Materiality of Plants," 801.
11. van der Veen, 801.
12. Most notably, the heraldic badges in the Wars of the Roses and the unifying Tudor rose designed in the aftermath.
13. Ruwe, "Charlotte Smith's Sublime," 123–24.
14. Curtis, "Oxalis Acetosella. Wood-Sorrel," in *Flora Londinensis*, n.p.
15. Woodville, *Medical Botany*, 57. In fact, the oxalis family thrives everywhere except the polar regions, and various species are indigenous to nearly every country and region on Earth.
16. See Koerner, *Linnaeus*.
17. Daston and Galison, *Objectivity*, 86.
18. Marder, *Plant-Thinking*, 5.
19. Marder, 12.
20. Kranz, "Language of Flowers," 195.
21. Farina, "Vegetal Continuity," 421.
22. See Sagal, *Botanical Entanglements* (2022).
23. For a start, see Baluška, Mancuso, and Volkmann, *Communication in Plants* (2006); Mancuso and Viola, *Brilliant Green* (2005); and Mancuso, *Revolutionary Genius of Plants* (2018).
24. See Hall, *Plants as Persons* (2011); Gagliano, *Language of Plants* (2017); and Gagliano, Ryan, and Vieira, *Mind of Plants* (2021).
25. Darwin, *Phytologia*, 132, 136.
26. As an import from Asia or Africa, ivory was a rare luxury and a pricey investment—an unusual set of associations to pile on an indigenous plant.

Bibliography

Bailes, Melissa. "Linnaeus's Botanical Clocks: Chronobiological Mechanisms in the Scientific Poetry of Erasmus Darwin, Charlotte Smith, and Felicia Hemans." *Studies in Romanticism* 56, no. 2 (2017): 223–52.

———. *Questioning Nature: British Women's Scientific Writing and Literary Originality, 1750–1830*. Charlottesville: University of Virginia Press, 2017.

Baluška, Frantisek, Stefano Mancuso, and Dieter Volkmann. *Communication in Plants: Neuronal Aspects of Plant Life*. Berlin: Springer-Verlag Berlin Heidelberg, 2006.

Bellamy, Liz. *The Language of Fruit: Literature and Horticulture in the Long Eighteenth Century*. Philadelphia: University of Pennsylvania Press, 2019.

Curtis, William. *Flora Londinensis, or, Plates and Descriptions of Such Plants as Grow Wild in the Environs of London: With Their Places of Growth, and Times of Flowering, Their Several Names According to Linnæus and Other Authors*. London: B. White, 1777.

Darwin, Erasmus. *Phytologia; or, the Philosophy of Agriculture and Gardening*. London: J. Johnson, 1800.

Daston, Lorraine, and Peter Galison. *Objectivity*. New York: Zone Books, 2010.

Duhamel du Monceau, M. *A Practical Treatise of Husbandry: Wherein are contained, many useful and valuable experiments and observations in the new husbandry . . .* London, 1762.

Farina, Lara. "Vegetal Continuity and the Naming of the Species." *Postmedieval: A Journal of Medieval Cultural Studies* 9, no. 4 (2018): 420–31.

Gagliano, Monica, John Ryan, and Patricia Vieira, eds. *The Mind of Plants: Narratives of Vegetal Intelligence*. Santa Fe: Synergetic Press, 2021.

Hall, Matthew. *Plants as Persons: A Philosophical Botany*. Albany: SUNY Press, 2011.

Hill, John. *The Vegetable System: or, a Series of Experiments, and Observations Tending to Explain the Internal Structure, and the Life of Plants; Their Growth, and Propagation; the Number, Proportion, and Disposition of their Constituent Parts; with the True Course of Their Juices; the Formation of the Embryo, the Construction of the Seed, and the Encrease from that State to Perfection*. Vols. 1–5. London, 1759–63.

Koerner, Lisbet. *Linnaeus: Nature and Nation*. Cambridge, MA: Harvard University Press, 1999.

Kranz, Isabel. "The Language of Flowers in Popular Culture & Botany." In *The Language of Plants: Science, Philosophy, and Literature*, edited by Monica Gagliano, John C. Ryan, and Patricia Vieira, 193–214. Minneapolis: University of Minnesota Press, 2017.

Mancuso, Stefano. *The Revolutionary Genius of Plants: A New Understanding of Plant Intelligence and Behavior*. New York: Atria Books, 2018.

Mancuso, Stefano, and Alessandra Viola. *Brilliant Green: The Surprising History and Science of Plant Intelligence.* London: Island Press, 2005.

Marder, Michael. *Plant-Thinking: A Philosophy of Vegetal Life.* New York: Columbia University Press, 2013.

Marshall, William. *Rural economy of the west of England: including Devonshire; and parts of Somersetshire, Dorsetshire, and Cornwell. Together with minutes in practice.* London, 1796.

A New Medicinal, Economical, and Domestic Herbal: Containing a Family and Accurate Description of Upwards of Six Hundred British Herbs, Shrubs, Trees, &c. . . . Blackburn, 1808.

Pascoe, Judith. "Female Botanists and the Poetry of Charlotte Smith." In *Revisioning Romanticism: British Women Writers, 1776–1837,* edited by Carol Shiner Wilson and Joel Haefner, 193–209. Philadelphia: University of Pennsylvania Press, 1994.

Powell, Rosalind. "Linnaeus, Analogy, and Taxonomy: Botanical Naming and Categorization in Erasmus Darwin and Charlotte Smith." *Philological Quarterly* 95, no. 1 (Winter 2016): 101–24.

Ruwe, Donelle R. "Charlotte Smith's Sublime: Feminine Poetics, Botany, and Beachy Head." *Prism(s): Essays in Romanticism* 7, no. 1 (1999): 117–32.

Sagal, Anna K. *Botanical Entanglements: Women, Natural Science, and the Arts in Eighteenth-Century England.* Charlottesville: University of Virginia Press, 2022.

Smith, Charlotte. "Beachy Head." In *Charlotte Smith: Major Poetic Works,* edited by Claire Knowles and Ingrid Horrocks, 203–10. Peterborough, ON: Broadview Press, 2017.

Tayebi, Kandi. "Undermining the Eighteenth-Century Pastoral: Rewriting the Poet's Relationship to Nature in Charlotte Smith's Poetry." *European Romantic Review* 15, no. 1 (March 2004): 131–50.

van der Veen, Marijke. "The Materiality of Plants: Plant-People Entanglements." *World Archaeology* 46, no. 5 (2014): 799–812.

Woodville, William. *Medical Botany: Containing Systematic and General Descriptions, with Plates, of All the Medicinal Plants, Indigenous and Exotic, Comprehended in the Catalogues of the Materia Medica.* London: James Phillips, 1790–93.

Spicy Forests and Amboyna Burl

Dryden and the Ecology of Disaster

RAJANI SUDAN

John Dryden's "Indian plays" have been the object of postcolonial interest for some time now, with the figures of an Aztec king, a Moghul emperor, or an Indian queen (*The Indian Emperour, Aureng-Zebe, The Indian Queen*, respectively) reflecting the English fascination with the East. Certainly, the idea of Moghul emperors and queens bolsters an equally powerful attraction to the goods and commodities such places yielded, as many critics have argued for the past forty years. More recently, scholars have turned their attention to the nuances of "worlding," the term Gayatri Spivak introduced in 1985 with the publication of "The Rani of Simur: An Essay in Reading the Archive."[1] Briefly, Spivak's term addresses the transformative moment when the native land is experienced as a colonized space by natives, a space that is methodically remapped by its colonial masters through writing, cartography, and surveillance or simply traveling over the colonized land. Hence the necessity of the entourage: the train of natives accompanying the often lone English traveler in order to (re)experience the land as it is imagined, recorded, and controlled by the English master. "He [the Englishman] is engaged in consolidating the self of Europe by obliging the native to cathect the space of the Other on his home ground"; the story of how conquest is less the result of brute force and more a consequence of Michel Foucault's panopticon, or the process of surveillance.[2]

Worlding, however, presupposes a specific dynamic: the power relation between the metropole and the colonized space. Spivak's account of this dynamic privileges the idea of European sovereignty; recent scholarship questions the adamancy of power relations, focusing on the exchange of techne that shapes both colonized space and the metropole.[3] Worlding is recalibrated as worldmaking, and the very elements that counted as metonymies of new spaces are collected and synthesized into a legible conceptual framework.[4] I would argue that those very elements folded into a synchronized understanding of "global" are also where fissures erupt the legibility of a global framework. So, for example, a padauk tree may function as a metonymy of the New World or Aztec culture more specifically, but what happens when this same tree turns up in places where it doesn't "belong"? Does this displacement question early modern European accounts of a "world"? Richard Grove argues that a European recognition of a "world flora" by 1550 is an important prerequisite to the new ways in which botanical gardens as collections brought the world together and, as such, "acquired a meaning as symbols of an economic power capable of reaching and affecting the whole biological world."[5]

I would like to draw on Grove's identification of the profound changes of perspective in human relations to the environment caused by the circumstances of colonial expansion as a way of thinking through Restoration senses of world-making. Grove identifies three dominant ideas that defined relations between humans and nature from classical times to early colonial expansion: the idea of a designed earth, the preoccupation with environmental influences on cultural development, and the notion of humans as geographical agents. The first developed largely from theology, mythology, and philosophy; the second originated from medical theory and pharmaceutical lore; and the third, less prominent in antiquity, emerged from quotidian activity like cultivation, carpentry, and timber-cutting.[6] I want to consider Grove's argument in a reading of Dryden's *Amboyna, or the Cruelties of the Dutch to the English Merchants*, in particular a reading that directs new attention to the characters of Perez (a Spanish captain), Julia (wife of Perez), and Ysabinda (an "Indian Lady" betrothed to Gabriel Towerson, the English captain). In so doing, I hope to complicate the ecological theater of this drama and uncover the complex web of epistemological systems at work in Dryden's polemic against the Dutch. Robert Markley identifies *Amboyna* as a play in which Dryden deploys national martyrdom as a strategy for self-definition. He contends that Dryden transforms English abjection as "the martyrdom of national virtue, liberty, and nobility." But such a transformation is only achieved by "*simplifying* a complex history of international rivalry in Southeast Asia," allowing Dryden to project the image of heroic English mercantilism "that mystifies the sources and nature of Anglo-Dutch conflict."[7] And while I agree with this argument, I also believe there are many material genealogies embedded in this literary representation of global trade that are worth considering.

A lot happens in the wood of *Amboyna*, particularly in act 4. In this "wild woody walk," husbands are lost, brides are found and lost, and the loyalties of the heroic East India Company merchant Gabriel Towerson are sorely taxed. But the wood itself provides a stable counterpoint to the political and sexual turmoil: the "fragrant . . . eastern groves" of the "rich Molucca isles" yield a fabulous profit to the Dutch East India Company (VOC) in cloves and nutmeg, while the "poor Englishman" remains firmly locked out of the prosperous compound of Dutch enterprise, not even coming in for "a third part of the merchandise."[8] Frustrated by their impotence when the Dutch governor imprisons them, the English merchant Beaumont frets, "If we deserve our tortures, 'tis first for freeing such an infamous nation, that ought to have been slaves, and then for trusting them as partners, who had cast off the yoke of their lawful sovereign."[9] What would it mean for such an "infamous nation" to be "slaves" rather than "partners" for the English? The "yoke"—a word used at least twice in the play—refers to the Spanish subordination of the Dutch. But the word *yoke* is a cognate of Old Dutch, and most of the first meanings of this word are associated with harnessing draft animals for agricultural purposes: the clearing of land and plowing of fields to conform a savage place to the idea of a designed earth, an improvement of the land from wild wood to fruitful plantation. The implication of this cast-off yoke, then, as a rebellion against the "lawful" sovereign's pastoral work and, as the play develops, the replacement of a cleared field with a fragrant grove signifies different cultural responses to ecological anomalies—in particular, the necessity for preserving natural groves for the cultivation of cloves and nutmeg. Grove describes the profound influence Herman Boerhaave had in the medical school of Leiden, which was attended by many students from England, but in particular, the "establishment of Dutch power in Cochin on the decline of Portuguese power in Malabar was marked in botanical terms by the preparation of the *Hortus indicus malabaricus* as a personal project initiated by Hendrik van Reede tot Drakenstein."[10] So it would seem that such English grumblings as Beaumont articulates suggest an ironic alignment with the Spanish against the superior horticultural and botanical knowledge of the Dutch.

This wood, perhaps the displaced "spicy forests" of *Annus Mirabilis,* would seem to be fragrant from the cloves and nutmeg groves that yield "luxuries" from their "o'erflowing bounties." These luxuries appear to defy the logic of Hobbesian scarcity during a historical period when the fantasy of mercantile proliferation seemed to overcome the problems of seventeenth-century environmental and ecological crises such as timber shortages, plagues, and various infestations of vermin that were the unfortunate results of the Little Ice Age.[11] Daniel Defoe's South Sea fiction, *The New Voyage around the World,* is one such example, but even his nonfiction treatises argue for the possibilities of trade in the South Seas in spite of the financial riot the Bubble wreaked.[12] While the Dutch may have been pioneers of botanical knowledge and, more specifically, of the spice trade—although the Spanish also had a wealth of pharmaceutical lore in their materia medica of the New World—their desire for an exclusive market for these fabulously valuable spices was as great as anyone else's, and these spice groves were jealously guarded to the extent to which they would rather burn their groves than allow for the possibility of transplantation.[13] More interestingly, what *did* grow on Amboyna without human cultivation, whether Dryden knew it or not, was the padauk tree, genus *Pterocarpus indicus*. There are many species of this tree, abundant in Southeast Asia and Africa, but an intriguing fact is that the species specific to Amboyna is also found on the American continent, particularly in the part that was historically known in Europe at this time as New Spain. The wood from this tree was well known in early modern Europe for its peculiar ability to change the color of water into opalescent shades, especially blue. Cups carved from the tree were prized gifts given to royalty not only because of the beauty of the water but because of the substance *Lignum nephriticum*, the traditional diuretic derived from the wood and the primary agent for its color transformation. Known in Nahautl as coati, coatli, or coatl, this substance was also familiar to the Aztecs, who used it to cure kidney problems.

The historian Daniela Bleichmar reminds us of the importance of New World materia medica to the emergence of early modern botanical knowledge, in particular the Sevillian physician Nicolas Monardes's sixteenth-century *Historia medicinal de las cosas que se traen de nuestras Indias Occidentales que sirven en medecina* (1565, 1569, 1574), a publication that established him as Europe's foremost authority on New World materia medica.[14] Although he never left Spain, he established an important partnership in 1553 with Juan Nunez de Herrera, who was settled in Tierra Firme (Central America). In addition to this partnership, four of his eight children immigrated to Tierra Firme, thus ensuring a steady exchange of medicinal products from the New World to the Old. Published in Monardes's *Historia medicinal* was one of the first European descriptions of the medicinal powers of *Lignum nephriticum*. Although this compendium of materia medica lost its epistemological value later in the eighteenth century, it was extraordinarily well received when it first appeared and was translated into Latin, English, French, and German, and a Flemish revision (1574) by the naturalist Carolus Clusius ensured its position as the authoritative source of information on New World materia medica. So it would seem that unlike Beaumont's bitter remarks against Dutch greed and Spanish tyranny, the real problem was English ignorance. After all, it wasn't until a hundred years later that Robert Boyle explained the phenomenon of *Lignum nephriticum* as dependent on pH balance. A year later, Isaac Newton mentioned the cups made of the wood in his treatise "Of Colours"; word of this phenomenon was disseminated in England just a few years before Dryden's play was staged, endowing this piece of pharmaceutical lore with a false contemporaneity.

At the conclusion of the first part of *Historia medicinal,* Monardes asks,

> How many trees and plants with great medicinal virtues are there in our Indies . . . leaving no need for the spices from the Moluccas, and the medicines from Arabia and Persia, given that our Indies yield them so spontaneously in the untilled fields and in the mountains. It is our fault not to investigate them, not to look for them, not to be as diligent as we should be in order to profit from their marvelous effects. And I hope that time, which is the discoverer of all things, and diligence and experience will demonstrate them to our great profit.[15]

Part of the "imperfect workings of nationalism in Dryden's play," as Markley argues, is the "inconsistencies, gaps, and anxieties that constitute the fiction of an essential, transhistorical, national identity."[16] But Monardes writes from the secure standpoint of a highly organized state and scientific discipline that originated in cosmography, a set of theories, language, grammar, and practices that found an institutional home in Spanish universities and later became part of a scientific enterprise with a mission deployed solely for the benefit of the state, specifically the Habsburg monarchy. Even if his musing sets a precedent for English fantasies of inexhaustible troves of treasure somehow produced through the notion of a designed world, eluding the economy of scarcity and supply, the idea that the Moluccas are unnecessary to commercial profit—that New Spain can furnish the same spices, the same materia medica available in "*our* Indies"—suggests a degree of national confidence missing in Dryden's play (emphasis mine).[17]

Until, that is, we take a closer look at Captain Perez's character and role in the play. Granted, this Spanish character is the principal subject of an elaborate cuckolding scheme that renders his role humiliating, and he is killed off at the play's conclusion, but he proves himself capable of the heroic loyalty that makes this play a "mercantile morality play."[18] The alliance between Towerson and Perez seems crucial. Perez embodies access to another knowledge of navigation—witnessed by the perils at sea of Harman Junior (the governor of Amboyna's son) and by Towerson's own account of his voyage to Amboyna: "myself to have escaped the storm that tossed me long, doubling the Cape, and all the sultry heats, in *passing twice the Line*" (emphasis mine).[19] Maria Portuondo writes about Spanish navigation manuals,

> One such problem was the deviation of the compass needle from north or magnetic declination. The problem had vexed sailors since Columbus noted the phenomenon during his first voyage. The erratic behavior of the compass needle was known as the *nordesteare* and the *noruesteare*. In long transatlantic voyages, the phenomenon added a new dimension of difficulty in charting a course at sea, since the deviation from north seemed to vary according to one's meridian. Navigation manuals became a forum for advocating different ways of coping with the problem, either by introducing instruments that "corrected" the compass reading or by trying to use the needle's deviation as an indicator of longitude. When travelling from east to west, as in the route taken by Spanish galleons sailing to the Indies, the compass needle deviated in a manner that suggested the deviation was proportional to longitudinal distance. This apparent correlation was used by generations of Spanish cosmographers in attempts to construct an instrument that would determine longitude at sea.[20]

What Portuondo argues is that far from allowing the "erratic behavior" of the compass needle to thwart their charted courses, the Spanish used the margin of error to address and supplement a serious seafaring problem—accurately determining longitude—and then continued to use this "correlation" to develop a reliable instrument. The problem of determining longitude at sea had vexed many English sailors, who were often forced to share shipping lanes with rivals and pirates, although they didn't publicly address the problem until the Longitude Act of 1714. Towerson's rescue of the governor's son would very much have depended on instruments of location, even in Dryden's fanciful representation. But perhaps even more importantly, Perez's importance is articulated by the character of Van Herring (a Dutch merchant), in consultation with the Fiscal, as he declares, "Dispatch him; he will be a *shrewd witness against us* if he returns to Europe" (emphasis mine).[21]

Even Perez's humiliation is of consequence. In the second act, the English merchant Beaumont encounters the Fiscal in close relations with Perez's wife, Julia:

> BEAUMONT. Now, Mr. Fiscal, you are the happy man with the ladies. . . . You've the Indies in your arms, yet I hope a poor Englishman may come in for a third part of the merchandise.

FISCAL. Oh, sir, in these commodities, here's enough for both; here's mace for
   you, and nutmeg for me, in the same fruit . . .
JULIA. My husband's plantation is like to thrive well betwixt you.[22]

Sharing mace and nutmeg—or the outer and inner parts of the ovary[23]—both Beaumont and the Fiscal misread Julia's body as a metonym of the highly profitable commodity enriching Dutch merchants and providing the hope for a share to English ones. Julia describes herself jokingly as the "plantation" whose ground, when properly tilled, should thrive. The female body collapses into exotic produce, both a fruit and a spice but, most importantly, a commodity that's capable of hemispheric migration.

There is another plant, however, not as portable a commodity as nutmeg yet found with equal abundance in the East and West Indies. In addition to the *Lignum nephriticum* yielded by the *Pterocarpus indicus* that grows so abundantly in Southeast Asia is something known as Amboyna burl. The burls of the padauk tree produce a peculiar and highly figured wood that was much prized for its beauty and rarity (as it still is today). The wood cups given to Ferdinand III, emperor of the Holy Roman Empire, in 1637 were carved from this burl, as the Swiss botanist Johann Bauhin described in his 1650 *Historia planetarum universalis*. These cups were, in fact, part of the native industry of the Luzon area of the Philippines before the Spanish arrived; after their conquest, these cups were imported into Mexico through the Manila-Acapulco Galleon, a shipping line started in 1565 (and that remained active until 1815), and from there into Europe.

While yielding richly grained wood, however, burls on trees are sites of stress caused by injury or disease (virus or fungus). As residual scar tissue, they tend to have a denser grain than that of the rest of the tree and are most often found on the underground roots as a type of malignancy—a natural disaster, as it were. In other words, the site of trauma furnishes some of the most expensive and valuable ornamental wood. To return to the play, Julia figures herself as a "plantation" (albeit her husband's), an agricultural term devoted to the cultivation of a specific crop, a term more popularly associated with the West Indies, and a term very much bound up with Grove's account of the dominant ideas about "nature" and the ways in which human agency shapes this world, most importantly the means of "improvement" meted out by Reformation logic.[24] Like the indentured servitude and chattel slavery exploited to make the plantations of "our Indies" profitable, Julia is the fertile ground bound up with Spanish colonial power, with the less powerful but no less covetous English interest embodied by Beaumont's "poverty" ("a poor

Englishman"). India and the Caribbean are thought of in the same frame, from either more English ignorance or pragmatic arrogance, given the protoimperial crossing between the "two Indies" of people, plants, objects, and ideas. "The Indies" function as a mentality, not a mistake, as Ashley Cohen argues, and even as early as the Restoration, Dryden is aware of what constitutes a colonial model of world-making.[25]

Ysabinda, however, is a different story. The perils of that "wild woody walk" are great for her; in its shades Harman Junior confronts her, blinded by his lust: "You are a woman; have enough of love for him and me; I know the plenteous harvest is all his: He has so much joy. That he must labour under it. In charity, you may allow some gleanings for a friend."[26] This is an example of the eloquence with which he purports to win her, but naturally, she finds his rhetoric hardly persuasive, not the least because she has just married Towerson. Frustrated by her unwillingness to give him her "gleanings," Harman Junior binds her to one of the padauk trees and rapes her. The problem is that Harman Junior has misread Ysabinda's role on the island. Not only is she already tied to the English merchant Towerson, but also, unlike Julia, she is not part of a plantation economy. Her "husband's plantation" will not "thrive" from the ministrations of a shady Englishman and even more suspect Dutch merchant because her husband's mercantile virtue will not allow for underhanded transactions. Far from offering Harman Junior a share of the "plenteous harvest," Ysabinda has to be raped, which positions Harman Junior as a thief. She cannot yield either the "plenteous harvest" or any "gleanings" because her ecological ties are not to plantation bounty but to Amboyna burl, the site of trauma. She is a product of the grove, not the plantation, and her rape is part of the ecological disaster that underwrites this mercantile tragedy.

Obviously, this scene is traumatic on a number of levels, but what I find most interesting is the way in which Ysabinda is forcefully bound to the indigenous tree. Unlike the flirtatious relations Julia encourages between the Dutch Fiscal and the English merchant during Perez's cuckolding, Ysabinda wards off Harman Junior's advances until she's incarcerated by his lust. Yet she retains her virginal virtue—an intangible, immaterial something Towerson notes when he finds her and declares her "still as fragrant as [her] fragrant groves ... without the gross allay of flesh and blood."[27] Ysabinda is part of the ecology of the East Indian "grove," situated in Dutch-controlled Amboyna, unable (as it turns out) to be commodified like nutmeg and mace and therefore transported to Europe despite her conversion to Christianity and her marriage to Towerson. As a "native" of Amboyna, she is tied to that soil, and she demonstrates this in her final lines: "O'er the green turf, where my love's laid, there will I mourning sit, and draw no air but from the damps that rise out of that hallowed earth; and for my diet, I mean alone my eyes shall feed my mouth."[28] Sitting on Towerson's grave, she gradually breathes the diseased "damps" that "rise" from the ground, watered by her tears.[29] Julia, on the other hand, is an abstract "plantation" to be bargained for by her English and Dutch lovers but, importantly, *not* her Spanish husband. I'll return to this shortly.

Bridget Orr, Heidi Hutner, and Robert Markley have all published convincing arguments regarding Ysabinda's role in Dryden's play and the implications her role has for early modern colonial expansion. Orr and Hunter argue that the figure of the Indian queen and the violence perpetrated upon her person represent the politics of colonial aggression crucially redirected to *individual* stories of virtue.[30] Conquest is justified through the erotic alliance such figures negotiate with their English partners, and her beauty is metonymically connected to New World riches, something that seems unproblematic to English merchants. Markley reads Ysabinda's role similarly as a transplantation of "the Pocahontas myth on to the Spice Islands."[31] Elizabeth Maddock Dillon, on the other hand, reads the torture of English merchants by the Dutch as part of a self-conscious political strategy by English playwrights: she argues, for example, that in William D'Avenant's 1653 *Proposition,* the "staging of the torture of English merchants by the Dutch . . . [was] an ideal subject for the theater of moral instruction," as D'Avenant himself proposed "to Cromwell's Council of State."[32] I find these readings compelling but want to offer a supplement, something I'm identifying as an ecology of disaster. Resisting the economy of the spice trade, Ysabinda is the product not of Dryden's strangely figured plantation but of the indigenous "grove" of padauk trees. Furthermore, she still retains virtue: even ravished and purportedly ruined for any kind of marital success, Towerson figures her as untouched: "Your breast is white, and cold as falling snow . . . your whole frame as innocent and holy as if your being were all soul and spirit."[33] Because Ysabinda is a metonym for Amboyna burl—the ecological product of a natural disaster—she plots her death as a return to her native soil.

I'm not suggesting that Dryden knew this (although there's also plenty of evidence that this information was available), but because of his *lack* of knowledge of the world beyond English shores, both East and West Indies get abstracted from their local contexts and become to him interchangeable. The idea of a flawed postlapsarian world was becoming less attractive to Iberians (and to the Dutch and English, but later than the Spanish) largely because seafaring technological developments made travel, particularly to the extraordinary variety and richness of the tropics, much more viable, so there was new interest in the profits that the world could offer. As Grove argues, "Furthermore, the tropical island had become a focus for understanding natural processes and a metaphor for handling new ideas about nature, 'new worlds' and social Utopias."[34] Ysabinda's return to the land while wielding a virtuous upper hand is far less profitable to Dutch and English merchants than Julia's sexualized abundance. The simplification "of a complex history of international rivalry in Southeast Asia" that Markley levels at Dryden is the simplified standpoint of an English poet and playwright who had no ecological knowledge of the world, no understanding of medical theory and pharmaceutical lore, and a limited command of environmental influences affecting cultural and political development.[35] The discourse of a designed earth, emerging from theology, mythology, and philosophy, structures his representation of the mystified "Indies."

But the larger implications of Dryden's simplification are interesting. Captain Perez exits the stage to go to his death; Julia's parting words are "Farewell, my dearest! I may have many husbands, but never one like thee." As the "plantation," Julia's history is firmly fixed in the west, moving back and forth under both English and Dutch mastery. But the Spanish still retain much more power than either the English or the Dutch, and even Dryden's narrowly informed representation makes a vague allusion to that power. At the time of Dryden's play, England had had the East India Company and some trading relations at some ports but not much else in the East Indies. Their colonies in Virginia (Jamestown), Plymouth Colony, and Bermuda were still struggling; they had Barbados as a main colony, but this was a risky business because of Barbadian trade with the Dutch, and the parliamentary act passed to limit trade solely to English vessels precipitated the First Anglo-Dutch War. Really the only form of profit came from attacking Spanish galleons and robbing them, acts that hardly were in line with Dryden's projection of virtuous English merchants. Despite Dryden's representation, England wasn't invested in one Indies over the other, and Spain was still very powerful in spite of the VOC's claims to the Moluccas. England had no serious investment in the East until they lost their purchase in the west, which didn't effectively occur until the mid-eighteenth century with the American Revolution and the loss of the thirteen colonies. I'm not suggesting that England didn't struggle for a part of the lucrative spice trade—there were many Anglo-Dutch wars that attest to British conflict with the Low Countries—but Spain was still very powerful.

Of course, rivalry with the Dutch and the VOC characterized most of England's bid for power in developing trade.[36] Grove reminds us how, in the wake of religious Reformation, the VOC played a pivotal role in establishing the authority of church and state. Here, again, the Dutch prevail, particularly in a new painterly tradition. Grove argues,

> The rapid accumulation of convertible capital by a large urban class at a stage when Amsterdam had become the financial capital of the world helped to accelerate the process of aesthetic transition. These developments help one to understand the very striking early portrayals of often "wild" landscapes, the accurate depictions of "Little Ice Age" conditions and the growing readiness to move away from the stereotyped image of the Italian pastoral. Such readiness can well be equated with a growing willingness to describe tropical island landscapes in less stereotyped and more dynamic terms. . . . All of these developments led to a more empirical and more informed image of the far tropics, particularly on the part of the Dutch, to which was added their particular interest in the careful husbanding and control of scarce and hard-won resources.[37]

The English were also slow to recognize the importance of environmental control. Clearing, drainage, and timber loss—part of the puritanical impetus to "improve" the land—also led to soil erosion and deforestation, hardly the conditions to sustain a belief in the providential bounty available to reformed Christians. On the other hand, the enormous projects of land reclamation exercised by the Low Countries led to a heightened awareness of geographical space and climate that eventually emerged as a highly developed form of environmental control.[38]

The growing volumes of capital that Grove suggests structured the market economies of France, Britain, and the Netherlands in the late sixteenth century were accompanied by another development, one that was as crucial to the "transformed reality of [Dutch] artificial and often planned cities, ports and drained landscapes" as the influx of capital.[39] Timothy Reiss identifies this development as "a gradual disappearance of a class of discursive activity, a passage from what one might call a discursive *exchange within* the world to the expressions of knowledge as a reasoning *practice upon* the world."[40] Reiss argues that analytico-referential discourse is primarily a sixteenth-century European creation that emerged from an earlier discursive form where words were presented as the image of things. This discourse of resemblance or patterning, while placing humans in an ordered relationship to the world, situates epistemology in the confines of exchange *within* the known world. The move toward epistemology as a practice *upon* the world adds another crucial dimension to the effects of joint-stock companies not just on the relations between town and country and the consolidation of state power but on emerging discourses of world-making. The early modern cultural project of synthesizing new global experiences, as Ayesha Ramachandran argues, is part of the process of reconfiguring the "particular against the universal,"[41] thus turning discourse into something that provides an external and objective knowledge of the real structure of the world, the "expressions of knowledge as a reasoning *practice upon* the world."[42]

One of the earliest beliefs that structured human relations with nature was the concept of a designed earth, emerging, as Grove has suggested, from mythology, theology, and philosophy. Grove writes, "It is assumed that the earth is designed for people alone, as the highest beings in creation, or for the hierarchy of life, with man at the apex. This conception presupposes the earth or certain parts of it to be a fit environment not only for life but for high civilization."[43] Here, then, are the conditions out of which the discourse of patterning and resemblance emerges, although, as Reiss hastens to point out, medieval circumstances of the slowness of communication, the separation of the educated clergy from the dominant politics of the feudal aristocracy, and the separation of both powerful classes from village life all inevitably contributed to other emergent discourses. The important point to keep in mind is that this variety still occurs under the sway of a single discursive dominance, and this discourse put humans in an ordered relationship to the world by presenting words as the reflection of things. Divine design, therefore, makes human engagement with the environment a fixed, somewhat immobile affair, encased within the confines of a discursive exchange *within* the world.

Grove identifies two other concepts that define human relationships to nature: (1) their preoccupation with environmental influences and (2) their conception of people as geographical agents in themselves. Although these ideas appear largely from a Hippocratic standpoint to the European mind, they still presuppose the notion of travel; the third idea establishes humans in control of the transformation of their environments, improving and cultivating a world created solely for them. Movement and travel are implicated in the discursive moment Reiss identifies, as if the stasis of a discourse based on resemblance and pattern is too confining for European mercantile culture. The emergence of discourse as a semiotic system and the fluidity of meaning that words now convey not only reflect a world larger than the urban and rural enclaves. Especially in the case of the Dutch, the middle classes began to observe transformed environments of drained lands and ports and were thus able to appreciate this transformation as the work of geographical agents. It's no wonder that it is the Dutch in *Amboyna* who are capable managers and custodians of the groves that yield nutmegs and cloves. As Emma Spary argues, "The 'botanist' as a figure of epistemological expertise was thus a retrospectively fashioned identity, reflecting the outcomes of local disputes. This was particularly true in colonial situations since travel disrupted individuals' patronage relations and thus caused the sources of their scientific expertise to be destabilized."[44] Thus, the capacity of analytico-referential discourse to provide an epistemology as a reasoning practice *upon* the world privileges the anthropogenic importance of humans to the environment.

Sixteenth-century England was behind Europe in terms of maritime trade, but in spite of this handicap, Dryden demonstrates a loose familiarity with the rules of the game. Using virtue as a necessary supplement for lack of economic or navigational skills, the tortured and murdered English merchants suggest an equally complex conflation of the East and West Indies.[45] Dryden levels the differences between Amboyna and Barbados; they are abstractions to him, and their peculiar ecologies lose local definition to become interchangeable. Of course, the creation of the British West Indies took place in successive eras of colonial expansion, and the integrated plantation was both the sign of prosperity and the economic foundation of the first British Empire. The first phase, starting in the early seventeenth century, began with the English claim to Barbados and Jamaica, the former of which became the richest colony by 1625. There are different historical accounts of this process of colonization in which the role of the English is either as reckless sugar planters who wielded chattel slavery to their advantage or as a perfectly sober people eager to transplant and maintain social order as they knew it: the legal, religious, and governmental institutions that constituted life in England. In either case, the notion of transplantation was tied to the circulation of commodities, even if the early English settlers in Barbados were not profligate settlers made wealthy through the proceeds of chattel slavery.[46]

The spicy forests, wood, and Amboyna burl in Dryden's play are uprooted from their local conditions and conflated with the plantation that Julia embodies, and yet this plantation is one that surrenders the fruits of the "spicy forests"—mace, nutmeg, and cloves—that grow without human intervention, unlike the sugar plantations of the West Indies. Dryden is deploying the idea that the garden or grove representing a symbolic relocation of paradise is as available for cultivation and harvest as agrarian husbandry, hence Julia's sly reference to her "husband's plantation" to the Fiscal and Beaumont, who "betwixt" them (the English merchant and Dutch accountant) will "thrive well." In fact, the Fiscal alludes to the commodities when sharing the same fruit: "Oh Sir, in these Commodities here's enough for both, here's Mace for you, and Nutmegg for me in the same Fruit." Dryden abstracts trees from their ecology and "fruit" from their groves because, as Grove argues, "tropical islands had by the mid seventeenth century acquired a very specific role as the subject of a discourse based in large part on archetypal Utopian and Edenic precepts, many of them with eastern roots."[47] Edenic islands were useful to Dryden's optimistic vision for English mercantilism; the British Isles, associated with the Protestant understanding of "improvement," as Grove defines it, could look forward to global movement under the new conditions of the market economy.[48] Operating as a parallel, Dryden's collapse of East and West provides a sanguine future for prospective English merchants even while addressing the horrors of their torture and murders in *Amboyna*.

But this confidence is misplaced, and the ecology of disaster dramatized by Ysabinda's demise remains stubbornly indelible in Dryden's play. To clarify the importance of disaster in Dryden's theatrical representation, it may be useful to consider the ways in which the politics of early modern monoculture operated and the effects that agricultural husbandry and extraction had on the larger colonial project. Ramachandran opens her study on world-making with a description of the prized Dutch *Atlas,* also called the *Representation of the Universal World,* in which this "world emerges into view not through divine revelation but by dint of human effort."[49] James C. Scott also addresses the importance of cartography to understanding a global scope or, at the very least, a comprehension of "state": "Much of early modern European statecraft seemed . . . devoted to rationalizing and standardizing what was a social hieroglyph into a legible and administratively more convenient format."[50] Discussing forest management in early modern Prussia and Saxony, Scott describes the importance of uniformity and neatness to comprehending the availability of timber for shipbuilding, urban development, and profit, but uniformity also became a dominant aesthetic. Thus, the "geometric, uniform forest was intended to facilitate management and extraction," but the "*visual* sign of the well-managed forest . . . came to be the regularity and neatness of its appearance" (emphasis mine).[51] Because this form of monocrop—conifers—takes at least eighty years to grow, it wasn't until the second cycle of planting that the problems started to become visible. Without the complex process of soil building, nutrient uptake, and the symbiosis among fungi, insects, mammals, and flora, all of which had been removed in the interests of aesthetics and more efficient management, these plantations—if that's the right word—started to die, coining a new phenomenon known as *Waldsterben,* or forest death. Thus, the cartographic abstraction of the global contours that putatively "make visible the global whole—now understood as the 'universal world,' has serious ecological and environmental consequences."[52] The results of religious "improvement," now primarily the province of human endeavor, were manifest primarily by clearing. As Grove argues, the importance of clearing land was largely political: it was (like the uniformity and neatness of managed forests) the most visible sign of ownership, something of tantamount importance to the colonial project. To clear was to subjugate the land to human hegemony, although this was certainly not limited to the land itself; an adequate labor force was also necessary to control cultivation.

The many acts that privatized land through its enclosure and engrossment, timber loss, and deforestation were acute environmental problems for seventeenth-century England, especially as the prospect of colonizing the West Indies and the Americas became exponentially more attractive and shipbuilding assumed greater importance. New seventeenth-century joint-stock companies—the Dutch, English, and French East India Companies engaged in transforming landscapes—the Low Countries' drainage schemes, or drainage projected on the East Anglian fens, both in Europe and in the East and West Indies, were largely responsible for soil erosion and deforestation, and the transformation of market economies with the influx of capital had massive environmental effects. Many scholars have written in great detail about these environmental issues. The intervention I want to make is to think about these events as a crucial intersection between material geography and the psychosocial formation of national identity and present a few notes that link terror, trauma, trade, and the "global" position early modern England navigated. To return to Dryden's drama, how does the trauma of Amboyna burl get rewritten as a history of trauma?

It might be useful to recall Cathy Caruth's contention concerning trauma and the possibility of history, where she argues that in the "bewildering encounter with trauma . . . we can begin to recognize the possibility of a history that is no longer straightforwardly referential." So perhaps the burls formed on the base of padauk trees, burls that embody visible signs of trauma, are transformed into valuable commodities that contribute to an increasingly crucial materia medica for the English that depended on global travel, commodification, and capital. These elements all contribute to another epistemology shaped by pharmaceutical lore, cartography, and an awareness of a "universal world" that is accessible through primarily human means (and not divine will). As Caruth speculates, "Through the notion of trauma . . . we can understand that a re-thinking of reference is not aimed at eliminating history, but at resituating it in our understanding, that is, of precisely permitting history to arise where immediate understanding may not."[53] Ysabinda's trauma—her conversion to Christianity, her marriage to Towerson, and her bondage and rape by Harman Junior, or in short, the ways in which her body is used and abused by the English and the Dutch—represents a nonlinear, nonreferential history whereby her return to indigeneity becomes her only refuge. By contrast, Julia—as "plantation" and specifically as the lucrative and portable "fruit," nutmeg—aligns with European navigations that define a "universal world." But such plantations come with their own ruinous cost. Grove argues that the demand for timber for urban development and shipbuilding, together with the expansion of arable agriculture, placed a great deal of pressure on the European environment. But more crucially, he explains that the "ecological pressures at the colonial periphery were felt more rapidly and catastrophically."[54] Dillon's reading of Dryden's theatricality focuses on a "new account of territorial sovereignty," one that is more in keeping with an emergent imperial ideology favoring the authority of territorial occupation "from below" rather than divine will.[55] I argue that Dryden's representation of English mercantilism in *Amboyna* collapses into an ecology of disaster in which the disavowal of violence perpetrated by colonial monoculture practices is rewritten as the traumatic torture and murder of virtuous English merchants. Thus, the promise of world-making, the "universal world" made visible through Dutch cartography, is a fictional fabric recovering environmental decimation. The logical outcome of "spicy forests" is the trauma of Amboyna burl.

Notes

1. Gayatri Spivak, "The Rani of Simur: An Essay in Reading the Archive," *History and Theory* 24, no. 3 (October 1985): 247–72.
2. Spivak, 253.
3. See, for example, Rajani Sudan, *The Alchemy of Empire: Abject Materials and the Technologies of Colonialism* (New York: Fordham University Press, 2016).
4. At least, this is part of Ayesha Ramachandran's argument in *The Worldmakers: Global Imagining in Early Modern Europe* (Chicago: University of Chicago Press, 2015).
5. Richard Grove, *Green Imperialism: Colonial Expansion, Tropical Island Edens, and the Origins of Environmentalism, 1600–1860* (Cambridge: Cambridge University Press, 1995), 75.
6. Grove, 24–25.
7. Robert Markley, *The Far East and the English Imagination, 1600–1730* (Cambridge: Cambridge University Press, 2006), 143.
8. John Dryden, *Amboyna, or the Cruelties of the Dutch to the English Merchants*, 2.1.497, Kindle. I am grateful to David Alff, who pointed out that the criticism of the Dutch in this scene is replicated in *Annus Mirabilis*, where Dryden admonishes the Dutch for their failed reciprocity:

> Trade, which, like blood, should circularly flow,
> Stopp'd in their channels, found its freedom lost:
> Thither the wealth of all the world did go,
> And seem'd but shipwreck'd on so base a coast.

9. Dryden, *Amboyna*, 5.1.1216.
10. Grove, *Green Imperialism*, 78.
11. For a fascinating study of vermin, literature, and environmental discourse and a much more thorough reading of my gloss, see Lucinda Cole, *Imperfect Creatures: Vermin, Literature, and the Sciences of Life, 1600–1740* (Ann Arbor: University of Michigan Press, 2016), 24–48.
12. I should clarify that Defoe still continued to point out two fundamental flaws in the EIC's charter: that the company was first and foremost invested in itself and that it precluded trade in the South Seas. For more about Defoe's issues with EIC, see Robert Markley, "Defoe and the Problem of the East India Company," in *Robinson Crusoe in Asia*, ed. S. Clark and Y. Yoshihara (Singapore: Palgrave Macmillan, 2021), https://doi.org/10.1007/978-981-16-4051-3_2. Robert Markley's chapter, "'So inexhaustible a treasure of gold': Defoe, Credit, and the Romance of the South Seas" is the most persuasive treatment of Defoe's novel I've read. See *Far East and the English Imagination,* 210–40.
13. My thanks to Danielle Spratt, who pointed out that Hans Sloane came under a lot of scrutiny regarding this matter, as the broader Scriblerian circle was extremely skeptical of the colonial dimensions of spice acquisition/distribution for "disinterested" purposes and steered me to William King's *Transactioneer* and *Voyage to Cajamai*. See E. C. Spary's argument about the Dutch and nutmeg in "Of Nutmegs and Botanists: The Colonial Cultivation of Botanical Identity," in *Colonial Botany: Science, Commerce, and Politics in the Early Modern World,* ed. Londa Schiebinger and Claudia Swan (Philadelphia: University of Pennsylvania Press, 2005), 187–203.
14. Daniela Bleichmar, "Books, Bodies, and Fields: Sixteenth-Century Transatlantic Encounters with New World *Materia Medica*," in Schiebinger and Swan, 83–99.
15. Quoted in Bleichmar, 90.
16. Markley, *Far East*, 44.
17. The concept of the "Indies" is one that defies cartographical certainty and becomes, in both Dryden's play and Monardes's treatise, a matter of an interesting geographical displacement. For Monardes, the Indies reflect New Spain, while for Dryden, mindful of the tragic outcome of the incident on Amboyna, the Indies is the province of rival Dutch merchants. It is only when notions of plantations and groves are conflated that he claims some purchase on this coveted entity.
18. Markley, *Far East*, 143.
19. Dryden, *Amboyna*, 1.2.258.

20. María M. Portuondo, *Secret Science: Spanish Cosmography and the New World* (Chicago: Chicago University Press, 2009), 52. Here one wonders what Dryden knew, although the European struggle to create instruments that could accurately determine longitude at sea without relying on the often wayward inclinations of dead reckoning and lunar distance was certainly no secret. The Spanish navigation manual was indebted to the Portuguese tradition of *roteiros*, the use of astrolabes to calculate latitude, and together with Martin Fernández de Enciso's *Summa de geographia que . . . trata largamente del arte de marear,* Iberian cosmographers had quite a reputation for controlling sea travel, something Dryden would certainly have known.
21. Dryden, *Amboyna*, 5.1.1123.
22. Dryden, 2.1.
23. I recognize that this is an overdetermined word to use in this context considering the conflation of the spice(s) and Julia's body, but it just so happens that the fruit of the nutmeg is called an ovary.
24. Although Dryden converted circa 1685, a shift in religious thinking that would clearly question the Protestant origins of a providential god, *Amboyna* was staged in 1673, well before the death of Charles II (Dryden's patron) and his deathbed conversion that probably influenced Dryden's own relationship to the church.
25. See Ashley L. Cohen, *The Global Indies: British Imperial Culture and the Reshaping of the World, 1756–1815* (New Haven: Yale University Press, 2020). Cohen focuses on strategies of world-making (or "reshaping the world") from the mid-eighteenth to the mid-nineteenth centuries. What's interesting is how her argument for the two Indies in fact has a much earlier historical precedent.
26. Dryden, *Amboyna*, 4.1.857.
27. Dryden, 3.3.959.
28. Dryden, 5.1.395.
29. There is some suggestion of reading this scene as an act of *suttee*, or widow sacrifice, with which Ysabinda, as "an Indian Lady," may have been familiar.
30. See Heidi Hutner, *Colonial Women: Race and Culture in Stuart Drama* (New York: Oxford University Press, 2001); and Bridget Orr, *Empire on the English Stage, 1660–1714* (Cambridge: Cambridge University Press, 2001).
31. Markley, *Far East*, 143.
32. Elizabeth Maddock Dillon, *New World Drama: The Performative Commons in the Atlantic World, 1649–1849* (Durham: Duke University Press, 2014), 68.
33. Dryden, *Amboyna*, IV.i.
34. Grove, *Green Imperialism*, 95.
35. Markley, *Far East*, 143.
36. The Levant Company trade was, however, relatively free of this rivalry; by the 1620s, Britain had become the Ottoman Empire's biggest trading partner

in Europe. For more on the Ottoman influence on British colonialism, see Gerald Maclean and Nabil Matar, *Britain and the Islamic World* (Oxford: Oxford University Press, 2011).
37. Grove, *Green Imperialism*, 53–54.
38. Grove, 54. The Dutch farmed out these technologies of environmental control to their holdings in the Moluccas. According to E. C. Spary, the Dutch "policed these crops [of nutmeg and cloves] with naval precision at the centers of their spice trade, the islands of Amboina and Banda, and exterminated plants growing wild elsewhere." Spary, "Of Nutmegs and Botanists," 189–90. Part of the ways in which the Dutch managed to establish a monopoly on the spice trade throughout most of the seventeenth century depended on judicious destruction (to keep prices high) and reconstruction: a prudent attention to the microclimatic conditions of the Banda Islands, where cloves and nutmegs flourished.
39. Grove, *Green Imperialism*, 52.
40. Timothy J. Reiss, *The Discourse of Modernism* (Ithaca: Cornell University Press, 1982), p. 30.
41. Ramachandran, *Worldmakers*, 7.
42. Reiss, *Discourse of Modernism*, 30.
43. Grove, *Green Imperialism*, 25.
44. Spary, "Of Nutmegs and Botanists," 203.
45. Cohen claims that the Seven Years' War constitutes a "watershed in the history of Indies pairing," which is the "jumping-off point for [her] study." As I've pointed out, the historical precedent for this pairing can be read in Dryden's drama. Cohen, *Global Indies*, 4.
46. See Trevor Burnard, *Mastery, Tyranny, and Desire* (Chapel Hill: University of North Carolina Press, 2004); and Larry Gragg, *Englishmen Transplanted: The English Colonization of Barbados, 1627–1660* (Oxford: Oxford University Press, 2003).
47. Grove, *Green Imperialism*, 72.
48. One could argue that Grove shackles "improvement"—or the capitalization of agriculture—to "the Protestant Reformation," a Whiggish argument that could lead to English Protestant triumphalism vis-à-vis "improvement." For more on trading networks, the commercial empire that proceeds governmental empire, see Jonathan Israel's essay "The Emerging Empire: The Continental Perspective, 1650–1713," in "The Origins of Empire," *The Oxford History of the British Empire*, vol. 1, ed. Nicholas Canny (Oxford: Clarendon Press, 1998).
49. Ramachandran, *Worldmakers*, 2.
50. James C. Scott, *Seeing like a State* (New Haven: Yale University Press, 1998), 3.
51. Scott, 18.
52. Ramachandran, *Worldmakers*, 2.

53. Cathy Caruth, "Unclaimed Experience: Trauma and the Possibility of History," *Yale French Studies* 79 (1991): 182, accessed January 9, 2022, http://www.jstor.org/stable/2930251.
54. Grove, *Green Imperialism*, 61.
55. Dillon, *New World Drama*, 7.

Geomythography
A Genealogy

TOBIAS MENELY

Might the new science have begun not only in experimentalist procedure but also in the critical interpretation of ancient myth? In *De sapientia veterum* (1609), Francis Bacon reads thirty-one classical myths as partial—"shadowed"—but prescient anticipations of the new learning.[1] "Poetical fables," for Bacon, borrowing Marcus Terentius Varro's influential periodization, come from the "mythic" period between "earliest antiquity," known only from scripture, and classical antiquity, from which we have written records (317). Stories circulated in a "pre-literate age" before they were recorded by Hesiod, Homer, and Ovid.[2] "These fables," Bacon writes, were "delivered down and related by those writers, not as matters then first invented and proposed, but as things received and embraced in earlier ages" (320). Myths are stories retold and remade, fragmenting and accreting over time so that "sometimes a piece of history or other things are introduced, by way of ornament; or, ... the times of the action are confounded; or, ... part of one fable be tacked to another; ... for all this must necessarily happen, as the fables were the inventions of men who lived in different ages, and had different views" (319–20). Bacon recognizes an affiliation between the new learning and the pagan fables, which unlike scripture are defined not by a single authoritative text but by an openness to revision and reconfiguration. The old myths are more like science, human knowledge negotiated as it circulates, than religion, divine knowledge revealed and passed down unchanged.

This very plasticity also explains the difficulty of interpreting myths. Since "fables" are "composed of ductile matter," any "witty talent" can attribute to them a range of invented but "plausible meanings" (317). Earlier mythographers, from Euhemerus of Messene to Niccolò Machiavelli, had projected historical or political meanings onto the pagan legends, reflecting their own allegiances. Correctly interpreting the "allegory" in these myths, according to Bacon, requires recognition that they serve "two different and contrary ends": "to instruct or illustrate" and to "wrap up and envelop" (321).³ In mythic tales, knowledge of the world was given a pleasing form. For Bacon, these two ends work together, and "allegory, metaphor, and allusion" remain "useful, and sometimes necessary in the sciences" (321–22). Bacon ranges widely in his exegeses, considering myths with civic meanings and others with cosmological significance. Stories recounting the birth of Pan, for example, reconcile two accounts of creation as beginning in the "divine word," embodied by Mercury, or "the confused seeds of things," represented by "Penelope and all her suitors" (335). The "Destinies," the sisters of Pan, stand for "the natures and fates of things . . . as the chain of natural causes links together the rise, duration, and corruption; the exaltation, degeneration, and workings; *the processes, the effects, and changes, of all that can any way happen to things*" (336; emphasis mine). While some scholars have asked how Bacon's valorization of ancient wisdom accords with his bias toward modern learning, Paolo Rossi identifies the specific intellectual commitment that motivates Bacon's recovery of mythic knowledge. Bacon saw classical myth as encoding a "pre-Socratic naturalism and Democritean materialism" that could counter scholasticism and provide inspiration for the new science as it sought to uncover nature's hidden secrets.⁴

In this essay, I explore the intellectual genealogy of a curious form of inquiry, geomythography, a history that unsettles familiar stories about discipline formation. I first turn to seventeenth- and eighteenth-century savants beginning to investigate Earth's history, natural philosophers who looked not only to rocks and shells and strata but also, following Bacon's lead, to what Robert Hooke called "the Mythologick Stories of the Poets."⁵ I focus on the development of a historicist method, which digs beneath surfaces, whether lithic or literary, based on the principle that formal layering, displacement, and unconformity provide evidence of change over time. There is a resemblance, or what Pratik Chakrabarti calls a "methodological consilience," between the allegorical approach to myth, according to which a narrative is defined by the relation between its manifest surface and hidden depth (especially when this break is, as for Bacon, itself an expression of historical change), and a corresponding idea in the earth sciences that planetary history can be read in the relation between strata.⁶ As my genealogy of mythography moves from the seventeenth century to the present, I highlight the emergence of a symptomatic disagreement about how to interpret the latent content of myth as having a planetary or a social referent. For Bacon, either interpretation is available: the "allegory" may be "turned" toward "natural philosophy" or "morality or civil policy" (320). Later thinkers, however, stabilize the allegory by assigning its latent content either a physical or social significance. In the theory of myth, we witness the emergence of Bruno Latour's "modern constitution," as the study of human history and planetary history forge their separate paths. I conclude, then, by suggesting that the task of geomythographic inquiry in the Anthropocene, as we confront the catastrophic inseparability of planetary and human history, is to learn to read in the stories we tell, and retell, a record of our vexed inhabitation on a planet with its own elemental energies and capacity for cataclysmic change. Perhaps inspired by the example of scientific geomythography, literary studies, in particular, might expand its understanding of the modes in which stories circulate and of the socioecological meanings they negotiate.

Many of the seventeenth-century savants forging new approaches to earth history followed Bacon in reading ancient myth as the first science, an anticipation of modern learning. William Poole notes the importance of Ovid's *Metamorphoses* in Bernhardus Varenius's influential account of geomorphological vicissitude in *Geographia Generalis* (1650).[7] At the end of *Prodromus* (1669), the Danish cleric Nicholas Steno cites pagan myth to support a history of geological change after the universal deluge. While, Steno observes, many ancient legends reflect the imperative to "celebrate the actions of renown'd Men," they also provide a record of the sort of "Mutations"—"Earthquakes, Eruptions of Fires, Inundations of Rivers and Seas"—that a philosopher may never experience in person: "Of that kind I there find divers things, whose falsity rather than verity seems dubious to me; such as are that the Mediterranean Sea was sever'd from the western Ocean; that there was a passage out of the Mediterranean into the Red Sea; the submersion of the Atlantis Island; And the Description of so many places in the Expeditions of Bacchus, Triptolomus, Ulysses, Æneas, and others, may be true, though it agree not with things as they are at this day."[8] That the mythic landscape fails to correspond with present-day geography attests not to the falsity of myth but to the transformations of Earth's surface that have occurred in the intervening time.

Hooke complained that Steno was unfairly credited with geological insights that Hooke himself had first proposed in his lectures given to the Royal Society between 1668 and 1693. It seems just as likely, though, that when Hooke began devoting attention to "the Fables of the Poets" (376) in his later lectures, he was inspired by Steno. Hooke was a polymath of unequaled range, known today for his *Micrographia* (1665) and his dispute with Isaac Newton over precedence in formulating the inverse square law of gravity. Paolo Rossi, Ellen

Tan Drake, and other intellectual historians have worked to recover Hooke's centrality to the emergence of earth science: his impressive record of correct if speculative hypotheses about planetary history (including sedimentary stratification and superpositioning, the extinction and evolution of species, and even tectonics) and his role in modeling a historicist approach to the Earth as an object defined by change through time.[9] The enigma at the center of Hooke's lectures, assembled and published in the 1705 edition of his *Posthumous Works*, was the presence of marine fossils in upland strata. Hooke argued for the organic provenance of these "figured" stones, a controversial topic among the virtuosi. Fossilized organisms might be extinct species, Hooke proposed, or even the ancestors of living species that have changed due to climatic shifts. Bodily metamorphoses follow alterations of the Earth, the "differing Latitude of places in differing Ages ... the differing Altitudes of Places ... and consequently of changing the Nature, Soil, Climate &c. of the superficial Parts of the Surface" (372). In accounting for changes to Earth's surface and so the conditions of life, Hooke accepted the role of Aristotelian "vicissitude" over extended temporal durations, but he emphasized sudden cataclysmic events. The central agents in producing the "Alterations and Changes of the superficial Parts of the Earth" (321) were earthquakes, an expansive category for Hooke that includes any lithic disturbance, tremors as well volcanoes, landslips, and even slow erosion.

Hooke's hypotheses—always verging on heterodoxy—elicited skepticism from his learned colleagues, so in lectures beginning in spring 1687, he turned to a new type of evidence. As he observes, "One of the most considerable Objections I have yet heard is, that History has not furnish'd us with Relations of any such considerable changes as I suppos'd to have happen'd in former Ages of the World" (372). Yet, he continues, ancient myth is full of such accounts. What is Plato's narrative of the destruction of Atlantis ("a true History and not a Romance") if not a record of a "prodigious Earthquake and Inundation which happened in a Day and a Night, the Earth cleaving swallow'd up all those War-like Men" (374)? Myth, in Hooke's account, provides such a rich archive because it integrates a "Cosmography" with an empirical record of geological and historical events, some of which have been forgotten and some of which match up with known chronologies (395). As he writes, "This Mythologick History was a History of the Production, Ages, States and Changes that have formerly happened to the Earth, partly from the Theory of the best Philosophy; partly from Tradition, whether Oral or Written, and partly from undoubted History" (384).

As for most early modern mythographers, Hooke's main source of pagan myth was the *Metamorphoses*.[10] In turning to the "Hypothesis in *Ovid*," Hooke speculates about the process of transmission and, in doing so, explains the obscurity of myth (377). One of Ovid's source texts, Hesiod's *Theogonia*, integrates records of "real and actual Catastrophies" that had been passed down from the Egyptians, the Phoenicians, or perhaps from Orpheus himself (394). What remains from this "dark and uncertain" premythic time before Hesiod and Homer are "fragments." Any "written" texts, "being committed to small and perishable Substances . . . have been more easily drowned and swallowed by time, or buried and overwhelmed with the Dust of Oblivion" (394). The absence of records of even earlier stories, other than as obscure fragments within later stories, supports Hooke's argument for the Earth's cataclysmic mutability. The ancient poets, who are to be "accounted modern" with respect to earlier bards, would have likely had better access to "these more ancient Histories or Traditions than what we now can find," integrating them into their works much as the ancients erected "great Buildings" with the ruins of earlier civilizations. A myth is a story that has been broken or worn away before being reclaimed, "like Structures made up and pieced of the Rubbish, Ruins, and Fragments of . . . Antiquities," its very allegorical "obscurity" testifying to the catastrophic history it also preserves (394).

Ovid's *Metamorphoses* integrates all the "Records of Antiquity concerning the Changes and Catastrophies that had happened to the Earth from the Creation unto his own time" (394). This is the geohistorical implication of Ovid's opening assertion that he will treat universal principles of change. Hooke, rather prosaically, translates Ovid's famous opening lines ("My mind is bent to tell of bodies changed into new forms") thus: "My design in the Book is to speak concerning the various alterations which the Bodies or superficial Parts of the Earth have, by the Divine Powers, undergone" (395).[11] The myth-making poet, in Hooke's reading, is a protoscientist. During his late lectures over the course of a half decade, Hooke developed geological "explications" of Ovid's most famous characters and stories, including the formation of the Earth, the four ages, and the primeval flood survived by Deucalion and Pyrrha as well as the tales of "*Daphne* turned into a Laurel by *Apollo*" ("nothing but the pleasant verdures the Sun produced upon the Earth" [394]), "the Rape of *Proserpina* by *Pluto*" ("some dreadful Earthquake that had formerly happened in Sicily" [402]), Perseus ("hot inflamed Air or Lightening" [397]), Andromeda ("an half drowned and Rocky Country, by turns overflowed with the Sea at High Water and covered with Sand" [401]), and Atlas (a "prodigious Mountain" [400]).

Beyond his somewhat fanciful readings, Hooke also contributed to the theory of myth, particularly in his discussion of personification, the metaphorical displacement of natural forces onto humanlike figures. Hooke observes that in the opening verses of the *Metamorphoses*, "the Poet gives us a short History of the formation of the Earth . . . nor has he yet Personated or Mithologized any thing" until he "begins calling the Winds Brothers. . . . The Sense of the rest is plain till the eighty second Verse, where he began again to personate Actions Mythologically" (378). Ovid, Hooke speculates, shifts from a literal to an allegorical treatment of the Earth, as person-like, once the Earth begins to undergo change in time: "He had spoken of things as Dead and Unactive Earth, but from hence he will describe the Earth as changed and clothed with the various shapes of Men and Persons, and so having described the Formation or first Generation of all things Physically and plainly, he comes next to tell the Age or Ages of the World, and what Periods of Life or Being it hath had, and the States it hath been during those several Periods" (379). "Personation" is a way of knowing an entity that is subject to alteration, persisting even as it undergoes change, acting even as it is acted upon. Hooke pushes further, suggesting that all of the various personified figures in the *Metamorphoses* are displaced figures of elemental forces: "All these Poetical Expressions, which the Author seemeth to speak, as of Men, and their Actions, and Enjoyments, I take to be significative of all acting Powers of the Earth" (379). Extravagant personifications, when properly interpreted, are revealed as records of an eventful, changeable, and variegated Earth.

Hooke characterizes the allegorical quality of myth as an expression of the catastrophic discontinuity of history but also as a way of narrating dynamic change through time and making such history memorable (396). He offers a sociological explanation as well, directed to his audience, the savants of the Royal Society. Myths reflect the emergence of a division of knowledge. In order to "conceal their Knowledge," he proposes, "Fraternities" of "Adepti" "contrived and digested" it "into fabulous Stories," which could "amuse and awe the Vulgar" while also serving "to instruct and inform" (394). Whatever the cause of its allegorical obscurity, for Hooke a myth passed down and reconfigured through time is much like a fossilized shell embedded in upland strata: an enigmatic historical object demanding interpretation. A "rotten Shell" may seem like a "trivial thing," but such objects are in fact "Records of Antiquity which Nature have left as Monuments and Hieroglyphick Characters of . . . Transactions of the Body of the Earth" (411). In reading the hieroglyphics of a fossilized shell or allegorical tale, the natural philosopher begins to map the layers of Earth's history. And these lithic and literary archives are best read in conjunction with each other: "And tho' it must be granted, that it is very difficult to read them, and to raise a *Chronology* out of them, and to state the intervals of the Times wherein such, or such Catastrophies and Mutations have happened; yet 'tis not impossible, but that, by the help of those joined to other means and assistances of Information, much may be done" (411).

Giambattista Vico's *New Science,* appearing in a series of editions in the 1720s, inaugurates a major development in the theory of myth. For Bacon and the early virtuosi, myth was specifically associated with pagan antiquity; the problem of myth is the problem of the relation of ancient wisdom to modern learning. With his references to Indian, German, and Native American myth, Vico redefines myth as a comparativist problem. In doing so, he claims the mythic imagination for a science of culture. Myths are not so much the first science as the first politics. Vico begins with the axiom that myth extrapolates from what is known most intimately: "When men want to create ideas of things of which they are ignorant, they are naturally led to conceive them through resemblances with things that they know. And when there is a scarcity of known things, they judge the things of which they are ignorant in accordance with their own nature. Hence, since the nature that we know best consists in our own properties, men attribute to things that are insensate and inanimate, movement, sense and reason, which are the most luminous labours of poetry."[12] Before we know the world, we know ourselves as sensate and animate beings. Myth is produced in the projection of this self-knowledge onto an enigmatic world. The "first fable of all" is that of some proto-Jove, a hurler of thunderbolts. Humans in the state of nature "expressed their passions by shouting, grunting, and murmuring." Since they were "ignorant of the causes of the thunderbolts," they imagined that "the sky was a vast, animate body which, by shouting, grunting and murmuring, spoke and wanted to communicate with them" (152). The axiom Vico takes from this example is that myths are "born . . . wholly ideal, in that the idea of the poet gives things all the being that they lack" (152). This is why myths ultimately always express "civil truths." Myths are the social body refracting its organizing principles because the social world is what we know as clearly as we know our own animacy. Vico identifies a categorical difference between the laws governing physical nature and those governing human societies, which develop according to autonomous principles reflected in his three ages of myth: the theological, the heroic, and the human (focusing on deities, on man as deity, and on man alone).

In the second half of the eighteenth century, the philosophe Paul-Henri Thiry, Baron d'Holbach, revisited and refined the approach to myth developed by the seventeenth-century virtuosi. In *Système de la Nature* (1770), he describes mythology as "the daughter of natural philosophy, embellished by poetry."[13] Whereas Vico expresses surprise that primitive philosophers would study natural conditions, from Holbach's perspective as a materialist, it is obvious that the natural world, the foundation of all human productive activity, would be the primary object of observation and reflection: "The observation of nature was the first study of those who had leisure to meditate: they could not avoid being struck with the phenomena of the visible world. The rising and setting of the sun, the periodical return of the seasons, the variations of the atmosphere, the fertility and sterility of the earth, the advantages of irrigation, the damages caused by floods, the useful effects of fire, the terrible consequences of conflagration, were proper and suitable objects to occupy their thoughts" (175). Early thinkers delineated the forces of nature according to "their own peculiar energies," characterizing them in terms of their risks and benefits to human projects. Myth expresses an imperative to apprehend natural forces by familiarizing them. According to Holbach, priests "began to make a distinction between Nature and the energy in nature," turning the elemental forces into "an incomprehensible being." These bardic priests, seeking to address "the imagination of their audiences," treated natural forces as "fictitious personages": "the igneous matter, the ethereal electric fluid.... The source of heat, was deified under the name of Jupiter" (181). Mythic stories were passed down from generation to generation until people forgot "it was nature ... that lay buried under a heap of allegories" (180). Religion supplants myth when its adherents forget that the deity is merely a "personation" of the "energy in nature." This forgetting, though, expresses a social function: the "interests" of the priests in accruing authority as interpreters. The "first institutors of nations ... only spoke to the people by fables, allegories, enigmas, of which they reserved to themselves the right of giving an explanation" (179). In addition to this process of forgetting, Holbach, like Hooke, describes the transmission of myth as a process that registers the Earth's catastrophic history: "Those who were able to escape from the ruin of the world, filled with consternation, plunged in misery, were but little conditioned to preserve to their posterity a knowledge, effaced by those misfortunes of which they had been both the victims and the witnesses: overwhelmed with dismay, trembling with fear, they were not able to hand down the history of their frightful adventures, except by obscure traditions" (176). The obscurity of myth testifies to the "universal ravages" that have affected the Earth. The dangerous dynamism that is the first subject of myth is also the force that explains why the real meaning of myths has been forgotten.

Even as they pioneered empirical methods of geological study and began extending planetary chronology far back into prehuman deep time, earth scientists did not altogether abandon geomythographic inquiry. Indeed, a century after Hooke, theories of rock formation had separated into two camps distinguished by mythic appellations: Plutonism (igneous formation) and Neptunism (aqueous formation). Colonial expansion, scientific expeditions, and early ethnographic research shifted attention from classical myth to the oral traditions encountered in Indigenous societies. In *Deep Time: A Literary History*, Noah Heringman recounts how Johann Reinhold Forster and Georg Forster, the father-and-son naturalists on the second James Cook voyage, integrated ethnographic findings with speculations about the human and geologic history of the Pacific. Johann Reinhold Forster notes that the "mythology" of Tahitians and Society Islanders includes a deified figure, Maui, "who in his anger shakes the earth." He cites this myth to support his conjecture that "these isles were raised by earthquakes and fire," and he speculates that the legend of Maui "dragging . . . the land through the sea" records the earthquake and flood that formed the archipelago.[14] The development of an "ethnographic attitude toward myth," in Heringman's words, provided a framework authorizing "its application to natural knowledge."[15] In *Inscriptions of Nature*, Pratik Chakrabarti extends this story of scientific geomythology into nineteenth-century India, where archaeologists, geologists, and colonial administrators cross-referenced myths, ruined monuments, and rock formations to reconstruct the human history of the subcontinent. Geology in the nineteenth century remained a "hybrid enterprise" in which ethnographic approaches to myth promised access to human prehistory, a rescaling of time that provided an imaginative precondition for geological deep time.[16]

Even as geology developed into a specialized discipline, geologists continued to read myth as an archive of planetary eventfulness. The American volcanologist Dorothy Vitaliano, credited with coining the term *geomythology*, published her groundbreaking *Legends of the Earth: Their Geologic Origins* in 1973. In 2007, an extensive collection, *Myth and Geology*, was published by the Geological Society of London. In an introductory essay, W. B. Masse et al. offer an overview of twentieth-century theories of myth, noting that "none" of the predominant accounts—psychological, sociological, structuralist, historicist—"is seemingly willing to suggest that a real observed natural process or event may lie at the core of myth storylines."[17] Taking up the project initiated by Hooke, the contributors to the collection explore a multitude of myths that seem to record geological events. The Klamath Indians tell of a battle between two chiefs who hurled rocks and flames at each other until the chief standing atop Mount Mazama was vanquished, which sounds like an account of the eruption 6,500 years ago that led to the formation of Crater Lake. Floods are a frequent subject of myth and a fertile subject for geomythography. Scientists searching for an event corresponding with the Noahic deluge have identified a possible catastrophic inflow of the Mediterranean into the Black Sea, a highly controversial hypothesis. In Australia, Aboriginal oral traditions describe objects falling from the sky and forming craters, possibly recording a 4,700-year-old meteor impact. Masse et al. observe that the "myth of the destruction of Atlantis"—Hooke's starting point—may be "a refracted image of the supereruption of the Thera volcano around 1625 BC" (16). *Myth and Geology* has been cited nearly 150 times, attesting to an active and ongoing scholarly conservation.

The geologist reads myth to "extract" records of unusual geological events.[18] Anthropologists face a more difficult interpretive task: to read narrative traditions—oral, written, and performative—as a privileged source of knowledge about the long history of human migration, settlement, and land-use practices. Stories, for the anthropologist, are a record of "geohistory," the inseparability of planetary eventfulness and human world-making. In *Yuganta*, for example, Irawati Karve developed a revisionary reading of the *Mahabharata* as recording the decisive role of deforestation and the displacement of forest-dwelling people for agriculture and grazing in the Gangetic plain.[19] In *Do Glaciers Listen?*, Julie Cruikshank, drawing on three decades of fieldwork with Tlingit and Athapaskan storytellers, describes how oral traditions record "geophysical changes associated with late stages of the Little Ice Age and European colonial incursions." While European scientists sought to distinguish the physical from the social world, she notes, Indigenous storytellers "explore the connections between nature and culture."[20]

Despite the tradition of geomythographic inquiry I have explored in this essay, myth is, today, a devalued term, defined in opposition to scientific knowledge, connoting false belief (e.g., the myth of the free market—a "myth," it should be noted, that itself has profound geohistorical implications). Yet, as Matthew Spellberg writes, myth is best understood as a form of collective knowledge that flourishes in oral societies, an "epistemology which uses stories for its procedures of reasoning, argument, and verification."[21] Or, in the words of Robert Bringhurst, a "myth is a theorem about the nature of reality, expressed not in algebraic symbols or inanimate abstractions but in animate narrative form."[22] The early geomythographers picked up on a crucial quality of myth emphasized also by Spellberg and Bringhurst—its reliance on retelling, its availability to revision and change—that underlies its affinity with science as a mode of knowledge. It is appropriate to keep this understanding of myth in mind, in tension with a definition of myth as ideological falsehood, if we are to begin to consider the geological significance of myth in the Anthropocene, particularly in instances where "Indigenous storywork" remains relevant in conflicts over sovereignty, knowledge, and practices of environment-making.[23]

In Aotearoa New Zealand, for example, the Te Awa Tupua (Whanganui River Claims Settlement) Act of 2017, recognizing the Whanganui River as a legal person, and the Te Urewera Act of 2014, recognizing a densely forested region of the North Island as a legal person, drew on Māori storywork—"ancestral cosmological ideas"—according to which the river and the forest are respected ancestors.[24] A recent article by Priscilla Wehi et al. explores oral traditions attesting to Polynesian exploration of Antarctic waters going back to the seventh century, a project of recovering narratives of Māori precedence and Māori principles—such as whakapapa (genealogy) and kaitiakitanga (guardianship)—with implications for claims of sovereignty and governance today.[25] A 2019 double issue in the *New Zealand Science Review* examined the difficult but productive relation between Western science and mātauranga Māori (Māori knowledge). Georgina Tuari Stewart describes her work integrating Māori cosmology into secondary science education, starting with "traditional accounts of Rangi and Papa (Father Sky and Mother Earth) and their many godly children . . . who act as guardians and metaphors for knowledge of the different elements and domains of the natural world."[26] Mātauranga Māori, which integrated the sophisticated ancestral knowledge of the skies and the ocean used by the Polynesian star navigators, developed over a millennium of exploring and inhabiting Aotearoa. It has relevance today for Māori cultural identity but also in environmental risk management and biodiversity conservation. Doug Jones et al. give the example of a highway that was redesigned to avoid encroaching on the lair of Karu-tahi (a taniwha, a dragon-like creature). A year after the completion of the highway, "a flood inundated the lair of Karu-tahi, but the redesign ensured that the expressway was not threatened."[27] In conflicts over science curriculum and land-use policy, such Indigenous storywork is generally not referred to as myth, given the dominant connotation of myth as either ideological falsehood or archaic stories. Indeed, the astronomer Rangi Mātāmua, in response to recent controversy over the status of Mātauranga as a science, asserts, "We did not navigate to Aotearoa on myths and legends. We did not live successfully in balance with the environment without science. Māori were the first scientists in Aotearoa."[28] In this essay, I have sought to unsettle this opposition between science and myth, showing how Earth science has recognized myth as a valuable record of planetary eventfulness and even as a prototype for scientific ways of knowing.

Compared with geology and anthropology, the discipline of literary studies has less of a place for geomythography—and, indeed, for myth itself. Scholars of modern literature define their primary objects of inquiry—the book, the text, reading, authorship, fictionality, realism, style, irony—in opposition to myth. As Spellberg observes, it is difficult to understand "an oral epistemology" when "our entire tradition of analysis is based on the conceptual language of the text."[29] It may be that a scholarly discipline organized around the study of literature rather than myth is fatally diminished—as an ecocritical method and as a decolonial praxis—insofar as it separates textual interpretation from a broader analysis of "storywork," the always ongoing social recirculation and remaking of old stories in new media but also in oral storytelling practices and ritual performance.

The twentieth-century critic who made the strongest case for the continuity between myth and literature was Northrop Frye. For Frye, literature recycles structural patterns such as the quest narrative and the trickster figure from the mythic past. As for many of the myth theorists I have discussed, what defines myth for Frye is its availability to retelling, a dialectic of recurrence and revision. Frye's account of myth, however, is thoroughly Viconian:

> Mythology . . . belongs to the world of culture and civilization that man has made and still inhabits. As a god is a metaphor identifying a personality and an element of nature, solar myths or star myths or vegetation myths may suggest something of a primitive form of science. But the real interest of myth is to draw a circumference around a human community and look inward toward that community, not to inquire into the operations of nature. Naturally it will draw elements from nature, just as creative design in painting or sculpture would do. But mythology is not a direct response to the natural environment; it is part of the imaginative insulation that separates us from that environment.[30]

Myth, for Frye, is passed down from oral societies and absorbed into literate civilization not as the first science, an attempt to know nature, but as culture, a symbolic accompaniment to the work of cultivating and forging a human world.

The shock of the Anthropocene—the disappearance of this supposed "insulation" between human culture and planetary environment—reveals the inadequacy of the Viconian theory of myth. Consider the insistent return of myth in recent climate change fiction, the reassertion of the resonance of myth and mythic storytelling in novelistic worlds representing climate breakdown: the Khembalung myth in Kim Stanley Robinson's *Green Earth* trilogy (2004–7), the Great Mystic Points in Nnedi Okorafor's *Who Fears Death* (2010), the Aboriginal myth in Alexis Wright's *The Swan Book* (2013), the "stone lore" in N. K. Jemisin's *Broken Earth* trilogy (2015–17), and the Manasa Devi legend in Amitav Ghosh's *Gun Island* (2019). In *Anatomy of Criticism*, Frye addresses the "presence of mythical structure in realistic fiction," which he defines in terms of the operation of "displacement."[31] Literary realism picks up on something necessary in myth, reflecting an inadequacy in realism, but displaces that mythic mode of representation to make it compatible with realism's probabilistic standards of verisimilitude and logics of causality. Perhaps, though, displacement works the other way, such that the dialogic reassertion of the mythic as a stratum of meaning within realist fiction facilitates representations of the Earth's elemental energy and cataclysmic potential. After all, as I have argued here, there is a long history of reading myth as precisely the narrative impulse to know a dynamic and dangerous Earth. As a final example, consider Octavia Butler's pioneering work of climate fiction, *Parable of the Sower* (1993), in which the young protagonist, Lauren Olamina, displaced from her Southern California home by climate-induced social collapse, develops a mythic narrative, Earthseed, in order to bring together a community of survivors. Earthseed begins "The only lasting truth / is Change."[32] Lauren responds to the catastrophic conjunction of human history and planetary history by telling a new story. Yet this story expresses the same impulse that led Ovid to retell old myths—"My mind is bent to tell of bodies changed into new forms"—in a manner that Hooke read as conveying the "Records of Antiquity concerning the Changes and Catastrophies that had happened to the Earth from the Creation unto his own time."

Notes

1. Francis Bacon, *Essays and Wisdom of the Ancients* (Boston: Little, Brown, 1884), 319; hereafter cited parenthetically in text.
2. Rhodri Lewis, "Francis Bacon, Allegory, and the Use of Myth," *Review of English Studies* 61, no. 250 (2010): 383.
3. Gordon Teskey reads a conscious "irony"—awareness "of the convenience of what one decides to believe"—in Bacon's distinction between the faulty readings of earlier mythographers and his own interpretations of the pagan myths as anticipations of the new science. Teskey, *Allegory and Violence* (Ithaca: Cornell University Press, 1996), 88.
4. Paolo Rossi, *Francis Bacon: From Magic to Science*, trans. Sacha Rabinovitch (London: Routledge & Kegan Paul, 1986), 84. See also Gerard Passannante, *Catastrophizing: Materialism and the Making of Disaster* (Chicago: University of Chicago Press, 2019), 154–57.
5. Robert Hooke, "Lectures and Discourses of Earthquakes, and Subterraneous Eruptions," in *The Posthumous Works* (London, 1705), 372; hereafter cited in text.
6. Pratik Chakrabarty, *Inscriptions of Nature Geology: and the Naturalization of Antiquity* (Baltimore: Johns Hopkins University Press, 2020); hereafter cited in text.

7. William Poole, *The World Makers: Scientists of the Restoration and the Search for the Origins of the Earth* (Oxford: Peter Lang, 2010), 99–101.
8. Nicolas Steno, *The Prodromus* (London, 1671), 108–9.
9. Ellen Tan Drake, *Restless Genius: Robert Hooke and His Earthly Thoughts* (Oxford: Oxford University Press, 1996). Paolo Rossi writes, "Slow, uniform changes, unforeseen catastrophes, the destruction and emergence of continents, the mutations within species, all spoke to Hooke of a history that is like a river of unknown course, which might even contain, in its past, treasures of wisdom now lost forever." Rossi, *The Dark Abyss of Time: The History of the Earth and the History of Nations from Hooke to Vico,* trans. Lydia Cochrane (Chicago: University of Chicago Press, 1984), 15. See also Gordon Davies, "Robert Hooke and His Conception of Earth-History," *Proceedings of the Geologists' Association* 75, no. 4 (1964): 493–98.
10. On Ovid's status as a touchstone in early science, see Samuel Galston, "Ovid's *Metamorphoses* and the Scientific Revolution" (PhD diss., Princeton University, 2016).
11. Ovid, *Metamorphoses* (Cambridge, MA: Harvard University Press, 1971), 1:3.
12. Giambattista Vico, *The First New Science,* ed. Leon Pompa (Cambridge: Cambridge University Press, 2002), 151; hereafter cited in text.
13. Paul-Henri Thiry, Baron d'Holbach, *The System of Nature,* trans. H. D. Robinson (Boston: J. P. Mendum, 1889), 178; hereafter cited in text.
14. Johann Reinhold Forster, *Observations Made during a Voyage round the World* (London, 1778), 1:158–59.
15. Noah Heringman, *Deep Time: A Literary History* (Princeton: Princeton University Press, 2023), 114.
16. Chakrabarty, *Inscriptions,* 19. According to Adeline Buckland, Lyell and Darwin called on the "geological imagination" to reveal the deep history and unified processes of the earth, but they limited such imaginative mastery to educated Europeans. Buckland, "'Inhabitants of the Same World': The Colonial History of Geological Time," *Philological Quarterly* 97, no. 2 (2018): 220.
17. W. B. Masse, "Exploring the Nature of Myth and Its Role in Science," in *Myth and Geology,* ed. L. Piccardi and W. B. Masse (London: Geological Society, 2007), 13.
18. Masse et al., 25. A. J. Nocek develops a sophisticated critique of this extractive hermeneutics in "Geology, Myth, Media," *SubStance* 47, no. 2 (2018): 84–106.
19. Irawati Karve, *Yuganta: The End of an Epoch* (New Delhi: Orient Longman, 1961).
20. Julie Cruikshank, *Do Glaciers Listen? Local Knowledge, Colonial Encounters & Social Imagination* (Vancouver: University of British Columbia Press, 2005), 10–11.
21. Matthew Spellberg, "Myth and Anarchy," *Yale Review* 107, no. 2 (April 2019): 93.

22. Robert Bringhurst, "The Meaning of Mythology," in *Everywhere Being Is Dancing: Twenty Pieces of Thinking* (Berkeley: Counterpoint, 2008), 63.
23. See Jo-ann Archibald et al., *Decolonizing Research: Indigenous Storywork as Methodology* (London: Zed Book, 2019).
24. See Anne Salmond, *Tears of Rangi: Experiments across Worlds* (Auckland: Auckland University Press, 2017), chap. 9. Salmond quotes the right-wing politician Rodney Hide on the settlement: "Who would have believed it? Singing a song can make a river yours" (298).
25. Priscilla M. Wehi et al., "A Short Scan of Māori Journeys to Antarctica," *Journal of the Royal Society of New Zealand* 52, no. 5 (2022): 587–98.
26. Georgina Tuari Stewart, "Mātauranga and Pūtaiao: The Question of 'Māori Science,'" *New Zealand Science Review* 75, no. 4 (2019): 66.
27. Doug Jones et al., "Weaving Mātauranga into Environmental Decision-Making," *New Zealand Science Review* 76, nos. 1–2 (2020): 51.
28. Quoted in "Scientists Rubbish Auckland University Professors' Letter Claiming Māori Knowledge Is Not Science," *New Zealand Herald*, July 28, 2021.
29. Spellberg, "Myth and Anarchy," 95.
30. Northrop Frye, *The Great Code: The Bible and Literature* (San Diego: Harcourt Brace Jovanovic, 1982), 37; emphasis mine.
31. Northrop Frye, *Anatomy of Criticism* (Princeton: Princeton University Press, 2000), 136.
32. Octavia Butler, *Parable of the Sower* (New York: Warner, 1995), 3.

CONTRIBUTORS

DAVID ALFF is Associate Professor of English at the University at Buffalo, State University of New York. He is author of *The Northeast Corridor* and *The Wreckage of Intentions: Projects in British Culture, 1660–1730*. He is currently finishing a monograph entitled *Rights of Way: A Literary Approach to Infrastructure*.

MELISSA BAILES is Associate Professor of English at Tulane University. She is the author of *Questioning Nature: British Women's Scientific Writing and Literary Originality, 1750–1830*, which won the British Society for Literature and Science Book Prize, as well as *Regenerating Romanticism: Botany, Sensibility, and Originality in British Literature, 1750–1830*.

FRANK BOYLE is an Associate Professor of English at Fordham University and co-chair of the New York Eighteenth-Century Seminar. He is the author of *Swift as Nemesis: Modernity and Its Satirist* and numerous articles on eighteenth-century topics. His current book on early neurology and literature is nearing completion.

LISA FORMAN CODY is Associate Professor of History at Claremont McKenna College. She is the author of *Birthing the Nation: Sex, Science, and the Conception of Eighteenth-Century Britons*, which won the Berkshire's Best First Book Prize among other awards. Her essays on the history of medicine, midwifery, gender, politics, visual culture, literature, and the law have received several article prizes, including the Walter D. Love Prize of the North American Conference on British Studies. Her most recent article, "'Marriage Is No Protection for Crime': Coverture, Sex, and Marital Rape in Eighteenth-Century England" (*Journal of British Studies*) received the James Clifford Award of the American Society for Eighteenth-Century Studies. Her current book on marital sensibility is nearing completion.

AL COPPOLA is Associate Professor of English at John Jay College, City University of New York, and the past chair of the Columbia University Seminar in Eighteenth-Century European Culture. His first book, *The Theater of Experiment: Staging Natural Philosophy in Eighteenth-Century Britain*, is a critical study of science in—and as—performance. His current book project, *Enlightenment Visibilities*, studies the eighteenth-century innovations that structure twenty-first-century modernity by bringing previously unimaginable or imperceptible phenomena into the domain of knowledge.

ERIC GIDAL is Professor of English at the University of Iowa. He is the author of *Ossianic Unconformities: Bardic Poetry in the Industrial Age*, *Poetic Exhibitions: Romantic Aesthetics and the Pleasures of the British Museum*, and a range of articles on English, Scottish, and French romanticism.

KRISTIN M. GIRTEN is Professor of English and Assistant Vice Chancellor of Research and Creative Activity at the University of Nebraska Omaha. She is the author of *Sensitive Witnesses: Feminist Materialism in the British Enlightenment* and coeditor (with Aaron R. Hanlon) of *British Literature and Technology, 1600–1830*.

AARON R. HANLON is Associate Professor of English and Chair of the Science, Technology, and Society Department at Colby College. He is the author of *A World of Disorderly Notions: Quixote and the Logic of Exceptionalism* and *Empirical Knowledge in the Eighteenth-Century Novel* and coeditor (with Kristin M. Girten) of *British Literature and Technology, 1600–1830*.

JESS KEISER is Associate Professor of English at Tufts University. He is the author of *Nervous Fictions: Literary Form and the Enlightenment Origins of Neuroscience*.

JAYNE LEWIS is Professor of English at the University of California, Irvine. The editor of the anthology *Religion in Enlightenment England*, she is the author of numerous articles and several monographs on eighteenth-century literature, religion, science, and culture, including *Air's Appearance: Literary Atmosphere in British Fiction, 1660–1794*.

RAMESH MALLIPEDDI is Associate Professor in the Department of English Language and Literatures at the University of British Columbia, Vancouver, and editor of *Eighteenth-Century Studies*, the official publication of the American Society for Eighteenth-Century Studies (ASECS). He is the author of *Spectacular Suffering: Witnessing Slavery in the Eighteenth-Century British Atlantic*. He has recently published articles and book chapters on Native American bondage, roads and transportation infrastructure, and land. His special issue of *The Eighteenth Century: Theory and Interpretation* on "Empire, Capital, and Climate Change" appeared in fall 2022. His current book on soil—*Expendable Lives, Disposable Lands*—is nearing completion.

TOBIAS MENELY is Professor of English at the University of California, Davis. He is the author of *The Animal Claim: Sensibility and the Creaturely Voice* and *Climate and the Making of Worlds: Toward a Geohistorical Poetics*, which received the Warren-Brooks Award and the Michelle Kendrick Prize. With Jesse Oak Taylor, Menely coedited *Anthropocene Reading: Literary History in Geologic Times*.

LAURA MILLER is Professor of English at the University of West Georgia and author of *Reading Popular Newtonianism: Print, the "Principia," and the Dissemination of Newtonian Science*. She is currently a co-investigator for Eighteenth-Century Libraries Online (ECLO), a library history database funded by the Arts and Humanities Research Council of the UK (AHRC). Her current book project, *Prescriptive Communities*, uses historical library records to reconstruct a history of preventive care and public health in the eighteenth century.

VIVIAN ZULUAGA PAPP is Doctoral Lecturer of English at New York City College of Technology, City University of New York. She is currently working on a book project that investigates vision, technology, the science of the novel, Enlightenment science writing, and AI.

ADELA RAMOS was, until recently, Associate Professor of Eighteenth-Century Studies at Pacific Lutheran University, where she focused on critical animal studies, ecofeminism, and the history of science. She is now Director of Partnerships and Communications at the Tacoma Tree Foundation, where she directs the science education and communications programs. She has published essays on Homero Aridjis, Henry Fielding, Jonathan Swift, and Mary Wollstonecraft. She just completed an essay on animals in Jane Austen adaptations for the forthcoming *Reading Jane Austen: An Introduction*, edited by Danielle Spratt.

ANNA K. SAGAL is Associate Professor of English and Creative Writing at Cornell College. She is the author of *Botanical Entanglements*, released by the University of Virginia Press in 2022. She has also written several articles on eighteenth-century literature and science, including articles on women and botany, marine ecology, women's periodicals, disability and the novel, and scientific epistles.

DANIELLE SPRATT is Professor of English and Director of Community Engagement at California State University, Northridge. With Bridget Draxler, she is coauthor of *Engaging the Age of Jane Austen: Public Humanities in Practice* and editor of the forthcoming *Reading Jane Austen: An Introduction*. She is also finishing a manuscript on theories of conception and reproduction in the literary imagination across the long eighteenth century.

RAJANI SUDAN is Professor of English at Southern Methodist University. She is the author of *Fair Exotics: Xenophobic Subjects in English Literature* and *The Alchemy of Empire: Abject Materials and the Technologies of Colonialism*. She is also coeditor of *Configurations*.

ANNE M. THELL is Associate Professor of English literature at National University of Singapore. She is author of *Minds in Motion: Imagining Empiricism in Eighteenth-Century British Travel Literature* and editor of Margaret Cavendish's *Grounds of*

Natural Philosophy. She is now coediting (with Lara Dodds) an edition of Cavendish's collected works for Oxford University Press.

HELEN THOMPSON is Professor of English at Northwestern University. She is author of *Fictional Matter: Empiricism, Corpuscles, and the Novel* and a range of articles or volume chapters discussing topics that include West African gold extraction, alchemy, and empiricist representation in long eighteenth-century scientific and fictional prose forms.

INDEX

Page numbers in italics refer to illustrations.

Abolitionist Society, 271–72
Abstract of the Evidence . . . before the House of Commons, 271–72
Addison, Joseph, 95
Aeschylus, 129
aesthetic education, 285
Aikin, John, *An Essay on the Application of Natural History to Poetry*, 291, 292
air pumps, 4, 52–55, 58–59, 111, 243, 247
Aït-Touati, Frédérique: *Fictions of the Cosmos*, 50; *Terra Forma* (with Alexandra Arènes and Axelle Grégoire), 49–50
Alaimo, Stacy, 241n34
alchemy, 165, 167, 169, 181, 182n5
Alff, David, 279
Amboyna burl, 311–13, 317, 319–20
analogy. *See* figurative language; metaphors
Anand, Nikhil, 285
anatomy, Barker and, 209, 210, 215–18
anemones, 295, 298, 300–301, *300*
animals: for agricultural labor, 261–62, 264–66, 273–75, 307; in metaphors for population, 188, 189, 192–94, 196–97, 200, 205n41; women and children associated with, 243–48, 253
animal spirits, 180, 186n81, 209, 218, 224n31
Anson, George, 107n28; *A Voyage Round the World*, 4, 89–105, *90*
Anthropocene, 7, 327, 334, 336

anthropology, 334–35
apparition novels, 54
Appel, Hannah, 285
Arbuthnot, John, 5, 170, 173–75, 182; *Essay concerning the Nature of Ailments*, 179; *Examination of Dr. Woodward's Account of the Deluge*, 173, 176; *Three Hours after Marriage* (with John Freind and John Gay), 175–78, 185n51
Aristotle, 53, 264
art: appeal of, 95–96; science in relation to, 91–92, 104–5; and the veracity of visual representation, 90–92, 94–96, 104–5
Asfour, Amal, 130, 131
Astell, Mary, *Defence of the Female Sex*, 209, 211
Athapaskan people, 334
Atlas (Representation of the Universal World), 318
atomism, 228, 232–33. *See also* corpuscles; micromatter; particles
Aubrey, John, 227
Austen, Jane, *Northanger Abbey*, 77–78
Australian Aboriginals, 334
authorship, 33–44

Backscheider, Paula, 210
Bacon, Francis, 4, 22–23, 126, 168, 217, 227, 229–34, 237, 325–27, 331, 337n3;

Bacon, Francis (*continued*)
  Advancement of Learning, 253; *De sapientia veterum*, 325; *Novum Organum*, 230
Bailes, Melissa, 293
Barbon, Nicholas, 190, 191, 193
Barker, Edward, 209, 211, 215–17, 219–20
Barker, Jane, 209–21; "A Farewell to Poetry with a Long Digression on Anatomy," 6, 210, 215–21, 223n30; "An Invitation to My Friends at Cambridge," 212–13; *The Lining of the Patch-Work Screen*, 223n19; "On the Death of my Brother. A Sonnet," 222n13; *A Patch-Work Screen for the Ladies*, 6, 209–11, 214, 216; *Poetical Recreations*, 209, 211, 211–21, 222n12; "A Virgin Life," 210–11, 211, 213–15, 223n19
Bartholin, Caspar, *Anatomicae Institutiones Corporis Humani*, 216
Battigelli, Ann, 241n46
Bauhin, Johann, 311
Bayley, William, 114–15, 117, 119–20
Beddoes, Thomas, 66, 75, 80n9
Behn, Aphra, 35
Bender, John, 93; *The Culture of Diagram* (with Michael Marrinan), 48–49, 60, 150–51
Bewick, Thomas, *History of British Birds*, 251
Bindman, David, 125
birds, women associated with, 243–54
Blair, William, *Essay on the Venereal Disease*, 75
Blake, Liza, 232, 241n34
Bleichmar, Daniela, 308
Bligh, William, *A Narrative of the Mutiny*, 120
Blith, Walter, *The English Improver Improved*, 262, *262*, *263*, 267
Board of Longitude, 110, 112–15, 121
Boerhaave, Herman, 178–80, 307
Borlik, Todd Andrew, 227
botany. *See* plants
Botiș, Florin, 59
Bouvier de Fontanelle, Bernard le, *Conversations on the Plurality of Worlds*, 50

Bowker, Geoffrey, 278, 285
Boyle, Robert, 4, 38, 46–47, 50–55, 59, 61, 62n19, 128, 165–70, 179, 182n5, 221, 309; *New Experiments Physico-Mechanicall*, 47, 50–55; *The Sceptical Chymist*, 165, 169
Braudel, Fernand, 265
Breton, André, 78
Brett, Peircy, 90–105, 107n28, 107n42; "The Burning of the Town of Payta on the Coast of Santa Fee," 96, 102, *103*; "Cape Virgin Mary," 101; "View of Patagonia," 101; "A View of Streights Le Maire between Terra del Fuego and Staten Land," 102, *102*; "A View of the Commodore's Tent at the Island of Juan Fernandes," 98, *99*; "View of the Watering Place at Tinian," 96, 103
Bridgens, Richard, *West India Scenery . . . from sketches taken during a voyage to, and residence of seven years in . . . Trinidad*, 264
Bringhurst, Robert, 334
Brown, John, 191
Browne, Thomas, 78
Brydges, James, Duke of Chandos, 143
Buchan, William, *Domestic Medicine*, 71
Buckland, Adeline, 338n16
Burgess, Will, 178, 183n16
Burnet, Thomas, *Sacred Theory of the Earth*, 171
Burton, Robert, *Anatomy of Melancholy*, 216
Butler, Marilyn, 254
Butler, Octavia, *Parable of the Sower*, 337
Byron, George Gordon, 42, 43–44

Cantillon, Richard, 188, 189, 193–94, 199
cartography. *See* maps
Caruth, Cathy, 319
Catholicism: Barker's conversion to, 210, 211; and placebos, 71–73; and population, 194, 195
Cavendish, Margaret, 6, 38, 227–38, 245; *The Blazing World*, 236; "A Condemning Treatise of Atomes," 228; *Grounds of*

Natural Philosophy, 229; *Observations upon Experimental Philosophy,* 229, 232, 236–37; "Observation X. Of a Butterfly," 234; "Of the Senses and Brain," 236; *The Philosophical and Physical Opinions,* 228; *Poems and Fancies,* 228, 229; *The Worlds Olio,* 229, 236–37
Centlivre, Susanna, 33–44; *A Bold Stroke for a Wife,* 3, 33–44
Chakrabarti, Pratik, 326, 333
Chandler, Anne, 246
Chardin, Jean-Baptiste-Siméon, 60
charlock, 293–97, 301
Charlotte, Queen, 143
Chaucer, Geoffrey: *Merchant's Tale,* 71, 72; *Parson's Tale,* 78
Chico, Tita, 1, 56–57, 112, 231, 249
chronometers, 110
Churchill, Charles, 34
Cipriani, Giovanni Battista, 126
Clark, William, *Digging Holes for Planting Sugar Cane,* 270
cloves, 307–8, 316–17, 323n38. See also spice trade
Clucas, Stephen, 228
Clusius, Carolus, 319
Codrington, Christopher, 269
Cody, Lisa Forman, 223n29
Cohen, Ashley, 312
Coleridge, S. T., 42
College of Physicians, 166
colonial agriculture and technology, 6, 261–75, 318–20
colonial world-making, 305–6, 311–12, 315–20
Complete System of Geography, A, 93
Conduitt, Catherine Barton, 5, 125, 134, 135, 136–37
Conduitt, John, 5, 125, 128–37
Congreve, William, 34–35, 44, 129
Cook, James, *The Journal of Captain Cook's Last Voyage to the Pacific Ocean,* 120
Coppola, Al, 2, 185n51
copyright, 140–41, 148–49, 157. See also intellectual property

Corner, James, 286
coronas. See glories
corpuscles, 168–82. See also atomism; particles
Cottegnies, Line, 228
Cowley, Abraham, 216–17, 223n23; "Ode upon Doctor Harvey," 216–17; "To the Royal Society," 3, 14, 21–25
Coyer, Abbé, 89
Crayle, Benjamin, 212
Creake, Bezaleel, 140–50, 155–57, 160n19
Cruikshank, Julie, 334
Cullen, William, 69–71, 73; *First Lines of the Practice of Physic,* 69; *Materia Medica,* 69
culture, nature in relation to, 18–25, 220–21, 306
Curtis, William, *Flora Londinensis,* 295–96, *296, 300*

Dale, Amelia, 223n19
Darwin, Charles, 247, 338n16
Darwin, Erasmus, 246, 299–300; *The Botanic Garden,* 292; *The Loves of the Plants,* 299; *Phytologia,* 299; *Zoonomia,* 299
Daston, Lorraine, 151
D'Avenant, William, *Proposition,* 313
David, Jacques-Louis, 136
Davidson, Donald, 27n5
Davis, Vivian, 39–40
Davy, Humphry, 66
Dawson, Catherine, 161n27
Dawson, Paul, 141, 145–48, 156, 161n27
Defoe, Daniel, 57, 321n12; *A Journal of the Plague Year,* 4, 46–48, 50–61; *The New Voyage around the World,* 307–8; *Serious Reflections During the Life and Surprising Adventures of Robinson Crusoe,* 51
De La Beche, Henry, 271
demography. See population
Deneys-Tunney, Anne, 285
Desaguliers, John Theophilus, 5, 140–57; *A Course of Experimental Philosophy,* 144, 155; illustrations associated with lectures of, *150, 152, 153, 154*

Descartes, René, 166
description. *See* nondescription
diagrams, 150–55
Diderot, Denis, *Encyclopedie,* 49, 60
digestion, as transmutation, 178–79
Dillon, Elizabeth Maddock, 313, 320
dividing engine, 113
Donald, Diana, 243, 247
Donne, John, 215
Dooren, Thomas van, 244
draft animals, 261–62, 264–66, 273–75, 307
Drake, Ellen Tan, 328–29
Dryden, John, 5, 34–36, 44, 129, 130, 133, 322n24; *Amboyna, or the Cruelties of the Dutch to the English Merchants,* 7, 306–14, 316–20, 321n17, 322n24; *Aureng-Zebe,* 305; *The Indian Emperour,* 125, *127,* 129–30, 132, 305; "Indian plays," 305; *The Indian Queen,* 305; *Mac Flecknoe,* 35; "To My Dear Friend Mr. Congreve," 34–35
Dutch East India Company (VOC), 307, 314–15, 319. *See also* East India Company
Dutch trade and colonialism, 306–20, 323n38
Dymock, Cressey, 274

earth, history of, 171–78, 327–34, 337
East India Company (England), 314, 319, 321n12. *See also* Dutch East India Company; French East India Company
Edgeworth, Honora, 246
Edgeworth, Maria: *Belinda,* 6, 243–54; *Practical Education,* 243–49, 253
Eleazar, Albin, *A Natural History of Birds,* 251
embodiment, 5–6; alchemical epistemology and, 167, 181; Barker and anatomy, 209, 210, 215–18; Cavendish's sensitive witnessing and, 237–38; as metaphor for population, 191–92
empiricism: Cavendish and, 227, 229–38; challenges for, 4; critiques of, 227; figurative language in service of, 19–21; and imperceptible particles, 166–75, 184n48; mirror-knowledge paradigm and, 166–68; myth in relation to, 7; sentimental, 65, 66. *See also* atomism; epistemology; materialism; visual verification
enargeia, 20
Engraver's Copyright Act (1735), 129
Enlightenment, 3, 66, 126–27, 151, 229
environment. *See* nature
Epicureanism, 227–29, 232–35, 238, 240–41n31
epistemology: myth and, 325–26, 335–36; and the Woodward pamphlet wars, 165–82. *See also* empiricism; mirror-knowledge paradigm; truth
Equiano, Olaudah, *Interesting Narrative,* 112, 119–20, 121, 271
Euhemerus of Messene, 325
Evans, Chris, 268
experimental imagination, 112

Falconer, William, *Dissertation on the Influence of the Passions upon Disorders of the Body,* 73
Fara, Patricia, 126, 128
Farina, Lara, 298
Felski, Rita, 142
feminism, 246
Ferdinand III, Emperor, 311
Fielding, Henry, *Shamela,* 181
figurative language: ambivalence as a feature of, 13; applied to population, 188–200; creative/productive uses of, 1, 3, 13–14, 21–25, 26n5; dangers of, 15–18; instrumental/illustrative uses of, 1, 3, 13–14, 18–21, 26n5; in scientific writing, 3, 13–14, 18–25, 26n3. *See also* language; metaphors
Finch, Anne, 224n34
Fitzmaurice, William, 273
flood, biblical, 171–78
Forster, Georg, 333
Forster, Johann Reinhold, 333

Forster, Robert, 272
fossils, 171–78, 328, 330
Foucault, Michel, 305
Franklin, Benjamin, 4, 188; *Observations concerning the Increase of Mankind*, 194–97
Franklin report, 65, 67
Frederick, Prince, 143
Freind, John, 170, 180; *Three Hours after Marriage* (with John Arbuthnot and John Gay), 175–78, 185n51
French East India Company, 319
Freud, Sigmund, 74
Frye, Northrop, 336
Fussell, G. E., 267

Gagliano, Monica, 299
Galison, Peter, 151
Garrick, David, 126
Gay, John, 170; *Three Hours after Marriage* (with John Arbuthnot and John Freind), 175–78, 185n51
gender: and authorship, 35–37, 42–43; Barker's life and writing and, 209–21; birds and, 243–54; equality/insignificance of, 209, 212, 217, 219, 221, 238, 246, 250; modesty associated with, 237; Royal Society and, 237; subjectivities and, 5–6. *See also* misogyny; women
Geoghegan, Bernard, 74
Geological Society of London, 333
geology, 171–78, 327–30, 332–34, 338n16
geomythography, 7, 326–37
Gheban, Dan, 59
Ghosh, Amitav, *Gun Island*, 336
glories (coronas, halos), 103–4
Godwin, William, 4, 43, 65–67
Goethe, Johann Wolfgang von, *Sorrows of Young Werther*, 43
gonorrhea, 75
gothic literature, 66–67, 77–79
Grainger, James: *An Essay on the More Common West-India Diseases*, 271; *The Sugar-Cane*, 292
Graunt, John, 190–92, 199

Gray, Simon (pseudonym), 198
Grove, Richard, 306–7, 311, 314–16, 318–20, 323n48
Gupta, Akhil, 285
Gurney, John, 149

Hadley, John, 115–16
Hall, Matthew, 299
halos. *See* glories
Hanlon, Aaron, 233
Hanway, Jonas, 194
Haraway, Donna, 212, 221, 237
Harrington, Thomas, 192
Harrison, John, 110, 129
Harrison, Robert Pogue, 283
Hartlib, Samuel, 265, 267, 274
Harvey, Penny, 265
Harvey, William, 191, 216–21, 224n31, 224n34
Hauksbee, Francis, 143
Haygarth, John, 73–74, 77, 83n41
Hazlitt, William, 43, 188, 198–200, 207n79
Hellegers, Desiree, 224n34
Helmont, Joan Baptista Van, 165
Heringman, Noah, 333
Hesiod, *Theogonia*, 329
Hill, John, *The Vegetable System [. . .]*, 302n7
Hippocrates, 73, 192
Hoadly, Benjamin, 132
Hobbes, Thomas, 55, 166
hoes, 6, 261–62, 265–66, 268, 270–73, 275
Hogarth, Mary, 137
Hogarth, William, 91–92, 95–97, 104, 125–37; *The Analysis of Beauty*, 95, 105, 130, 131, 134; *Before and After*, 133; *The Conquest of Mexico*, 130; "The Country Dance," 105; *The Four Stages of Cruelty*, 134; *The Harlot's Progress*, 125, 133, 136; *Marriage à la Mode*, 131, 136; *Portrait of a Family*, 132; *A Rake's Progress*, 39, 125, 131, 133; *A Scene from "The Indian Emperour," or "The Conquest of Mexico by the Spaniards,"* 5, 125, 127
holing, 270–73, *270*

Hooke, Robert, 7, 52, 60, 112, 127, 229–31, 233, 235, 326, 327–31, 333–34, 337, 338n9; *Micrographia,* 49–50, 116, 229–30, 328
Hooper, Robert, *Quincy's Lexicon-Medicum,* 68
Hovey, Sylvester, 261, 271
Hunt, Leigh, 43
Hunter, John, *Treatise on the Venereal Disease,* 75
Hunter, J. Paul, 184n48
Hutchinson, Lucy, 240–41n31
Hutner, Heidi, 313
Huysmans, Jacob, 133–34
hypochondria, 69–70

Iliffe, Robert, 128, 130
images: in Desaguliers's unauthorized *A System of Experimental Philosophy,* 140–57; epistemological value of, 49–50, 53–54, 60; text in relation to, 97–100, 104–5
imagination: and the body, 73; Boyle and, 54, 61; Defoe and, 50–51, 56, 59, 61; experimental, 112; figurative language and, 15, 26n3; limitations of, 100; Newton versus Conduitt on, 128–29; nondescription and, 48–51, 56–57, 59, 61; placebos and, 65, 72–74; science and, 48, 50, 57, 65–66
Indies, 309–20, 321n17
infrastructural inversion, 278–80, 285–86
infrastructure, 278–86
inheritance, literary, 3, 34–44
innovation, 3
intellectual property, 140–57

Jacson, Maria, *Botanical Dialogues,* 292
Jarvis, J. Ereck, 7
Jay, Mike, 66, 80n9
Jefferson, Thomas, 69, 70
Jeican, Ionut Isaia, 59
Jemisin, N. K., *Broken Earth* trilogy, 336
Johns, Adrian, 158n5
Johnson, Samuel, 192; *Dictionary,* 179

Jones, Doug, 335
Jones, R. F., 25n1, 26n2
Jonson, Ben, 35
Jope, E. M., 266
Justman, Stewart, 71

Kahan, Benjamin, 211
Karve, Irawati, 334
Kassell, Lauren, 181
Keats, John, 42
Keenleyside, Heather, 72, 248
Keill, John, 142–44, 146–47
Keiser, Jess, 210
Keller, Eve, 227, 230, 233, 238
Keynes, Milo, *Iconography of Sir Isaac Newton,* 126, 132
Killigrew, Anne, 35
King, Kathryn, 210, 222n12, 223n30, 224n33
Klamath Indians, 333
Kneller, Godfrey, 130, 132
Kranz, Isabel, 298

Lady's Revenge, The (play), 194
Lamb, Jonathan, 98
landscape urbanism, 286
language: limitations of, 47, 98–100; nature as apprehended through, 16–25; Sprat's theory of, as communication, 15–17, 28n15; transparency and opacity as features of, 16–25; visual verification in relation to, 46–47. *See also* figurative language; rhetoric
Lanser, Susan, 210
Latour, Bruno, 142, 327
Law, James, 198
lectures. *See* science lectures
Lesage, Alain-René, *Devil upon Two Sticks,* 78
Levine, Joseph, 173
Lewis, Jayne, 4, 54
Lewis, Matthew, 79, 261; *The Monk,* 4, 78–79
Licensing Act (1662), 148
Licensing Act (1694), 140

Lieberman, Jennifer, 113
Lignum nephriticum (coati, coatli, coatl), 308–9, 311
Ligon, Richard, 268
Linnaeus, Carl, 292, 297, 298, 301
literary technologies, 111–16, 156
Littleton, Edward, 269
Locke, John, 28n15, 46, 48, 166, 182n10; *An Essay concerning Human Understanding,* 49, 251–52
Long, Edward, 261, 272–73
longitude, means of determining, 110–12, 310, 322n20. *See also* sextants
Longitude Act (1714), 111, 117, 310
Lower, Richard, 216, 219–21, 224n34
Lucretius, *De Rerum Natura,* 227, 232, 238, 240–41n31. *See also* Epicureanism
Lupton, Christina, 2
Lyell, Charles, 338n16
Lynall, Gregory, 117
Lynch, Deidre, 67

Machiavelli, Niccolò, 325
Macklin, Charles, *Love à la Mode,* 148–49
Macklin v. Richardson (1770), 149–50
Malthus, Thomas Robert, 188–89, 197–200, 282
Mandeville, Bernard de, 188
Mann, Rachel, 210
Māori people, 7, 335
maps, 49–50, 318
Marder, Michael, 297, 298
Markley, Robert, 8, 187n93, 306, 309, 313–14
Marrinan, Michael, *The Culture of Diagram* (with John Bender), 48–59, 60, 150–51
Martin, Benjamin, *An Explanation of a New Construction and Improvement of the Sea Octant and Sextant,* 115–16
Martin, Joshua Lover (J. L.), 115
Martin, Samuel, 261, 265, 274–75
Marvell, Andrew, "The Garden," 212
Marx, Karl, 263
Masse, W. B., 333, 334
mātauranga Māori, 335

materialism, 6, 66, 176, 180, 227–29, 232–34, 238, 326; and myth, 331–32. *See also* empiricism
materia medica, 308, 319
Mattern, Shannon, 279
Mazzeo, Tilar, 37, 40
McCormick, Ted, 273
McGrath, Alice Tweedy, 210, 224n38
McNeill, J. R., 265, 274
Mead, Richard, 170
Mears, W., 160n19
medicine, and placebos, 65–76
Meillassoux, Quentin, 46, 61n1
Menely, Tobias, 248
mercantilism, 7, 191–92, 195, 318, 320
Mesmer, Franz Anton, 4, 65–66
metaphors: applied to population, 188–200; argumentative use of, 190; devolving into clichés, 189; role of, in science and empiricism, 19–21; time-bound meanings of, 189. *See also* figurative language
micromatter, 166–68, 175, 182–83n10. *See also* atomism; corpuscles; particles
microscopes, 6, 53, 112, 227, 229–30, 235–36
midwifery, 74, 83–84n48
Mikhail, Alan, 3
Milton, John, *Paradise Lost,* 43, 213
mind, in relation to the world, 16–17, 28n15
Mintz, Sidney, 272
mirror-knowledge paradigm, 166–68. *See also* epistemology
misogyny, 39
modest witnessing, 4, 6, 156, 209, 212, 221, 231–34, 237, 249–50
Modyford, Colonel, 273
Molyneux, William, 172, 173
Monardes, Nicolas, *Historia medicinal de las cosas que se traen de nuestras Indias Occidentales que sirven en medecina,* 308–9, 321n17
monstrosity: biological, 34, 37–38, 43; gendered, 35; literary, 34, 40–43; moral, 38
Montagu, Charles, Earl of Halifax, 134

Montaigne, Michel de, 227
Montesquieu, Charles Louis de Secondat, Baron de La Brède et de, 193
Moran, Richard, 26n5
Mossop, Elizabeth, 286
Motherby, George, *Medical Dictionary*, 68
Mottley, John, 36–37
Mukerji, Chandra, 283
Murray, John, 43
myth, 7, 325–37; anthropology and, 334–35; cultural approach to, 331, 336; epistemology and, 325–26, 335–36; geology and, 327–30, 332–34; literary studies and, 335–37; science and, 325–27, 330–31, 335, 337n3

Naish, James, 89; copy of Anson's *A Voyage Round the World*, 90
Natural History of the Frutex Vulvaria, The, 289
natural philosophy. *See* science
nature: as accessible to direct experience, 17–18, 20, 23, 29n30, 166; Barker's poetry and, 211–14, 217, 219–21; Cavendish and, 232–33; culture in relation to, 18–25, 220–21, 306; design of, 282–86, 306, 314–15, 316, 318; as hidden/obscure, 24, 29n28, 29n30; humans in relation to, 306–20; language as means of apprehending, 16–25; mind's relation to, 16–17, 28n15; myths about, 331–32, 334
Navigantium atque Itinerantium Bibliotheca, 93
navigational aids, 310. *See also* quadrants; sextants
Neptunism, 333
New Medicinal, Economical, and Domestic Herbal, A, 293
new science: and agricultural technology, 262, 265, 267, 273–74; Bacon and, 227, 229–33; Barker and, 210, 211, 216–21, 223n29; Cavendish's critique of, 227–38; contradictions of, 5; and myth, 325–27, 331, 337n3; optical technologies used by, 229–30; satirical treatments of, 117, 121; scholarship on, 2–3; sex as a subject for, 223n29; Woodward and, 142; writing style suited to, 13, 17. *See also* science
Newton, Isaac, 5, 116–17, 125–36, 143, 221, 309; *Opticks*, 127, 176; *Principia*, 116, 117, 127–28, 197
Newton, John, 328
Newtonian philosophy, 140–43
New Zealand Science Review (journal), 335
Nicholls, Sutton, 151
Nietzsche, Friedrich, 27n5
nondescription, 46–60; authenticity associated with, 47; Boyle and, 4, 50–55; Defoe's *A Journal of the Plague Year* and, 4, 47–48, 50–61; features of, 48, 57; imagination engaged by, 48–51, 56–57, 59, 61
Nulman, Lin, 136
Nunez de Herrera, Juan, 308
nutmeg, 307–8, 311–12, 316–17, 320, 323n38. *See also* spice trade

objectivity: Cavendish's critique of, 233; critiques of, 227; truth associated with, 91–92, 94; in visual representation, 91–92, 94–97, 101–3. *See also* modest witnessing
Okorafor, Nnedi, *Who Fears Death*, 336
optical technologies. *See* microscopes; telescopes; visual verification
optics, 4
originality, 33–34, 36–37, 40–42, 44
Orr, Bridget, 313
Ovid, *Metamorphoses*, 129, 327, 328–30, 337
oxalis. *See* wood sorrel
Oxalis acetosella, 296

padauk tree, 306, 308, 311–13, 319
Paige, Nicholas, 279
Paracelsus, 167, 187n93
Parks, Lisa, 284
particles, 166–75. *See also* atomism; corpuscles; micromatter
Pascoe, Judith, 292–93
Paulson, Ronald, 131

Pelizza, Annalisa, 285
perception: art and, 90–92; Cavendish's critique of, 230, 235–36; challenges of, 4. *See also* empiricism; visual verification
Perera, Suvendrini, 254
Perkins, Benjamin, 83n41
Perkins, Elisha, 83n41
Peters, John Durham, 278, 279
Petty, William, 188–89, 193, 199
Philips, John (pseudonym?), *An Authentic Journal*, 93
Philips, Katherine, 212
Phillips, Adam, 76
Philosophical Transactions of the Royal Society (journal), 38, 115, 121, 143, 166–67, 172
Pierce, Robert, *History and Memoirs of the Bath*, 72
placebos: belief and, 66, 68, 73, 78; defining, 67–68; and "experience of nothing," 4; gothic literature and, 66–67, 77–79; and imagination, 65, 72–74; medicalized, 65–77; religion and, 71–73, 78
plagiarism, 33–34, 35, 40–41
plague. *See* Defoe, Daniel: *A Journal of the Plague Year*
plants: agency of, 291, 298–300; growth of, 165–66, 168–69; humans in relation to, 291–301; in metaphors for population, 195–97, 205n53; moral rhetoric's use of, 289–90; Smith's "Beachy Head" and, 290–301; taxonomy of, 7, 290–301; transpiration of, 166, 169, 171–72; violence linked to, 289, 295, 298; weeds, 293–94; Woodward's experiments with, 168–71
Plato, 328
plows, 6, 261–67, *262, 263,* 271–75
Plutarch, *Lives,* 43
Plutonism, 333
Polwhele, Richard, *The Pursuits of Literature,* 289
Poole, William, 327
Pope, Alexander, 34, 71, 72, 126, 170
population: animal metaphors and analogies for, 188, 189, 192–94, 196–97, 200, 205n41; birth versus migration as source of, 190–91; botanical metaphors and analogies for, 195–97, 205n53; city versus country concerning, 191; corporeal metaphors for, 191–92; labor associated with, 193–94; mathematical rhetoric applied to, 197–200; mercantilism and, 191–92, 195; metaphors applied to, 188–200; personification of, 198; race and, 194–96; Rousseau's *Julie* and, 282; sexual activity and, 192–94, 197, 199–200; slavery and, 195–96, 206n57
Portuondo, Maria, 310
Powell, Robert, 190
Powell, Rosalind, 293
Power, Henry, 233, 235
Practical Treatise of Husbandry, A, 293–94
Priestley, Joseph, 66
print media: and intellectual property, 140–57; the medical field and, 66, 68–69
Pritchard, Sara, 279

quadrants, 120, 122n1. *See also* sextants
queerness: Barker and, 210, 212, 221, 224n38; Woodward and, 175–78
Quine, Willard, 189

race: and colonial technology, 261; meanings of, 195; and population, 194–95
Radcliffe, Ann, 77–78; *The Italian,* 78; *The Mysteries of Udolpho,* 78
Ramachandran, Ayesha, 316, 318
Ramsden, Jesse, 113–15, 117
Ray, John, *Historia Plantarum,* 49
Reede tot Drakenstein, Hendrik van, 307
Rees, Emma, 228
Reiss, Timothy, 315–16
religion, and placebos, 71–73, 78
reproduction: biological, 3, 34, 36; literary, 3, 34–44
Reynolds, John Hamilton, *Peter Bell,* 42
rhetoric: and *enargeia,* 20; Malthus's use of, 197–99; science in relation to, 2–4, 13–21, 26n3; visual, 47, 49–50, 60–61. *See also* figurative language; language; metaphors; scientific writing

Richardson, Jonathan, *Essay on the Theory of Painting*, 106n21
Roberts, Justin, 271
Robins, Benjamin, 89, 93, 104
Robinson, Kim Stanley, *Green Earth* trilogy, 336
Rogers, John, 111–12, 241–42n52
Rorty, Richard, 166–68, 182n10
Rosenthal, Caitlin, 263
Rosenthal, Laura, 33, 35
Rossetti, Christina, *Goblin Market*, 289
Rossi, Paolo, 326, 328–29, 338n9
Roubiliac, Louis-François, 132
Rousseau, Jean-Jacques, 99, 246, 247, 252, 282; "Essay on the Origin of Languages," 283; *Julie, ou La Nouvelle Héloïse*, 6–7, 278–86
Rowden, Frances Arabella, *A Poetical Introduction to the Study of Botany*, 292
Royal College of Physicians, 224n34
Royal Society, 1, 3, 5, 14–25, 29n28, 38, 114–16, 118, 128, 132, 137, 140, 142–43, 157, 166–67, 172, 216, 217, 221, 229, 230, 232, 233, 237, 328, 330
Rush, Benjamin, 69, 73
Ruwe, Donelle, 295
Rysbrack, Michael, 126, 132

Sackfield, John, 140–50, 155–57, 160n19
Sarasohn, Lisa, 234
Schaffer, Simon, 53, 55, 62n19, 111, 117, 121, 156, 166–67, 181, 184n48, 221, 231
Schiebinger, Londa, 228
Schiller, Friedrich, 285
Schleifer, Ronald, 80n11
Schotte, Margaret, 122n1
science: art in relation to, 91–92, 104–5; Barker and, 209–11, 215–21; criticisms of, 39; Edgeworth and, 243–54; experiments as essential to, 17; and imagination, 48, 50, 57, 65–66; and instruments for the determination of longitude, 110–21; literary technologies and, 111–16; medicine and, 67–69; modern versus ancient and medieval, 17–18; and myth, 325–27, 330–31, 335, 337n3; placebos and, 65–79; rhetoric in relation to, 2–4, 13–21, 26n3; Sprat's theory of, as disclosure, 17; women and, 243–46, 249–50; women and children associated with, 254. *See also* empiricism; new science
science lectures, 140–44, 148, 150, 155
science studies, 1–2, 14
scientific illustration, 140–41, 150–57
scientific writing: and figurative language, 3, 13–14, 18–25, 26n3; literary status of, 112; plain style of, 3, 13, 15–18, 25n1, 26n2, 92; about the sextant, 115–16
Scilly naval disaster, 110–11
Scott, James C., 318
Scriblerians, 5, 170, 175, 181–82, 321n13
scurvy, 75–76
Seeman, Enoch, 132
Severinus, Petrus, *Idea Medicinae Philosophicae*, 170
sextants, 4; common cross-genre representations of, 112–16; credibility of, 114, 116–21; human usage of, 117–20; literary portrayals of, 111–22; promise and shortcomings of, 110, 112, 113–21; quadrants compared to, 120, 122n1; satirical treatments of, 116–19, 121
sexuality: Barker's poetry and, 212–15; gendered views of, 199–200; plants associated with, 289; population linked to, 192–94, 197, 199–200; virginity as alternative to, 209–21. *See also* queerness
Shackelford, Jole, 170
Shadwell, Thomas, 35; *The Virtuoso*, 175–76
Shakespeare, William, *Love's Labours Lost*, 290
Shapin, Steven, 53, 55, 62n19, 111, 117, 121, 156, 166–67, 181, 184n48, 221, 231
Shaw, Peter, 68
Shelley, Mary: *Frankenstein*, 3, 34, 42; "Giovanni Villani," 43
Shelley, Percy, 42, 43
Shovell, Cloudesley, 110–11
Silver, Sean, 2, 167
silver experiment, 171–72, 174

Simon, Julia, 282–84
Siskin, Clifford, 146, 285
slavery: activism for the abolishment of, 271–72; and colonial agriculture and technology, 6, 263–65, 270–75; and population, 195–96, 206n57
Sloane, Hans, 321n13
Smellie, William, *Treatise on [. . .] Midwifery*, 74
Smith, Adam, 197
Smith, Charlotte, 292–93; "Beachy Head," 7, 290–301; *Marchmont*, 292; *Minor Morals*, 292; *Rambles Farther*, 292; *Rural Walks*, 292
Smith, Courtney Weiss, 13, 14, 167
Smith, Pamela H., 166, 181
Smollett, Tobias, *Roderick Random*, 118–19
Sneed, Adam, 2
Society Islanders, 333
soil erosion and depletion, 268–71, 315
Spary, Emma, 316–17, 323n38
Spellberg, Matthew, 334, 335–36
Spencer, Jane, 245, 247
spice trade, 308–9, 313–14. *See also* cloves; nutmeg
Spivak, Gayatri, 305
spontaneous generation, 53
Sprat, Thomas, 1, 7–8, 14–25; *History of the Royal Society*, 3, 7–8, 14–25, 166, 232, 242n60
Star, Susan Leigh, 278, 285
Starobinski, Jean, 285
Starosielski, Nicole, 278
Stathas, Thalia, 37
Stationer's Company, 148
Statute of Anne (1710), 140, 148
Steele, Richard, 143–47, 160n27
Steno, Nicholas, *Prodromus*, 327
Stephens, James, 270
Stern, Simon, 148, 149
Sterne, Laurence, *Tristram Shandy*, 118, 210
Steuart, James, 197
Stewart, Georgina Tuari, 335
Stewart, Larry, 1, 143
Stubbe, Henry, 3

Sutherland, Alexander, *Attempt to Revive Ancient Medical Doctrines*, 72
Swift, Jonathan, 38; *Gulliver's Travels*, 112, 116–18, 121; *Tale of a Tub*, 180
system, concept of, 140–57

Tahitians, 333
taxonomy, of plants, 7, 290–301
Tayebi, Kandi, 293
Te Awa Tupua (Whanganui River Claims Settlement) Act (New Zealand, 2017), 334
telescopes, 6, 127, 227, 230
Temple, William, 197
Terry, Richard, 34
Teskey, Gordon, 337n3
Te Urewera Act (New Zealand, 2014), 334
text, and image, 97–100, 104–5
Thiry, Paul-Henri, Baron d'Holbach, *Système de la Nature*, 331–32
thistle, 293–97, 301, 302n7
Thomas, Pascoe, *A True and Impartial Journal of a Voyage to the South-Seas*, 93
Thompson, Helen, 54
Thomson, James, "A Poem Sacred to the Memory of Sir Isaac Newton," 135
Thornhill, James, 129, 132, 134, 136, 137
Thornhill, Jane, 134
Three Hours after Marriage (play), 5
Tlingit people, 334
transmutation, 165–66, 168–70, 178–79
trauma, and history, 319–20
Trotter, Thomas, *Observations on the Scurvy*, 75–76
truth: about Anson's *A Voyage Round the World*, 93–94; objectivity associated with, 91–92, 94; social determinants of, 166–67. *See also* epistemology; visual verification
Tryon, Thomas, 268–69
Tull, Jethro, 267

uncanny, 74

Vannetta, Jerry, 80n11
Varenius, Bernhardus, *Geographia Generalis*, 327

Varro, Marcus Terentius, 264, 325
Veen, Marijke van der, 294
vegetation. *See* plants
Vesalius, *De humani corporis fabrica*, 134
Vickers, Brian, 26n2
Vico, Giambattista, *New Science*, 331, 336
Virgil, *Georgics*, 266–67
virginity, 209–21
visual verification: art and, 90–92, 94–96, 104–5; Brett's engravings for Anson's *A Voyage Round the World*, 90–105; Cavendish's challenge to, 227, 229–30, 235–36; images and, 49–50, 53–54, 60; and the invisible, 46–61, 166–75, 229; language in relation to, 46–47; objectivity as factor in, 91–92, 94–97, 101–3; rhetoric of, 47, 49–50, 60–61; trustworthiness of, 46–48, 50–51, 55–61. *See also* empiricism; truth
Vitaliano, Dorothy, 333
VOC. *See* Dutch East India Company

Walaeus, Joannes, 216, 218–19, 221, 224n33
Wall, Cynthia, 46
Walpole, Horace, 89
Walter, Richard, 89–105
Walzer, Arthur, 197
Watt, James, 3
weeds, 293–94
Wehi, Priscilla, 335
Westfall, Richard S., *Never at Rest*, 128
Wharton, Joanna, 246, 249
Whig ideology: freedom, 36, 40–42; meritocracy, 40; politics, 5, 33; property, 3
Wickman, Matthew, 77
Williams, Eric, 263
Williams, Raymond, 279
Willis, Thomas, 186n81, 216, 218–21, 224nn33–34
Wilmot, John, Earl of Rochester, 133
Wilson, Catherine, 59
Wistar, Caspar, 69
witnessing: and Anson's *A Voyage Round the World*, 89, 91–92, 97, 100–103, 105; art and, 92, 97, 100–103, 105; and empiricism, 4; gender and, 6; of the invisible, 54, 55, 105; literary technologies as means of virtual, 111, 114–15, 117, 121, 156; modest, 4, 6, 156, 209, 212, 221, 231–34, 237, 249–50; scientific, 59, 156; sensitive, 6, 229, 234–38; Wright's *An Experiment on a Bird in the Air-Pump* and, 243
Wollstonecraft, Mary, 245; *A Vindication of the Rights of Woman*, 247
women: animals associated with, 243–48, 253; and authorship, 35, 37, 42–44; birds associated with, 243–54; celibacy of, 209–21; education of, 246–52; excluded from Royal Society, 237; intellectual capacity of, 209, 218–21, 237–38; and marriage, 250–51; and population, 190–94, 199–200; and science, 243–46, 249–50, 254; and sensitive witnessing, 237–38. *See also* feminism; gender; misogyny
wood sorrel (*Oxalis acetosella*), 295–301, 296, 302n15
Woodville, William, *Medical Botany*, 296
Woodward, John, 5, 166–82; *An Essay toward a Natural History of the Earth*, 166, 170–73, 176; "Some Thoughts and Experiments concerning Vegetation," 168–69; *The State of Physick*, 166, 170, 178–81
Wordsworth, William: *Peter Bell*, 42; *The Prelude*, 128
world-making, 305–6, 311–12, 315–20
Wragge-Morley, Alexander, 20, 49, 186n83
Wright, Alexis, *The Swan Book*, 336
Wright, Joseph, *An Experiment on a Bird in the Air-Pump*, 6, 134, 243–44, 247
Wright, Thomas, *Passions of the Mind*, 73

Young, Edward, *Conjectures on Original Composition*, 36

Zarka, Yves Charles, 285
Zeuxis, 23

www.ingramcontent.com/pod-product-compliance
Lightning Source LLC
Chambersburg PA
CBHW031426230426
43668CB00007B/456